AGRICULTURAL BIOECONOMY

AGRICULTURAL BIOECONOMY

Innovation and Foresight in the Post-COVID Era

Edited by

CHETAN KESWANI
Group Leader,
Academy of Biology and Biotechnology,
Southern Federal University, Rostov-on-Don,
Russia

CRISTINA POSSAS
Professor, Fiocruz, Rio de Janeiro, Brazil; Senior Scientific Advisor, Bio-Manguinhos/Fiocruz, Rio de Janeiro, Brazil

EMMANUEL KOUKIOS
Professor Emeritus, Chemical Engineering School, National Technical University, Athens, Greece

DAVIDE VIAGGI
Full Professor in Agricultural Economics and Rural Appraisal, University of Bologna, Bologna, Italy

ELSEVIER

ACADEMIC PRESS
An imprint of Elsevier

Academic Press is an imprint of Elsevier
125 London Wall, London EC2Y 5AS, United Kingdom
525 B Street, Suite 1650, San Diego, CA 92101, United States
50 Hampshire Street, 5th Floor, Cambridge, MA 02139, United States
The Boulevard, Langford Lane, Kidlington, Oxford OX5 1GB, United Kingdom

ISBN: 978-0-323-90569-5

For information on all Academic Press publications
visit our website at https://www.elsevier.com/books-and-journals

Publisher: Nikky P. Levy
Acquisitions Editor: Nancy Maragioglio
Editorial Project Manager: Howi M. De Ramos
Production Project Manager: Bharatwaj Varatharajan
Cover Designer: Miles Hitchen

Typeset by STRAIVE, India

Dedication

Dedicated to the loving memory of Prof. Dr. Cahyono Agus, whom we lost to COVID during the preparation of this volume.

Contents

Contributors *xiii*
Author note *xvii*
Preface *xix*
Acknowledgments *xxi*

1. Sustainable agro-bioeconomy after Covid-19: Nineteen utopian and dystopian scenarios for the world and Turkey **1**

İrfan Kalaycı

1. Introduction 1
2. Covid-19 and sustainable agro-bioeconomy 1
3. Scenarios for agro-bioeconomy after Covid-19 10
4. Conclusion 21

References 25

2. Organic systems under major changes—A deep crisis in the management of the organic world of our planet—New tasks for foresight: An opinion paper **29**

Emmanuel G. Koukios and Lazaros S. Karaoglanoglou

1. Introduction and approach 29
2. Results and discussion 32
3. Concluding remarks 34

Acknowledgment 35
References 35

3. Universal access to vaccines in post-COVID bioeconomy: Redesign for variants of concern and patent protection **37**

Cristina Possas, Adelaide Maria de Souza Antunes, Alessandra Moreira de Oliveira, Suzanne de Oliveira Rodrigues Schumacher, and Akira Homma

1. Introduction 37
2. Universal access to vaccines and pandemic preparedness: Conceptual framework 39
3. COVID-19 vaccines: Global distribution and related variants 40
4. SARS-CoV-2 variants and genomic vaccines: Overview 42
5. BRICS' vaccines: Development and production capacity 51

6. DNA and RNA vaccines in clinical trials: Patent landscape 52
7. Discussion 66
8. Conclusion 67
References 68

4. Peptide production by molecular farming with antiviral effects 77

Masoumeh Kordi, Zeynab Borzouyi, Robab Salami, and Behnam Asgari Lajayer

1. Introduction 77
2. Molecular farming for production drugs and vaccines 78
3. Antimicrobial peptide and antiviral effects 78
4. Peptide heterologous expression host systems 80
5. Peptide production in plants 81
6. Purification of recombinant antimicrobial peptides 81
7. Challenges of using peptides in treatment 82
8. Peptide-based drugs 82
9. Conclusion 82
References 83

5. An assessment of smallholder farmers' status in Umzinyathi and Harry Gwala districts of Kwazulu Natal Province, South Africa 85

Phokele Maponya and Ikechi Kelechi Agbugba

1. Introduction 85
2. Methodology 89
3. Results and discussion 93
4. Conclusion and recommendations 99
Declaration statements 100
References 101

6. Using geospatial technologies to manage COVID-19 103

Karim Ennouri, Slim Smaoui, Olfa Ben Braiek, and Mohamed Ali Triki

1. Introduction 103
2. Principal sensors, characteristics and applications 105
3. COVID-19: Impact assessments with insights from satellite data 110
4. Remote sensing as a tool to survey endemic diseases 111
5. Conclusion 115
References 116

7. Potential constraint of rainfall availability on the establishment and expansion of agroforestry in the Joe Gqabi, Alfred Nzo and OR Tambo Districts, Eastern Cape in South Africa **121**

Phokele Maponya, Casper Madakadze, Nokwazi Mbili, Zakheleni Dube, Thabo Nkuna, Meshack Makhwedzhana, Takalani Tahulela, and Kgosi Mongwaketsi

1. Introduction 121
2. Methodology 123
3. Results and discussions 127
4. Conclusion and recommendations 139

Declaration statement 139
Acknowledgments 140
References 140

8. Agricultural biotechnology in Bangladesh: The way forward **143**

Abdullah Mohammad Shohael and Kathleen L. Hefferon

1. Introduction 143
2. Agriculture in Bangladesh 145
3. Adopting agricultural biotechnology for circular bioeconomy 148
4. Recent breakthroughs in the cultivation of GM crops in Bangladesh 149
5. Genome editing as an emerging technology for the agricultural improvement in Bangladesh 154
6. The regulatory framework for GMOs adoption in Bangladesh 159
7. Consumer acceptance and future regulation of new breeding technologies in Bangladesh 160
8. Conclusions 161

References 162

9. Strategic tropical natural resources management and household socio-economic resilience due to the Covid-19 pandemic **167**

S. Andy Cahyono, Cahyono Agus, S. Agung Sri Raharjo, Agung Wahyu Nugroho, and Yonky Indrajaya

1. Introduction 167
2. The effects of Covid 19 pandemic: From global to households 169
3. Covid-19: Disruption on socio-economic and nature 171
4. Innovation: A new paradigm of knowledge-based development in Covid-19 era 174

5. Integrated bio-cycles farming system in post Covid-19 era: The next level 175
6. Household social economic resilience in Covid-19 era 177
7. Conclusion 178
References 179

10. The healing and bioeconomic prospect of tropical ecosystems due to the Covid-19 pandemic 183

Cahyono Agus, Alue Dohong, Dian Yuanita Wulandari, Meilania Nugraheni, Catur Sugiyanto, Enggal Primananda, and Zahra Zafira Mutiara

1. Introduction 183
2. Covid-19 and La-Nina phenomenon in Indonesia 185
3. The healing of tropical ecosystem in Indonesia 187
4. Bio-economy of nature-based resources management 192
5. Agricultural supply chain management 194
6. Precision and smart agriculture 195
7. Conclusions 198
References 198
Further reading 204

11. Towards the fifth innovation era in agricultural innovation 205

Per Frankelius and Matilda von Rosen

1. Introduction 205
2. Challenges of world agriculture 205
3. Challenges as drivers of innovation 212
4. Historic reflections: The first four innovation eras 214
5. The new fifth innovation era starting now 217
6. Enablers and disablers in the fifth innovation era 226
7. Conclusion 226
References 228
Further reading 229

12. The role of collaboration and entrepreneurship in strengthening the participation of primary producers in the bioeconomy 231

Nunzia Gabriella Fasolino, Matteo Zavalloni, and Davide Viaggi

1. Introduction and overview 231
2. Methodology 233

3. Results 235
4. Discussion and conclusions 241
Acknowledgments 243
References 243

13. Biofuel co-products for livestock feed 245
Ugo De Corato and Egidio Viola

1. Introduction 245
2. Structure of the chapter 249
3. Biofuel for a sustainable market 249
4. Biofuel co-products for livestock feed 264
5. Conclusions and future market trends 278
References 281

14. Education and training in agriculture and the Bioeconomy: Learning from each other 287
Yaprak Kurtsal, Giacomo Maria Rinaldi, Marco Maria Grande, and Davide Viaggi

1. Introduction 287
2. Bioeconomy education needs and approaches 288
3. Agricultural education and the epistemology of inter- and trans-disciplinarity 293
4. Transforming the AKIS framework toward collaborative learning and innovation: Can we extend it further to cover Bioeconomy 297
5. Discussion: Lessons learnt and the way forward 304
6. Concluding remarks 308
References 308
Further reading 313

15. Urban agriculture and agro-environment: A case study of homestead food gardeners in the City of Tshwane, Gauteng Province, South Africa 315
Phokele Maponya and D. Oelofse

1. Introduction 315
2. Methodology 318
3. Results and discussion 320

4. Conclusions and recommendations 333
Declaration statement 334
Acknowledgments 335
References 335

Index *337*

Contributors

Ikechi Kelechi Agbugba
Department of Agricultural & Applied Economics, Faculty of Agriculture, Rivers State University, Port Harcourt; School of Agribusiness Management, Rome Business School (Nigeria Campus), Ikeja, Lagos State, Nigeria

Cahyono Agus
PKBTS Tamansiswa Alumni Board; Faculty of Forestry, Universitas Gadjah Mada, Yogyakarta, Indonesia

Behnam Asgari Lajayer
Department of Soil Science, Faculty of Agriculture, University of Tabriz, Tabriz, Iran

Olfa Ben Braiek
Laboratory of Transmissible Diseases and Biologically Active Substances (LR99ES27), Faculty of Pharmacy, University of Monastir, Monastir, Tunisia

Zeynab Borzouyi
Agriculture Department, Islamic Azad University of Sabzevar, Sabzevar, Khorasan Razavi, Iran

S. Andy Cahyono
Research Center for Ecology and Ethnobiology, National Research and Innovation Agency (BRIN), Cibinong, Indonesia

Ugo De Corato
Department of Bioenergy, Biorefinery and Green Chemistry (TERIN-BBC-BIC), Italian National Agency for New Technologies, Energy and Sustainable Economic Development (ENEA), Bari, Italy

Alessandra Moreira de Oliveira
School of Chemistry, Federal University of Rio de Janeiro, Rio de Janeiro, Brazil

Suzanne de Oliveira Rodrigues Schumacher
School of Chemistry, Federal University of Rio de Janeiro, Rio de Janeiro, Brazil

Adelaide Maria de Souza Antunes
School of Chemistry, Federal University of Rio de Janeiro; National Institute of Industrial Property, Rio de Janeiro, Brazil

Alue Dohong
The Environmental Science, Post Graduate Programme at Palangka Raya University, Palangkaraya; Ministry of Environment and Forestry RI, Jakarta, Indonesia

Zakheleni Dube
University of Mpumalanga, Faculty of Agriculture and Natural Science, Mbombela, South Africa

Karim Ennouri
Laboratory of Amelioration and Protection of Olive Genetic Resources, Olive Tree Institute, University of Sfax; Digital Research Centre of Sfax, Technopark of Sfax, Sfax, Tunisia

Nunzia Gabriella Fasolino
Department of Agricultural and Food Sciences, University of Bologna, Bologna, Italy

Per Frankelius
Agtech 2030, Department of Management and Engineering, Linköping University, Linköping, Sweden

Marco Maria Grande
Department of Agricultural and Food Sciences, University of Bologna, Bologna, Italy

Kathleen L. Hefferon
Department of Microbiology, Cornell University, Ithaca, NY, United States

Akira Homma
Bio-Manguinhos, Oswaldo Cruz Foundation, Rio de Janeiro, Brazil

Yonky Indrajaya
Research Center for Ecology and Ethnobiology, National Research and Innovation Agency (BRIN), Cibinong, Indonesia

İrfan Kalaycı
Department of Economics, Faculty of Economics and Administrative Sciences, İnönü University, Malatya, Turkey

Lazaros S. Karaoglanoglou
Research Group BIOTOPOS, Laboratory of Organic Technologies, School of Chemical Engineering, National Technical University of Athens, Athens, Greece

Masoumeh Kordi
Department of Cell & Molecular Biology, Faculty of Life Sciences & Biotechnology, Shahid Beheshti University, Tehran, Iran

Emmanuel G. Koukios
Research Group BIOTOPOS, Laboratory of Organic Technologies, School of Chemical Engineering, National Technical University of Athens, Athens, Greece

Yaprak Kurtsal
Department of Agricultural and Food Sciences, University of Bologna, Bologna, Italy

Casper Madakadze
University of Pretoria, Faculty of Agriculture and Natural Science, Pretoria, South Africa

Meshack Makhwedzhana
Agricultural Research Council-Vegetable, Industrial and Medicinal Plants, Pretoria, Gauteng, South Africa

Phokele Maponya
Agricultural Research Council-Vegetable, Industrial and Medicinal Plants, Pretoria, Gauteng, South Africa

Nokwazi Mbili
University of KwaZulu-Natal, College of Agriculture, Engineering and Science, School of Agricultural, Earth and Environmental Science, Durban, South Africa

Kgosi Mongwaketsi
South Africa Forestry Company Limited (SAFCOL) Research, SAFCOL (Pty) LTD, Sabie, South Africa

Zahra Zafira Mutiara
Deutsche Gesellschaft für Internationale Zusammenarbeit (GIZ), Jakarta, Indonesia

Thabo Nkuna
Agricultural Research Council-Vegetable, Industrial and Medicinal Plants, Pretoria, Gauteng, South Africa

Meilania Nugraheni
Research and Development, Sinarmas Forestry, Riau; Faculty of Economics and Business, Universitas Gadjah Mada, Yogyakarta, Indonesia

Agung Wahyu Nugroho
Research Center for Ecology and Ethnobiology, National Research and Innovation Agency (BRIN), Cibinong, Indonesia

D. Oelofse
Agricultural Research Council-Vegetable, Industrial and Medicinal Plants, Pretoria, Gauteng, South Africa

Cristina Possas
Bio-Manguinhos, Oswaldo Cruz Foundation, Rio de Janeiro, Brazil

Enggal Primananda
Research Center for Plant Conservation and Botanic Gardens, Indonesian Institute of Sciences, Bogor, Indonesia

S. Agung Sri Raharjo
Institute for Implementation Standard of Environment and Forestry Instrument, Surakarta, Indonesia

Giacomo Maria Rinaldi
Department of Agricultural and Food Sciences, University of Bologna, Bologna, Italy

Robab Salami
Department of Cell & Molecular Biology, Faculty of Life Sciences & Biotechnology, Shahid Beheshti University, Tehran, Iran

Abdullah Mohammad Shohael
Department of Biotechnology and Genetic Engineering, Jahangirnagar University, Savar, Bangladesh

Slim Smaoui
Laboratory of Microbial, Enzymatic Biotechnology and Biomolecules (LBMEB), Center of Biotechnology of Sfax, University of Sfax, Sfax, Tunisia

Catur Sugiyanto
Faculty of Economics and Business, Universitas Gadjah Mada, Yogyakarta, Indonesia

Takalani Tahulela
South Africa Forestry Company Limited (SAFCOL) Research, SAFCOL (Pty) LTD, Sabie, South Africa

Mohamed Ali Triki
Laboratory of Amelioration and Protection of Olive Genetic Resources, Olive Tree Institute, University of Sfax, Sfax, Tunisia

Davide Viaggi
Department of Agricultural and Food Sciences, University of Bologna, Bologna, Italy

Egidio Viola
Department of Bioenergy, Biorefinery and Green Chemistry (TERIN-BBC-BIC), Italian National Agency for New Technologies, Energy and Sustainable Economic Development (ENEA), Bari, Italy

Matilda von Rosen
Agtech 2030, Department of Management and Engineering, Linköping University, Linköping, Sweden

Dian Yuanita Wulandari
Faculty of Agriculture, Universitas Gadjah Mada, Yogyakarta, Indonesia

Matteo Zavalloni
Department of Agricultural and Food Sciences, University of Bologna, Bologna, Italy

Author note

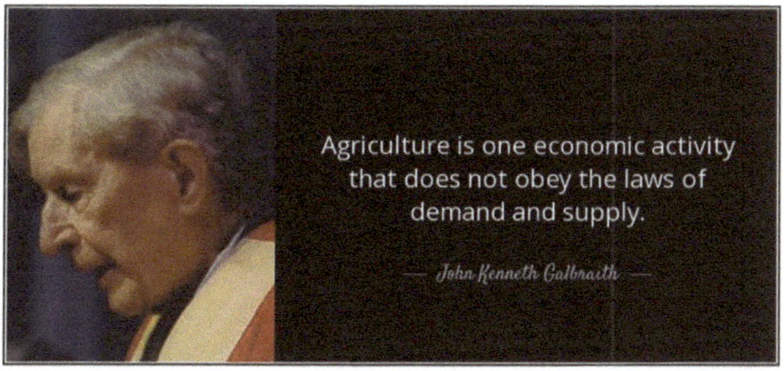
Agriculture is one economic activity that does not obey the laws of demand and supply.
— John Kenneth Galbraith —

Preface

The increasing demand in the major agricultural sectors of economy will increase the consumption of fossil energy sources, agricultural land, and fresh water. This demand could lead to irreversible changes in climatic patterns with unpredictable consequences. A possible direction to address this challenge sustainably is increasing the efficiency of currently used processes and displacing fossil fuel energy sources by production of useful biomass. Over the past three decades, the landmark discoveries in molecular biology have revolutionized multidimensionally. Contemporary bioeconomic strategies also support an integration of science, technology, economy, environmental issues, rural and industrial developments, regulatory processes, and social sciences initiatives. This is a great opportunity for molecular and engineering sciences to map and engineer the uncharted territories of molecular transformations that are the key to economic growth.

Sustainability in agro-based bioeconomy will evolve by incorporating concepts such as smart production with less impact on the environment as well as by reducing and recycling biowaste into its objectives. Finally, it should be emphasized that biological resources offer unique new functions and properties in comparison with nonbiological resources so that in all contexts, particularly considering the economy, the aspect of innovation must be highlighted. Excellent innovative and mission-oriented research requires long-term support and encouragement from all stakeholders in science, industry, and society to develop smart sustainable bioeconomies. Without this, there will be no "fuel" for the bioeconomy to maintain its innovativeness. The foundations and future of the agricultural bioeconomy are based on strong science and technology pillars, and a dynamic and innovative approach that responds to the constant demands for new scientific and technological knowledge needs to be created.

Furthermore, bioeconomy requires transforming basic knowledge into successful industrial production and agriculture, including food, novel bioproducts, and resilient germplasms. To ensure this, it will be vital to strengthen and stimulate mechanisms for creating an entrepreneurial atmosphere, whether from the public sector or from the private sector. Agro-based bioeconomy requires a subtle balance between public sector-supported science push and market and social pull, encouraged by industry and investors as well as policymakers and society.

This volume proposes an integrated approach for sustained innovation in various areas of agro-based bioeconomy, especially in the post-COVID perspective. Moreover, the contributors have tried to underline the upcoming challenges and present the most viable options for translating commercially viable ideas into easily affordable agricultural products and technologies.

Editors

Chetan Keswani
Cristina Possas
Emmanuel Koukios
Davide Viaggi

Acknowledgments

Working on this project has been an overwhelming experience, with lockdowns, multiple COVID waves, and loss of friends, family members, and collaborators. Among all the chaos, we are grateful to all the contributors who came out to share their valuable ideas for sustainable agricultural systems in the post-COVID era.

The editors express enormous gratitude to Mr. Howell Angelo M. De Ramos (from Elsevier) for his supreme patience, sustained support, and efficient management of this project. In addition, we are also thankful to Ms. Nancy Maragioglio, Senior Acquisitions Editor, for giving us the opportunity to edit this volume.

Lastly, Chetan Keswani gratefully acknowledges the Ministry of Science and Higher Education of the Russian Federation project on the development of the Young Scientist Laboratory (No. LabNOTs-21-01AB).

CHAPTER 1

Sustainable agro-bioeconomy after Covid-19: Nineteen utopian and dystopian scenarios for the world and Turkey

İrfan Kalaycı
Department of Economics, Faculty of Economics and Administrative Sciences, İnönü University, Malatya, Turkey

1. Introduction

The world economy and one of its major components, Turkey, continue to give serious human and economic losses because of the Covid-19 pandemic. The world will not be the same when this pandemic, which has a cumulative fatal and devastating effect, is over. The agricultural sector, which represents our source of life, and the agro-bioeconomy, which aims to increase food welfare, will also take its share. The sustainability goal in this area will be stuck between various utopias and dystopias. Dystopias give pessimism with the effect of bad marks while the utopias built here for the post-pandemic promise hope to humanity.

Within the scope of this study, utopian and dystopian scenarios come to mind at the intersection of the global epidemic and agro-bioeconomy. In these opposite scenarios, all the world's economies including Turkey will take part in different forms and will be affected at different levels.

2. Covid-19 and sustainable agro-bioeconomy

Like every global epidemic, Covid-19 has made the most basic needs more visible and urgent. The current need for our planet is sustainable agro-bioeconomy.

Agricultural Bioeconomy
https://doi.org/10.1016/B978-0-323-90569-5.00006-8

2.1 Covid-19 pandemic process

The Covid-19 pandemic resembles a "black swan"[a] incident, which was initially unpredictable, then surprised by its effects and made opposite statements about it. It is thought that the events that leave their mark in every aspect of life are black swans. From the very beginning, this pandemic was a dream the world never wanted to see because of the horrible loss of life and property it caused; because it is already living the truth. This pandemic has caused societies within themselves and countries to close or distance themselves from each other.

The Covid-19 pandemic which started as a global health crisis and turned into an economic crisis, and the globalization event which is supposed to make humanity happy, move in parallel. The pandemic has brought them back with drastic measures against globalization that eliminates distances, The agricultural sector also gets its share from this. In the ongoing pandemic period, the poverty and despair of societies and countries over food access has increased, depending on their current income levels and the welfare policies of governments. The bioeconomy may be an outlet for the whole world, especially if it contributes to repairing the disrupted supply chain in the agricultural sector.

2.2 Sustainable agro-bioeconomy: Basic concepts

Agriculture and bioeconomy are not opposites but complementary sectors, and to the extent that they support each other, one can speak of a sustainable agro-bioeconomy. This is needed to feed the growing world population in an orderly and equitable manner.

2.2.1 Sustainability

Sustainability, which has become popular with the sustainable development goal of the United Nations (UN), is the ability to remain productive for an indefinite period and is defined as the realization of current needs without directly or indirectly compromising the needs of future generations in terms of economic models (Vanhamaki et al., 2019).

[a] The "Black Swan" argument developed by the finance professor N. Taleb has also been the subject of a report of NASDAQ: WLTW (NASDAQ: WLTW (Willis Towers Watson), 2020), a global consulting firm. Accordingly, COVID-19 has three essential characteristics (conditions) of the "black swan": Unpredictable; great potential effective; late understanding of its importance after the event.

2.2.2 Agriculture 4.0

Agriculture refers to the cultivation of plants and animals for the benefit of humanity, to obtain products from them, to increase the quantity and quality of the products, to protect and market them under appropriate conditions. Agriculture, which is accepted as the art and science of tillage, crop and animal breeding; if it were not, there would not food, cotton, wool, leather, fabric, wood, paper, etc. there would be no products and people could not live.

As a result of the fourth phase of the industrial revolution, agricultural industrialization increased and "Agriculture 4.0" entered the agenda of modern economies with the label "green revolution" focused on innovations and healthy nutrition. Agriculture 4.0 is a term for the next big trends facing the industry, including a greater focus on precision agriculture, the internet of things (IoT) and the use of big data to drive greater business efficiencies in the face of rising populations and climate change. In 2018, the World Government Summit published their report called "*Agriculture 4. 0 – The Future of Farming Technology*," in collaboration with Oliver Wyman. According to the report, agriculture will be under the pressure of four main negative developments in the near future: increasing population, scarcity of natural resources, climate change and food waste (Proagrica, 2020). At this point, the bioeconomics will have a word to say and an action to do.

In the context of Agriculture 4.0, modern farms will primarily operate on sensors, robots, drones, GPS and information technology. These technologies will make agricultural businesses more profitable, efficient, safe and environmentally friendly. All fields no longer have to depend on water, fertilizers and pesticides, while farmers will use minimum quantities or even remove them from the supply chain. Even in arid areas, crops can be grown with the help of renewable energies and seawater (De Clercq et al., 2018).

2.2.3 Bioeconomy

Bioeconomy includes all kinds of industrial and economic sectors that produce, manage and distribute biological resources (e.g., agriculture, food, forestry, fisheries and other bio-based industry) to consumers for the purpose of high prosperity.

The "bioeconomy" has several related terms, such as "bio-based economy," "green economy" and "circular economy." There are clear synergies between especially the bioeconomy and circular economy concepts, such as cascading use of biomass (Biomonitor, 2020). This interaction is illustrated by the Venn diagram (see Fig. 1).

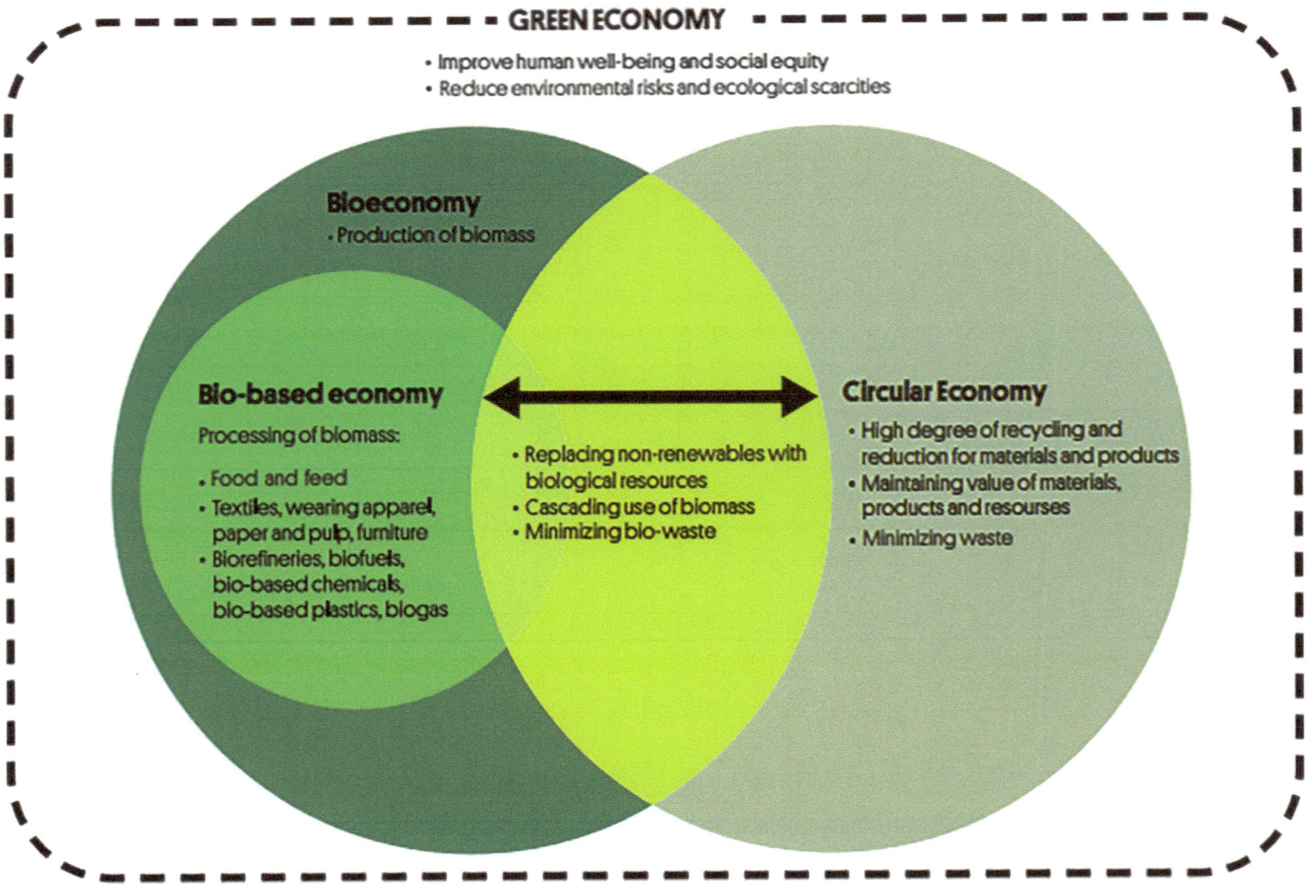

Fig. 1 VENN diagram of Bioeconomy, bio-based economy, green economy and circular economy (Kosir et al., 2021).

It is also emphasized that the bioeconomy is formed by the joint efforts of various technological platforms in the strategic document titled White Paper which describes the EU's (which Turkey is a candidate member) 2030 vision: A smart, sustainable and inclusive European Bioeconomy is targeted thanks to the social vision in sectors such as agriculture, animal husbandry, fishery, forestry, food, aquaculture, chemical industry, material production and energy (EC, 2008).

The key features of the three visions of bioeconomy can be identified. Table 1 focuses especially on implications in terms of overall aims and objectives, value creation, drivers and mediators of innovation, and spatial focus.

An increasing number of global, regional and national initiatives are working to define and achieve a sustainable bioeconomy. A total of 49 countries, including 23 countries from the Global South, already have a national policy or strategy on the bioeconomy as shown in Map 1 (UNOSSC, 2019).

2.2.4 Agro-bioeconomy

The Organisation for Economic Co-operation and Development (OECD, 2006) supposes the bioeconomy to be the aggregate set of economic operations in a society that use the latent value incumbent in biological products and processes to capture new growth and welfare benefits for citizens and nations: These benefits are manifest in product markets through productivity gains (agriculture, health), enhancement effects (health, nutrition) and substitution effects (environmental and industrial uses as well as energy).

According to OECD's (2006) visionary report focused on 2030, unsurprisingly, strategic interest is growing in the biosciences in both OECD and non-OECD countries. Indeed, the bioeconomy is growing faster in China, India and Singapore than in many OECD countries and the lead position occupied by OECD countries in most fields is being challenged.

Agro-bioeconomy basically aims to provide adequate and balanced nutrition for the rapidly growing world population. Agro-biotechnology, which is a component of bioeconomy, includes the following functions: Examining the cell, embryo, organ cultures, genetic structures of plants and animals; removing gene sequences; generating transgenic genotypes by direct gene transfer; finding microorganisms that produce enzymes that are used in a wide range of fields from the food and feed industry to the fertilizer, textile and detergent industry, especially in different fields of agriculture; genetically identifying them; practices to be carried out entirely in the laboratory environment such as genetically modified to increase efficiency.

Table 1 Key characteristics of the bioeconomy visions (Bugge et al., 2016).

	The bio-technology vision	The bio-resource vision	The bio-ecology vision
Aims and objectives	Economic growth and job creation	Economic growth and sustainability	Sustainability, biodiversity, conservation of ecosystems, avoiding soil degradation
Value creation	Application of biotechnology, commercialization of research and technology	Conversion and upgrading of bio-resources (process oriented)	Development of integrated production systems and high-quality products with territorial identity
Drivers and mediators of innovation	R&D, patents, technology transfer offices (TTOs), Research councils and funders (Science push, linear model)	Interdisciplinary, optimization of land use, include degraded land in the production of biofuels, use and availability of bio-resources, waste management, engineering, science and market (interactive and networked production mode)	Identification of favorable organic agro-ecological practices, ethics, risk, transdisciplinary sustainability, ecological interactions, re-use and recycling of waste, land use (circular and self-sustained production mode)
Spatial focus	Global clusters/central regions	Rural/peripheral regions	Rural/peripheral regions

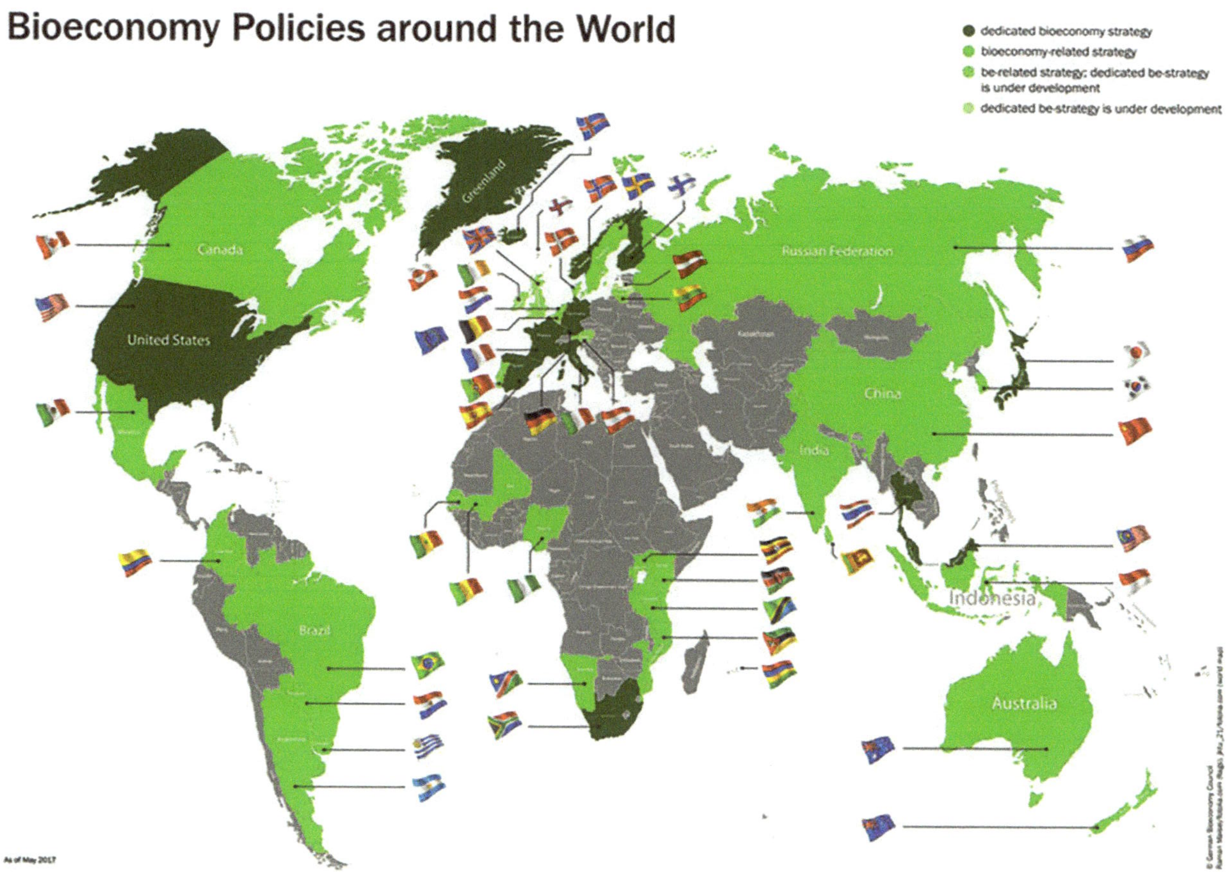

Map 1 Bioeconomy policies around the World (UNOSSC—United Nations Office for South-South Cooperation, 2019).

Wesseler (2015) lists the five main reasons that the bioeconomy beyond the agriculture sector should be taken seriously as follows: (i) Advances in biological sciences; (ii) An increase in horizontal and vertical integration in agricultural supply chains; (iii) An increase in inter- and intra-industry trade; (iv) Advances in information and communication technologies; (v) An increase in globalization. Later, Wesseler, Banse, and Ziberman (2015) added the challenges caused by climate change to this list. In addition, there is no harm in adding to these items the new development needs created by the Covid-19 pandemic in the context of bioeconomy.

As stated by the European Investment Bank (EIB, 2018), an important bank in the EU, both agriculture and bioeconomy sectors should invest in new business opportunities, adapt to challenges and innovate; therefore, especially long-term financing should be accessed. The EIB sustainably promotes the rural economy to benefit entrepreneurs and society as a whole.

2.2.5 Sustainable agricultural bioeconomy

The increasing need for food of the world population has brought the processing of renewable agricultural raw materials to the agenda. Because of this need, bioeconomy will soon contribute to sustainable prosperity by forcing the possibilities of recycling in agriculture as in industry.

According to Arnold's (2018) prediction, food and nutrition systems globally are facing a "perfect storm": (i) By 2050, 9 to 10 billion people will have to be fed, and it has been estimated that food production must increase by 60%. (ii) Today, there are more obese than undernourished people and yet 33% of food produced is wasted. (iii) Food systems in overall terms use 70% of freshwater resources, 30% of energy and produce 25% of GHG emissions worldwide. Also, as he stated again, while agriculture, food, fisheries and aquaculture represented 75% of Europe's bioeconomy turnover in 2013, these sectors employ 20% of the EU workforce and account for 6% of European GDP.

As calculated in a report of the OECD (2009); in 2003, the share of agricultural biotechnological R&D expenditures in total OECD enterprises was 4%; in 2030, the estimated potential share of the total biotechnology gross value added in the OECD region will increase to 36%. This suggests that sustainability for the agricultural bioeconomy is on the right track.

2.3 Relationship between the Covid-19 pandemic and sustainable agro-bioeconomy

The Covid-19 pandemic did not directly affect the agricultural sector since the virus is not transmitted from humans to plants and food animals,

However, the effects of this pandemic on the agricultural sector were indirect, due to the fact that farmers and workers had to stay home or work in a limited way. Agricultural industries were also affected to a certain extent by this negative development. Improvements in the agro-bioeconomic sector may have zigzagged according to countries or development regions due to the priority given to vaccination and other studies.

Yet regardless of the pandemic, the development of the bioeconomy is affected by biological and digital revolutions. As highlighted at the Global Bioeconomy Summit (2020), both revolutions have a wide range of applications in crop and livestock production and can therefore increase the productivity and environmental sustainability of producing bio-based resources for the bioeconomy. The extent to which such innovations are implemented depends on how societies perceive these options and what policies governments implement.

The Food and Agriculture Organization of the UN (FAO) points out that the world's food systems are deteriorating and this deterioration is accelerating due to the global pandemic(FAO, 2019). At this point of danger, the need for agro-bioeconomic governance consisting of social sensitivities and urgent government policies comes to the fore. The triple trivet of this governance is certainly based on less food waste, more efficient and affordable food production through biotech investments, and a guarantee of food access for future generations. A stable agro-bioeconomic governance that can be established in each country will able to problematize the national and/or global food urgency and solve it over time.

FAO (2020) demands USD 1.3 billion in initial investments that will provide an agile and coordinated global response to provide nutritious food for everyone during and after the pandemic. The "Covid -19 Response and Recovery Program," funded with this modest money, will enable donors to leverage the organization's collector power, real-time data, early warning systems and technical expertise for direct support where and when it is needed most. As FAO believes and expects, together, we can help protect the world's most vulnerable and poor people, prevent further crises, increase shock resilience, and accelerate the rebuilding and sustainable transformation of our food systems.

As stated by the International Labour Organization (ILO, 2020), Covid-19 deepens agricultural poverty and threatens food security. While many agricultural workers work selflessly to nourish the world, they are unable to free themselves from poverty and food insecurity. According to ILO's observations, as the pandemic spreads, the continued functioning of food supply chains is crucial in preventing a food crisis and reducing its negative impact on the global economy. Coordinated policy interventions are needed to support agricultural trade and the livelihoods and working conditions of millions of agricultural workers in line with relevant international labor standards. As ILO can accept, modern biotech production option instead of traditional production for more employment and less unemployment in agriculture can be a good and sustainable part of these policies.

3. Scenarios for agro-bioeconomy after Covid-19

There is a philosophical conflict in the world in the form of utopia and dystopia, and this conflict is reflected in the economy as well: optimists imagine that happy people (utopia), but pessimists imagine that unhappy people will increase (dystopia). It can be said that dystopia is more dominant in this cycle where the bad causes and consequences of climate change (famine, hunger, poverty, unemployment, destruction of nature, state armament, individual armament, terror, war, imperialism and fascism, etc.) are more common. Preventing dystopia from becoming a reality depends on strong and determined inter-communal and international cooperation to exert pressure on governments. As the global climate deteriorates, more precisely, human beings who change the climate must change their exploitation mentality against nature and their own kind. There is a dialectical relationship between human development and its widening ecological footprint (the burden we put on nature with our production and consumption activities), and if we cannot manage this relationship correctly, we will move away from utopia and approach a dystopia. Humanity, stuck between the automotive age and the internet age, seeks happiness in the shadow of carbon emissions and radiation emissions. What he finds are selfishness, violence, exploitation, inequality, injustice, economic and health crises institutionalized under the guidance of capitalism (Kalaycı, 2020).

There are many reasons for humanity to think more utopian against less dystopia in this post-Covid-19 era in which we have entered the age of agro-bioeconomy—as listed below on the sustainability axis.

3.1 Utopian scenarios for the world and Turkey

3.1.1 Scenario 1. The Covid-19 process has changed the working life and opened the door to increase in biotech value added and employment

During the global Covid-19 pandemic process, both the real sector and the financial sector switched to a flexible and partial working order due to mandatory quarantine and limited social life at home all over the world. This experience has shown that the daily working time can be reduced from 8 to 4 h, and part-time and homeoffice work can be done. It raises hopes that more time can be devoted to family and social relations that are longed for and a lower unemployment rate will be achieved, thanks to the new welfare-based wage and personnel regime to be established in cooperation with the unions-business-government. Socializing work life can become widespread. Meanwhile, as universities and research centers compete for biotech patents, these patents can create high added value along with the demand for new labor[b] in agricultural industrial factories.

According to the OECD (2019), there are high expectations that the bioeconomy can make a significant contribution to the sustainable development of the agro-food system through the creation of new business and innovation opportunities. Many bioeconomic strategies, which have the potential to promote more productive and sustainable production systems, consider they can make a significant contribution to the agricultural sector or primary production (agriculture and forestry) in general. Some strategies also refer to more "flexible" and "resource efficient" production systems (EU and France) or to "sustainable intensification" of biomass production (Sweden, Norway and Spain).

3.1.2 Scenario 2. Robotic biotech will enter the agriculture sector more and more for the food supply chain

There is a possibility that the world will enter Covid-n (Covid-21, Covid-22, and more years) processes after Covid-19, depending on corona virus mutations and again new types. Against this unwanted possibility, robot workers and/or robot farmers can be engaged more rapidly and widely in order to minimize the risk of contamination and maximize human health so that the reproduction-working economy is not locked in the modern world (especially in the developed West). With this development, preparations for industry 5.0 will be initiated while continuing industry 4.0. The transition from labor-intensive to capital- and knowledge-intensive technologies can occur smoothly for the robot-intensive technology phase. This transition will be related to the depth and prevalence of biotechnology in the agricultural sector, which cannot be kept independent of robot technology to some extent. Of course, it is debatable whether this development is a utopia or not, but it is inevitable for humanity. The values of morality and

[b] However, biotech corporations also have the potential to increase technological unemployment along with efficiency and profit.

justice will determine the goodness of this utopia. The breaking point here is that the agricultural supply chain, which will be strengthened with the help of biotechnology, demonstrates its power to eradicate global hunger.

3.1.3 Scenario 3. Contemporary countries have to focus on "green budgets" that protect nature and improve agriculture

The main sources of this could be to grow the production economy, reduce public waste and military budgets. Green budgeting means using fiscal policy tools of public expenditure and taxes to help achieve environmental and climate goals.

According to Cremins and Kevany (2018), who stated that green budgeting means using the budget system to promote and achieve improved environmental outcomes, the budget process is not neutral, but reflects long-standing societal choices regarding how resources are allocated.

As the OECD (2020) points out, these tools which can be used by each country, should be based on the existing public financial management (PFM) framework, but may include:

- Green budget labeling—Classifying budget measures according to environmental and/or climate impacts;
- Environmental impact assessments—Requiring environmental impact assessments to accompany new budgetary measures;
- Ecosystem services, including carbon, pricing—Setting a price on environmental externalities such as greenhouse gas emissions, usually through taxes and emissions trading systems, to facilitate achieving national environmental and climate targets;
- Green perspective on expenditure review—Taking into account efficiency issues as well as the impact of measures on national environmental and climate targets;
- Green perspective in performance determination—Integrating performance targets related to national environmental and climate targets.

3.1.4 Scenario 4. After the Covid-19 epidemic, there is a need for "global agro-bioeconomic diplomacy" on the basis of justice and security

Aguilar and Patermann (2020) say "Biodiplomacy is the new frontier of the bioeconomy." They define biodiplomacy as a new natural resource governance that addresses global challenges and is based on sustainable circular bioeconomies around the World. Hence, in addition to the authors, it can be that FAO and universities can lead agro-bioeconomic diplomacy with project outputs. Such diplomacy will be the most effective method against waste of agricultural resources such as military wars. It will also include the free transfer of food from wealthy countries to scarce countries. The hunger in Africa for world peace does not suit the West, which prides itself on its lavish wealth and selfish prosperity. When biotechnology creates the abundance of agricultural products, it will facilitate the systematic dedication that the rich West must make to poor Africa and other food-poor regions. If the West wants to do a good historical job, it must teach the poor countries how to fish, including the use of biotechnology methods, rather than feed them little by little.

3.1.5 Scenario 5. The foundations of "agro-bioeconomic citizenship" can be laid with good global governance

Thus, it can pave the way for every citizen to benefit from every product grown in the country. All the products planted and harvested in a country's territory essentially belong to the citizens of that country. Excess of those products may be the subject of foreign trade or international cooperation. Bioeconomy can contribute to the consolidation of national and global citizenship feelings for world peace and non-exploitative wealth sharing. Although imperial powers do not like it, it must be a pleasant utopia to expect this consciousness to turn into a global epidemic, just like Covid-19.

3.1.6 Scenario 6. Bioeconomy could play a potential role in the growth of the collective food sector based on planning and partnership

Because biotechnology, the technological pillar of bioeconomy, will increase productivity by reducing input costs and increasing output in agriculture. When the amount of food becomes plentiful, food prices will decrease and the purchasing power of poor families may increase while other variables are constant (ceteris paribus). Duncan et al. (2020), seeing collective food production as the key to the future, argue between optimism and pessimism, emphasizing that the future of food is uncertain. They propose six principles for regenerative food systems conceived in conjunction with the collective food sector: to accept and include various forms of knowing and being; looking after people, animals and the planet; going beyond capitalist approaches; commoning the food system; encouraging accountable innovation; and long-term planning and rural-urban relationships. As they say, regenerative approaches take into account not only farming systems, but also farm families, rural communities, landscapes and regions, and ecosystems.

3.1.7 Scenario 7. The world is entering an era that will reduce the need for child labor, especially in the agricultural sector, through biotechnology

There is a great need for child labor[c] in areas with dense rural population. Child labor in the agricultural sector will be limited to the extent that biotechnological practices make labor intensive production unnecessary. When children complete the age of play and education and become adults, the level of happiness of rural families can rise. As ILO (2021) points out, family farm culture and tradition can make it difficult to distinguish between children's participation in light work activities and child labor. Participating in some farm activities can give children the opportunity to develop skills and develop a sense of belonging to the community.

However, this becomes a problem when farm duties interfere with school education and are dangerous in nature. As highlighted in an FAO (2019) report, in order for the efficient use of bioeconomy which is the new techno-economic paradigm of production and consumption, to be safe, creative, feasible and applicable to all kinds of agriculture and rural situations, necessary political, economic and environmental conditions must be met. These conditions should also include the child-friendly element, because children are our future generations.

3.1.8 Scenario 8. Neoliberalism may not be a threat if bioeconomy is supra-ideologies

Neoliberalism, which started as the privatization and deregulation movement in the 1980s and became the dominant ideology of our age, has a cold face as it blesses money, leaves the broke people to die, and rejects land ownership reforms that favor poor farmers. Neoliberalism (or "Washington Consensus" in the 1990s), which Karagiannis describes as "immoral, theoretically and politically unhealthy, even barbarian liberalism" has institutionalized on waves of financialization and hyperglobalization that depend on flexible markets and free trade and capital flows (Karagiannis, 2020). So there is no other way but to wait for the bioeconomy to soften this bad trait of that ideology as it mediates the transport of vital products to poor homes, starting with agriculture. If limited and autocratic state regimes turn into an interventionist and protective democratic regime, the bioeconomy can get rid

[c] According to ILO data, 60% of all child workers (about 100 million) in the age group 5–17 worldwide work in agriculture, including farming, fishing, aquaculture, forestry and animal husbandry. The majority of child workers (67.5%) are unpaid family members. Approximately 59% of all children aged 5–17 working in hazardous jobs are in the agriculture sector (ILO, 2021).

of neoliberalism and the whole world can meet both a satisfying and friendly agricultural economy.

3.1.9 Scenario 9. If every country opens its agricultural bioeconomy sector to tourism, both agriculture and tourism sectors and therefore the whole world will win

This can be called a "win-win game." Agro-bioeconomic tourism [d] means accommodation in a farm, camp or hostel associated with the rural environment, crops and accommodation, direct participation in biotechnology-supported land and livestock production, educational trips, food and beverage, sale of products and handicrafts on the farm and other rational (homoeconomicus) is the totality of activities. It is clear that biotech farms will be preferred in this tourism. The aim here is to create awareness and sensitivity in agro-biotech production and consumption through tourism on the one hand, and to make the strong interaction between tourism and agriculture sustainable on the other hand.

3.1.10 Scenario 10. Provided that not only the new coronavirus vaccine is invented, but all people are vaccinated in equality and justice, both the direct-overall socio-economic life and the private-indirect agricultural bio-economics sector will be relieved from the crisis

The Covid-19 vaccine was invented under different brands in many advanced multinational R&D laboratories around the world, mass production and distribution began. The persistent vaccine campaigns of the governments have made these vaccine brands a major global market. Vaccine manufacturers are making high profits in an oligopolistic structure for now. Thanks to the vaccination of people and the reduction of the risk of disease transmission, the return to the old welfare levels will accelerate in all countries and the agro-bioeconomy sector will also be able to make new and strong investments. This sector may confirm greater positive results from vaccines to the extent that it facilitates people to get enough and healthy food. An agro-bioeconomic sector that works efficiently in accordance with the food supply chain and embraces all humanity may also disrupt the possible "vaccine imperialism" games.

[d] According to Buonaccorsi (2020), pure-traditional agro-tourism activities represent a fundamental integration of income and include: (a) the direct sale of a farm's agricultural products in fenced markets; (b) tastings; (c) wine tourism; (d) cooking lessons; (e) educational farms for schoolchildren; (f) pet therapy; and (g) religious ceremonies (weddings, baptisms, congregations, approvals).

3.2 Dystopic scenarios for the world and Turkey

3.2.1 Scenario 11. Governments can introduce new corona tax in every country affected by the Covid-19 pandemic

This new tax may not be regulated directly under the name of the "corona tax," but it can be applied by hiding existing taxes or by hiding them in other types of taxes to compensate for astronomical public health expenses incurred during the epidemic. This can be called "near-corona tax." The agricultural sector may also meet new taxes that could make production and consumption expensive. The corona tax can act as an effective financial weapon for governments looking to fund their increased epidemic spending. As Vernengo and Nabar-Bhaduri (2020) say, if economies deteriorate, calls for political resistance and austerity at varying levels increase in each country.

3.2.2 Scenario 12. Migration waves may increase from places where agricultural production and food are scarce to places where there is plenty

Cities will become crowded due to internal and external migration and population composition in those cities may deteriorate. Migrations can change direction and shape. In the past, migrations from rural areas to cities may now occur from cities to rural areas. If bioeconomic practices become widespread and create new job opportunities in agriculture and other subsectors, immigrant people in search of employment may be curbed.

3.2.3 Scenario 13. "Pandession!" If the pandemic spreads over many years, agricultural production will decrease and consumption will become expensive

The pandemic has triggered economic crises in many countries. Even strong economies recorded negative economic growth (recession) in "V" or "U" shape at the beginning of the pandemic, but then gradually began to enter into a process of positive growth (recovery) in "n" or "L" shape. Since the World Economic Crisis of 1929, no crisis has emerged other than Covid-19, which has made economies such deep and unstable cyclical fluctuations. Also in developing countries, including Turkey, economic growth again V (W) may take shape. If negative growth rates continue for a long time, especially in the agro-food industry and subsidiary industries, the economies will expect depression beyond recession. This means more impoverishment and stress. This special and dramatic conjuncture where pandemic and recession unite is called "Pandession" in the special words of Chouhan (2021). If adequate and effective policies cannot be put in place, even bioeconomic techniques may not be able to reduce the pandession in the agricultural sector.

3.2.4 Scenario 14. The pandemic is reflected in negative externalities such as environmental pollution

This appears to be the revenge of nature on the people who exploited nature and made the world a sphere of inequalities and injustices. People have been consuming masks and cleaning products more than ever since the pandemic started. Unfortunately, these products turn into mountains of garbage, and some of them are mixed into agricultural soils and fresh water basins in sizes that threaten our future. While these protection products provide positive benefits when used correctly, they cause harm when they pollute the environment excessively. If biotechnology is not transformed into a powerful and sustainable mechanism that mediates the recycling of these products, the net benefit for humanity could decrease (net harm could increase).

3.2.5 Scenario 15. Asymmetric information and alternative costs increase with the pandemic

It has been argued that some developing countries play with the positive number of cases and deaths, or raise doubts about not disclosing these numbers transparently, in order to get more loans and grants from the World Bank for the Covid-19 protocol (test, treatment and vaccine budget). When transparent statistics began to be published, it was observed that the number of cases was several times higher than before. This process indicates the formation of asymmetric information. Due to the wrong and bad expenditures within the scope of the fight against the epidemic, public costs are inflating and increasing budget deficits. On the other hand, some states do not stop military armament for various reasons. As the expenditures allocated from the budget to armament increase, spending to be allocated to agro-bioeconomic researches may decrease unless new sources of financing are found. This is an example of alternative costs that should be considered, as it reminds us of the fact that the world is armed. Asymmetric information and alternative costs will be perceived as dystopian as long as they disturb societies in terms of quantity and quality.

3.2.6 Scenario 16. Climate change called global warming (for some, global cooling) continues to disrupt the agricultural supply chain

Because the climate change problem will continue to exist after Covid-19.[e] Because of every +1 degree of warming in the climate, water resources evaporate and agricultural lands are irrigated less.[f] Thus, access to food products becomes difficult as agricultural production shrinks. As food products will become expensive due to the scarcity of production, demand inflation will also increase, although it will vary from country to country. In the near future, global warming may cause hyperinflation due to high food prices. We must rely on biotechnology's solutions to global warming.

3.2.7 Scenario 17. In the established global order, most natural resource-rich countries will remain poor

The theoretical source of this is the phenomenon called the "resource curse" or "abundance paradox." This paradox means that countries (regions) with abundant natural resources (particularly non-renewable energies such as oil and gas) tend to have less economic growth and worse development outcomes than countries with less of these resources. Natural resources are divided into two main categories: point resources (oil, gas, minerals, metals) and common resources (agriculture, forest). According to scientific tests carried out by Zagozina (2014), the former Soviet economies (Russia, Azerbaijan, Kazakhstan, Turkmenistan), which are highly dependent on natural

[e] As van der Ven and Sun (2021) recall, while COVID-19 and climate change are different types of crises, they are linked in several ways: Warming temperatures and deforestation create increased disease vectors that could lead to the next global pandemic. Attention to the climate crisis will play a key role in avoiding future pandemics and increasing agricultural bioeconomic production. More existential support from governments and citizens should be sought to highlight the growing urgency, visibility, proximity and universality of the climate crisis. Moreover, as UNOSSC (UNOSSC—United Nations Office for South-South Cooperation, 2019) emphasizes, relying solely on North–South development cooperation models will not be enough for developing countries to achieve the Paris Agreement and the bold sustainable development goals (SDGs). As developed countries (North) reaffirm their commitments to support developing countries (South) under the Paris Agreement and to achieve the 2030 Agenda, the importance and potential of new partnerships with and between developing countries through the South–South are increasingly recognized.

[f] Drought conditions in a country occur after little rainfall for several seasons. As many of the world's geography, Turkey is also in trouble with the drought. Satellite data from NASA (2021) documents this situation. The scenarios he drew up for the 21st century are not very pleasant. Having very large and fertile lands can only be meaningful thanks to high water reserves. How biotechnology will compensate for the deficiency in water resources is another important issue.

resources, are quite tends to develop slowly. However, a significant positive effect of point and diffuse resource exports on economic growth has been found. As long as this phenomenon, known as the "Dutch disease"[g] in the history, continues its existence in some underdeveloped or developing economies of Middle East and South America, exploitation, conflict and violence will not be missing in the world. Hopefully, agro-bioeconomic policies and practices will help neutralize this paradox. In this case, it will be understood that abundant natural resources and processed agro-biotech products are not a curse but a blessing, as proved by successful countries such as North America, Australia and Norway.

3.2.8 Scenario 18. The globesite problem frightens humanity, moving from farm to fork

The Covid-19 pandemic has widened the distance in social relations; friendships are now measured by asocial distances and Japanese type greetings are becoming common. But people are finding it difficult to put the same distance between themselves and their food. Obesity is the main health problem of the USA and the developed West which are especially welfare selfish, and is the result of malnutrition. Therefore, the world is faced with a globesity reality. As Arabska (2021) points out, a "Farm to Fork Strategy" has been designed in the EU, which can be considered together with the Common Agricultural Policy (CAP). This strategy, which aims to create a fair, healthy and environment friendly food system with an integrated food safety policy, can be relatively effective in combating globesity within the scope of bioeconomics. But it seems that the task of food production that will not make people obese will belong to agro-biotechnology.

3.2.9 Scenario 19. A new but creepy branch of science that is called "Corona+Economics" is developing

The economic consequences of the Covid-19 pandemic and the global size and depth of the financial and social damage associated with it cannot be

[g] As long as this phenomenon, known as the "Dutch disease" in the history, continues its existence in some underdeveloped or developing economies of Middle East and South America, exploitation, conflict and violence will not be missing in the world. Hopefully, agricultural bioeconomic policies and practices will help neutralize this paradox. In this case, it will be understood that abundant natural resources and processed agricultural biotech products are not a curse but a blessing, as proved by successful countries such as North America, Australia and Norway. (See also (Brahmbhatt et al., 2010; Dadgar and Orooji, 2020).)

denied. As Balogu (2020) strongly reminds, an unpleasant hybrid concept called "Coronomy" has begun to enter the agenda of the world. A. De Alwis derived "corona" from the combination of the terms "corona" and "economy," which describes the study of the effects of the corona virus on the economy. While humanity is waiting for the virus to be completely defeated with the invention of the vaccine, the effects of the virus on the prominent sectors of the economy will be constantly examined with "Coronomy" and the approach to reduce possible risks will be emphasized. But "Coronomics" appears to be an artificial branch of science; It should not be glorified enough to prevent growth economy which is still functional in humanity's radical struggle especially with agricultural poverty. Here, the field where producers and policy makers will work synergistically should be the agro-bioeconomy worthy of a utopia, not coronomy that evokes a dystopia.

3.3 Validity and sustainability of the scenarios

Here, it is time to describe the validity and sustainability conditions of utopian and dystopian scenarios that have been set up in terms of the global Covid-19 pandemic and the agro-bioeconomic future. But for this, first a summary statistics of the current food order of the world should be included.

3.3.1 Selected data of the world's current food order

Covid-19 once again demonstrated all kinds of inequality and injustice in the world. Inequality and injustice in access to food is the most chronic and dramatic. While this pandemic allowed the rich class to get richer, it reinforced the reasons that led the poor class to become poorer and even some governments to become autocratic. Regardless, democratic societies and governments cannot be desperate for the establishment of a sustainable new world order on the axis of equality and justice in agriculture. Every agro-bioeconomic project that will be put into practice will keep the hopes alive. El-Chichakli et al. (2016) talk about the "difficult balance" when they set up the coordination principles between bioeconomy and sustainable development goals and bring up the R&D projects on bioeconomic cities.

According to Oxfam (Oxfam International, 2021), almost a billion people go to bed hungry every night, not because there is not enough food for everyone, but because of the profound injustice in access to food. As can be seen in Table 2, attention can be drawn to some numerical data that require urgent establishment of a sustainable global food order.

Table 2 Some data that requires the establishment of a sustainable global food order (schematized by the author based on data from Oxfam International, 2021).

821 M	821 million people around the world face chronic food deprivation. This is a return to levels from almost a decade ago.
20 M	In 2017, over 20 million people were on the brink of starvation and in need of humanitarian assistance in East Africa due to the impact of climate change.
5 %	For products like Ecuadorian bananas or Indian tea, less than 5% of the price paid by consumers in Europe and the US reach small-scale farmers.
100 M	The World Food Program estimates that giving women farmers more resources could reduce the number of hungry people in the world by 100-150 million.
1 / 5	Indigenous Peoples and local communities legally own just one-fifth of the lands they manage collectively and have protected for centuries.

Each utopia and dystopia has a period of validity and a sustainable nature (see Table 3). The better a utopia is logically viable and sustainable, the worse a dystopia is valid and sustainable. Because in the first case there is the promise of prosperity and happiness, in the second there is chaos and crisis. If a dystopia can sustain, we will have justified reasons to be pessimistic about the future of the world.

4. Conclusion

The Covid-19 is a global problem, but the measures taken against this epidemic are generally national. This reality also applies to agricultural supply and demand and other markets affected by the epidemic. If we are to fear as humanity, we should not fear the virus, but the lack of food production, its unbalanced distribution, class hunger, and the breakdown and disruption of the supply chain. Food shortages and unfair distribution on earth are worse and scary than the transmission effect of the virus.

In fact Covid-19 has revealed just how weak developed Western economies (United States, United Kingdom, EU, Japan, etc.) can be in the face of an epidemic, and just how underdeveloped (most of Africa) and emerging economies (BRICS, MIST,[h] etc.) may also have to delay their development hopes.

All industrializing countries (including Turkey) have to take advantage of the advanced science and technology base of the industrialized countries

[h] BRICS: Brazil, Russia, India, China, South Africa. MIST: Mexico, Indonesia, South Korea, Turkey.

Table 3 Validity and sustainability conditions of utopias and dystopias after Covid-19.

		Subject/concept	Validity	Sustainability
Utopias	1	"Sufficient added value and employment increase in agriculture"	Biotechnological methods should be used for a significant part of agricultural production	Biotech agricultural production must be in line with sustainable development goals (SDGs)
	2	"Biotechnology for the food supply chain"	Biotech robots are needed against the cessation of agricultural production due to the high risk of virus transmission	The food supply chain should be strengthened with biotech production to ensure that people are adequately fed
	3	"Green budgets"	Central and local governments should put spending and tax items in their main budgets to protect the ecosystem	Savings in military spending should be transferred to green budgets for the continuity of agricultural production
	4	"Bioeconomic diplomacy"	Biodiplomacy should have a holistic approach with a socioeconomic focus for the effective management of agricultural resources	Bioeconomic goals must be humane by integrating with the SDGs; there must be hope for food-poor Africa and inspire all humanity
	5	"Agro-bioeconomic citizenship"	According to universal citizenship law, every person is the real owner of her/his land and products within the scope of bioeconomy	Biotech agricultural products in a country, after satisfying domestic consumption demand, should also be produced for international trade and solidarity in the context of world citizenship and global peace
	6	"Collective food sector"	Biotechnological planning will increase potential efficiency in collective agricultural production	Biotech agricultural products can be consumed cheaper thanks to the increase in productivity
	7	"Less child labor"	As biotechnology enters the agricultural sector, the need for child labor will decrease	Biotechnology will change labor-intensive production with child labor in agriculture toward knowledge-intensive production

	8	"Neoliberalism is no longer a threat"	The exploitative side of neoliberalism can be tampered with by using biotechnology for the benefit of human beings	Neoliberalism will be left out thanks to the policies of the governments to protect the rural-agricultural population and the biotech production model
	9	"Agro-tourism"	Agricultural regions where biotechnological production is carried out should be opened to a new generation of tourism	Agricultural biotech destinations will become vibrant and attractive with proactive tourism policies
	10	"Fair vaccine for all"	The vaccine against Covid-19 and its mutants must be invented and the vaccine supply must be able to meet the demand	Each country should include especially the producers' agricultural population in the vaccination schedule and implement their protective protocols without loosening
Dystopias	11	"Corona tax"	Governments can impose such additional taxes to finance increased medical spending due to the pandemic	These types of taxes are unpleasant; objections will begin if they are reflected in the poor class and made permanent; tax costs of the agro-bioeconomic sector will also increase
	12	"Reverse migration waves"	Migration from areas where agricultural poverty has increased to more ideal regions and countries may increase	Migration from industrial cities to rural areas could get out of control and the interregional development scissors could open further
	13	"Pandession"	When the pandemic dragged economies into recession, macro crises occurred; from now on, historical depressions will be giving way to pandessions	As the pandemic intensifies, more fiscal policy will be mandatory for governments and more monetary policy for central banks; All of these policies will have various side effects; disruption of agro-bioeconomic production is also possible
	14	"Negative externalities"	As a result of excessive industrial production and consumption from the past to the future, environmental	The pollution of soil and water resources and the concern that this danger will

Continued

Table 3 Validity and sustainability conditions of utopias and dystopias after Covid-19—cont'd

		Subject/concept	Validity	Sustainability
			pollution increases and harms the agricultural sector	never end endangers even agro-biotech production
	15	"Asymmetric information and alternative costs"	Not all parties could have the same information during the pandemic process. Incentives for agriculture sector decreased and investment costs increased due to wrong and bad government spending	While medical expenses increased in the fight against the pandemic, agricultural incentives and biotech R&D investments decreased. The unsustainability of this negative situation is obvious
	16	"Climate change"	Global warming evaporates water reserves and leaves agricultural lands to an arid climate, which will reduce welfare	It is alarming that biotechnologies are limited in what they can do to address the growing dangers of global warming.
	17	"The paradox or curse of abundance"	A country that cannot hold its current rich natural resources or the new resource reserves it discovers is actually doomed to remain poor	When the existing or potential rich natural resources in that country are not transferred to the whole economy through agriculture with the support of biotech, that country will fall into poverty trap
	18	"Globesity"	Obesity has turned into a serious global health and economic problem and seems to grow further	One pillar of obesity, at least as dangerous as the virus pandemic, is based on the extreme prosperity selfishness of the West's farm-to-fork strategy
	19	"Coronomy"	This hybrid branch of science emerged when economic research was also conducted on the corona virus pandemic	Even though it looks frightening, this branch of science will survive as long as the effects of Covid-19 continue; may even be called "coro-bio-nomy" in the future

and develop their national R&D strategies in order not to be afraid of exclusion from global biotechnology. Whether or not Covid-19 ends, globalizing economies will increase the degree of competition in agro-biotechnology this time. A proverb says; "Someone eats, someone looks, doom will break from him." All the lands of the world are common, and technologies used in agricultural production should also be common. Humanist competition is essential for greater partnership and sustainable cooperation to feed the world and not starve.

The global Covid-19 pandemic is neither like a hot war nor a cold war, but more suited to the warm war process. By saying that this pandemic is the result of distorted globalization led by hegemonic powers, we can never have mitigated the task of the agro-bioeconomy. Sustainability goals await achievement. Long after the pandemic, whether or not the utopia of a "World Government" will be realized, set in Isaac Asimov's science fiction novels on the war between galaxies and planets, or in Bertrand Russell's philosophical book "The Future of Industrial Society" is unknown. But it should be well known that sustainable agro-bioeconomy is aimed at all living things and must be humanized; otherwise, dystopia is inevitable.

References

Aguilar, A., Patermann, C., 2020. Biodiplomacy, the new frontier for bioeconomy. New Biotechnol. 59, 20–25. https://doi.org/10.1016/j.nbt.2020.07.001.

Arabska, E., 2021. From farm to fork: human health and well-being through sustainable agri-food systems. J. Life Econ. 8 (1), 11–27. https://doi.org/10.15637/jlecon.8.1.02.

Arnold, T., 2018. Bioeconomy, food and nutrition security. In: Bergeret, P., et al. (Eds.), Bioeconomy Challenges and Implementation: the European Research Organisations' Perspective. European Agricultural Research Initiative (EURAGRI), Versailles Cedex, pp. 11–22. https://www.euragri.aau.dk/digitalAssets/388/388000_chapter-1-bioeconomy- -food-and-nutrition-security- - -thomas-arnold.pdf. (Chapter 1).

Balogu, I.E., 2020. Coronomics and Global Economy: An Overview of the Impacts of Coronavirus on the Global Economy. SFC. https://dlc.dlib.indiana.edu/dlc/bitstream/handle/10535/10731/Coronomics%20and%20Global%20Economy.pdf?sequence=1&isAllowed=y.

Biomonitor, 2020. D1.1: Framework for Measuring the Size and Development of the Bioeconomy, Date of Document—09/2019 (10/1). European Union's Horizon 2020 research and innovation programme under grant agreement N° 773297 https://ec.europa.eu/research/participants/documents/downloadPublic?documentIds=080166e5ccddd248&appId=PPGMS.

Brahmbhatt, M., Canuto, O., Vostroknutova, E., 2010. Dealing with Dutch Disease. The World bank-Economic Premise. June, No: 16 https://openknowledge.worldbank.org/handle/10986/10174.

Bugge, M.M., Hansen, T., Klitkou, A., 2016. What is the bioeconomy? A review of the literature. Sustainability 8, 691. https://doi.org/10.3390/su8070691.

Buonaccorsi, A., 2020. Agritourism in Crisis During COVID-19: Italian Farms' Resilience and Entrepreneurial Strategies to face the Impacts of the Pandemic., https://doi.org/10.13128/bae-9554.

Chouhan, Y., 2021. Pandession: corona virus pandemic leading to the biggest economic crises since the great recession. Psychol. Educ. 58 (2), 803–812. https://doi.org/10.17762/pae.v58i2.1913.

Cremins, A., Kevany, K., 2018. An Introduction to the Implementation of Green Budgeting in Ireland. Staff Paper, Dep. of Public Expenditure and Reform Climate Change Unit, Dec https://igees.gov.ie/wp-content/uploads/2019/01/The-Implementation-of-Green-Budgeting-in-Ireland.pdf.

Dadgar, Y., Orooji, Z., 2020. Dutch disease, rentier state, and resource curse: a characteristic triangle and ultra challenge in the Iranian economy. Iran. Econ. Rev. 24 (1), 129–157. https://doi.org/10.22059/IER.2020.74477.

De Clercq, M., Vats, A., Biel, A., 2018. Agriculture 4.0: The Future of Farming Technology. https://www.oliverwyman.com/content/dam/oliver-wyman/v2/publications/2018/February/Oliver-Wyman-Agriculture-4.0.pdf.

Duncan, J.A.B., Carolan, M., Wiskerke, J.S.C., 2020. Regenerating food systems: a social-ecological approach. In: Routledge Handbook of Sustainable and Regenerative Food Systems. Routledge, pp. 1–11, https://doi.org/10.4324/9780429466823.

EC, 2008. The European bioeconomy in 2030: delivering sustainable growth by addressing the grand societal challenges. In: European Commission's Seventh Framework Programme (FP7). https://www.bioeconomy-library.eu/wp-content/uploads/2019/11/the_european_bioeconomy_brochure.pdf.

EIB—European Investment Bank, 2018. Agriculture and Bioeconomy: Unlocking Production Potential in a Sustainable and Resource-Efficient Way. EIB—European Investment Bank, Luxembourg. https://www.eib.org/attachments/thematic/agriculture_and_bioeconomy_en.pdf.

El-Chichakli, B., Braun, J., et al., 2016. Five cornerstones of a global bioeconomy. Nature 533 (14), 221–223. https://doi.org/10.1038/535221a.

FAO, 2019. The Outlook for Agriculture and Rural Development in the Americas: A Perspective on Latin America and the Caribbean 2019–2020. FAO, San Jose. http://repositorio.iica.int/bitstream/handle/11324/8214/BVE19040295i.pdf.

FAO, 2020. COVID-19 Responseand Recovery Programme. http://www.fao.org/3/cb0439en/CB0439EN.pdf.

Global Bioeconomy Summit, 2020. Track 1: Science & Innovation, How Bioeconomy is Transforming Agriculture. https://gbs2020.net/track1_agriculture/.

ILO, 2020. COVID-19 and the Impact on Agriculture and Food Security. https://www.ilo.org/sector/Resources/publications/WCMS_742023/lang- -en/index.htm.

ILO, 2021. Child Labour in Agriculture. https://www.ilo.org/ipec/areas/Agriculture/lang--en/index.htm.

Kalaycı, İ., 2020. Global indices and the economic-political future. In: Kalaycı, İ. (Ed.), Turkey's Economy: Theory, History and Policy. Divan Book, (Küresel Endeksler ve Ekonomi-Politik Gelecek. iç. Türkiye Ekonomisi: Kuram-Tarih-Politika), İstanbul, ISBN: 978-625-7263-00-9, pp. 837–880 (Chapter 23).

Karagiannis, N., 2020. Neoliberalism must die because it does not serve humanity. Real-World Econ. Rev. (94), 27–31. http://www.paecon.net/PAEReview/issue94/whole94.pdf.

Kosir, L.G., Sakellaris, G., Zimniewska, M., Varnik, R. 2021. BIOEAST Foresight Exercise: Sustainable Bioeconomies towards 2050. Ministry of Agriculture of Hungary Doi:10.13140/RG.2.2.31865.19049.

NASA, 2021. Turkey Experiences Intense Drought. https://earthobservatory.nasa.gov/images/147811/turkey-experiences-intense-drought.

NASDAQ: WLTW (Willis Towers Watson), 2020. Our Scenarios. 30 March. www.willistowerswatson.com›Insights›2020/03›Sc.

OECD, 2006. The Bioeconomy to 2030: Designing a Policy Agenda. OECD, Paris. http://www.oecd.org/science/emerging-tech/34823102.pdf.

OECD, 2009. The Bioeconomy to 2030: Designing a Policy Agenda: Main Findings and Policy Conclusions. OECD International Futures Project, Paris, https://doi.org/10.1787/9789264056886-en.

OECD, 2019. Joint Working Party on Agriculture and the Environment: Bioeconomy and the Sustainability of the Agriculture and food System: Opportunities and Polcy Challenges. COM/TAD/CA/ENV/EPOC (2018) 15/FIN, June 24, Paris https://www.oecd.org/officialdocuments/publicdisplaydocumentpdf/?cote=COM/TAD/CA/ENV/EPOC(2018)15/FINAL&docLanguage=En.

OECD, 2020. Collaborative on Green Budgeting Green Budgeting Framework—Highlihts. OECD, Paris. http://www.oecd.org/environment/green-budgeting/OECD-Green-Budgeting-Framework-Highlights.pdf.

Oxfam International, 2021. Food, Climate and Natural Resources. https://www.oxfam.org/en/what-we-do/issues/food-climate-and-natural-resources.

Proagrica, 2020. What is Agriculture 4.0? https://proagrica.com/news/what-is-agriculture-4-0/.

UNOSSC—United Nations Office for South-South Cooperation, 2019. South-South and Triangular Cooperation on the Bioeconomy: In Light of the Paris Agreement and the 2030 Agenda for Sustainable Development. https://www.unsouthsouth.org/wp-content/uploads/2019/12/Bioeconomy-Publication_visualization-for-website.pdf.

van der Ven, H., Sun, Y., 2021. Varieties of crises: comparing the politics of COVID-19 and climate change. Glob. Environ. Polit. 21 (1). https://doi.org/10.1162/glep_a_00590.

Vanhamaki, S., Medkova, K., Malamakis, A., Kontogianni, S., 2019. Bio-based circular economy in European national and regional strategies. Int. J. Sustain. Dev. Plan. 14 (1), 31–43. https://doi.org/10.2495/SDP-V14-N1-31-43.

Vernengo, M., Nabar-Bhaduri, S., 2020. The economic consequences of Covid-19: the great shutdown and the rethinking of economic policy. Int. J. Polit. Econ. 49 (4), 265–277. https://doi.org/10.1080/08911916.2020.1857589.

Wesseler, J.H., 2015. Agriculture in the Bioeconomy: Economics and Policies. Wageningen University. https://www.wur.nl/upload_mm/6/2/2/bd6f22fa-d0ea-4c96-b4c0-b25060553443_oration%20bookletJustus_Wesseler.pdf.

Wesseler, J.H., Banse, M., Zilberman, D., 2015. Introduction special issue: the political economy of the bioeconomy. German J. Agric. Econ. 64 (4), 209–211. http://www.gjae-online.de/news/pdfstamps/freeoutputs/GJAE-815_2015.pdf.

Zagozina, M., 2014. The Resource Curse Paradox: Natural Resources and Economic Development in the Former Soviet Countries (Thesis submitted for a M.Sc. degree in forest sciences and business). University of Helsinki, Dep. of Forest Sciences. https://core.ac.uk/download/pdf/33733135.pdf.

CHAPTER 2

Organic systems under major changes—A deep crisis in the management of the organic world of our planet—New tasks for foresight: An opinion paper

Emmanuel G. Koukios and Lazaros S. Karaoglanoglou
Research Group BIOTOPOS, Laboratory of Organic Technologies, School of Chemical Engineering, National Technical University of Athens, Athens, Greece

1. Introduction and approach

An early challenge regarding the assessment of organic systems has to do with their appropriate definition. In order to address the particular point, we have compiled the following list of definitions provided by a large number of dictionaries and other available sources. Therefore, according to this list, we can call "organic" something

- relating to or derived from living matter,
- referring to food produced without the use of chemical inputs and other artificial chemicals,
- referring to farming methods without the use of chemical inputs and artificial chemicals,
- being an organic chemical compound,
- being or coming from living plants and animals,
- characteristic of living plants and animals,
- relating to animal or plant constituents or products having a carbon basis,
- describes things that are natural or related to nature,
- in common usage, meaning "healthful" or "close to nature,"
- consisting of different parts that all fit together well,
- happening or developing in a natural and continuous process,
- connected or pertaining to the body's organs,

Agricultural Bioeconomy
https://doi.org/10.1016/B978-0-323-90569-5.00008-1

- serving as a means or instrument (from Latin *organicus* and from Greek *organon*),
- from the Greek *organon* coming from the root ERG, "work," with the linked meaning of implement and instrument,
- as well as, used in various metaphoric expressions involving the above listed definitions.

Another key challenge regarding the assessment of the recent growing interest in the large spectrum of organic phenomena has to do with their web-o-metric analysis. According to the first findings from such an analysis presented in Table 1, in the last 5 years the presence of the term "organic" in the web has almost doubled, whereas that of the term "inorganic" has been clearly reduced.

We should also note from Table 1 that the absolute and relative growth of "organic" is only in part due to the growth of the web appearance of "chemistry-" and "biology-" related terms.

A working hypothesis opening the way toward a useful map of the Organic Systems, based on the multiple Organic—Chemical—Biological—Other interactions, is the one provided by the list of Table 2.

The object of this chapter will be to assess the Global dynamics of major changes emerging in several key Organic Systems, listed in Table 3. The focus of the analysis to be summarized in Table 4 will be on growth barriers and other obstacles, which can in turn offer opportunities for appropriate foresight exercises.

Table 1 Relative importance and growth of the organic contents of the World Wide Web—Results of two Web searches: 2016 and 2021.

	Web search May 9, 2016		Web search May 18, 2021	
Web presence	**Count—Millions**	**Relative value**	**Count—Millions**	**Relative value**
Organic	484	100	847	100
Organic chemistry	87.5	18	280	33
Inorganic	43.1	9	55.1	6.5
Inorganic chemistry	14.1	3	13.9	1.6
Chemistry	327	66	518	61
Biology	281	58	481	57
Mathematics	242	50	368	44

Table 2 Main components of organic systems.

• Organic carbon (sp^3 hybridized carbon atoms)
• Organic molecules
• Organic macro-molecules
• Inorganic molecules and macro-molecules required for the organic systems operation
• Cell components
• Cells
• Tissues
• Organs
• Organisms (plants, animals, microbes)
• Communities of organisms
• Eco-systems
• Local and regional organic systems
• Global organic systems

Table 3 Main types of organic systems to be considered.

• Soil
• Plant farming
• Animal farming
• Agro-industries
• Forests
• Forest industries
• Fishing
• Aquaculture
• Energy
• Food
• Organic products
• Biotechnology
• Healthcare
• Quality of life
• Environment
• New services

2. Results and discussion

Table 4 Major change factors of global organic systems.

Organic system	Key feature	Growth barriers
Soil (Panagos et al., 2015; Eswaran et al., 2001; Egidi et al., 2021)	Organic capital reservoir	Erosion Fertility losses Toxicity Desertification
Plant farming (Levine, 2012; Smith et al., 2014; Wysokiński et al., 2020; Nair et al., 2013; Adejumo and Adebiyi, 2020)	Organic photosynthetic machines	Intensive use of inputs and nutrients Wasting energy and water Use of monocultures dependence upon plant protection and seed sources Unutilized residues and wastes
Animal farming (Committee on Considerations for the Future of Animal Science Research, 2015)	Organic flows upgrade	Low feed conversion rates to proteins (<1/10) Problems with feed supply and animal health Unutilized wastes—Sources of greenhouse gases
Agro-industries (Gontard et al., 2018)	Organic flows converters	Over-consumption of water and energy Unutilized wastes Organic pollution of aqueous reservoirs Limited developmental roles and socio-economic values
Forests and forest industries (Forest Europe, 2021)	Organic photosynthetic carbon storage	Great market pressure for consumption/land use change Acid rain—Forest fires Abandoning non-profitable forests Downgrade of developmental roles of long-term views
Fish farming (POLITICO, 2008; Joint Report AFD,	Aquaculture and fishing as organic process stars	Intensive use of inputs, feeds and drugs Accumulation of metals and

Table 4 Major change factors of global organic systems—cont'd

Organic system	Key feature	Growth barriers
EC, GIZ, 2017; European Commission, 2015)		other toxic elements Genetic "interventions" Land use conflicts, e.g., in coasts and touristic areas
Energy (Diamantopoulou et al., 2011, 2016; Dincer and Rayash, 2020)	Organic strategic toolbox	Dependence upon fossil C Pollution along the whole supply/use chains Loss of valuable fossil organic capital Biofuels and bioenergy vectors need of proper design
Food (EPA, 2021; Usubiaga et al., 2018)	Organic fuels for self-use and nutrition purposes	Wasting resources, Over-consumption—Dietary imbalances—Hunger vs Obesity—Health effects—Quality of life—Irrational transport—Market demands
Organic products (Paull, 2014; Colom-Gorgues, 2009)	Organic carbon "Traps"	Dependence upon fossil C Replacing/neglecting renewable carbon New market models Emphasis on consumption
Biotechnology (Jawed et al., 2019)	Organic hopes by microbial metabolics	Viruses and other novel threat vectors Applications outside natural systems Control by markets & cheap C sources New risks: Cloning, Huxley/Orwell Dystopias
Healthcare (Crown et al., 2017)	Feasibility optimization	Market/profit dominated New risks from new methods and drugs Emphasis on a disease vs Health approach Best practices by institutional solutions

Continued

Table 4 Major change factors of global organic systems—cont'd

Organic system	Key feature	Growth barriers
Quality of life (Siddiqui and Moghadasian, 2020)	Pursuit of happiness	Search for eternal youth (the Gilgamesh Myth) Culture of beauty Risks from new methods/ supplements and para-drugs
Environment (Everard and Mcinnes, 2013; Green Alliance, 2021)	Search for systemic solutions	Energy- and water-intensive waste treatments Shift to greater problems Neither old nor new systems Emphasis on public relations (PR)
New services (Koukios et al., 2020)	Making use of new knowledge	Low quality CHOICES of low-cost services Low level of knowledge integration Inadequate preparation of young people New education and training models needed

3. Concluding remarks

The preceding analysis of major growth factors and other key barriers illustrates an emerging multi-component Management Crisis of Global Organic Systems. The Organic Crisis has the following critical characteristics, which could become the object of a new generation of Foresight studies, making possible a Unified Approach to the multiple Sectoral Crisis:

- **(i)** Value-added generation by using "mining" vs "cultivating" approaches;
- **(ii)** Creation of multiple "external" cost sources (economic/social/ecological unaccounted);
- **(iii)** Exponential pressure on natural resources;
- **(iv)** The challenge of Organic Wastes;
- **(v)** New Risks—New pathogen outbreaks;
- **(vi)** Political, strategic and other non-economic domination of market forces;
- **(vii)** Socio-technical design neglecting or conflicting with ecosystem functions;

(viii) Limited/selective use of new knowledge;
(ix) Lack of a suitable frame of socio-economic values and ethics;
(x) Social acceptance of emerging socio-technical Disrupting Factors and Confusion.

Acknowledgment

This opinion paper was first presented at the November 26, 2017 Meeting celebrating the 100 Years of the Organic Technology Laboratory at the National Technical University of Athens, 1917–2017. With our sincere thanks to the participants of that workshop, and a view toward the next 100 Years of Organic System studies. A second version of this opinion paper was presented at the November 20, 2020 Annual Meeting of the Foresight Europe Network (FEN).

References

Adejumo, I.O., Adebiyi, O.A., 2020. Agricultural Solid Wastes: Causes, Effects, and Effective Management. https://www.intechopen.com/books/strategies-of-sustainable-solid-waste-management/agricultural-solid-wastes-causes-effects-and-effective-management. (Accessed 10 May 2021).

Colom-Gorgues, A., 2009. The challenges of organic production and marketing in Europe and Spain: innovative marketing for the future with quality and safe food products. J. Int. Food Agribus. Market. 21 (2–3), 166–190.

Committee on Considerations for the Future of Animal Science Research, 2015. Critical Role of Animal Science Research in Food Security and Sustainability. The National Academy Press.

Crown, W., et al., 2017. Constrained optimization methods in health services research—an introduction: report 1 of the ISPOR optimization methods emerging good practices task force. Value Health 20 (3), 310–319.

Diamantopoulou, L.K., Karaoglanoglou, L.S., Koukios, E.G., 2011. Biomass Cost Index: mapping biomass-to-biohydrogen feedstock costs by a new approach. Bioresour. Technol. 102 (3), 2641–2650.

Diamantopoulou, L., Papadaki, S., Karaoglanoglou, L., Koullas, D., Koukios, E., 2016. The new era of European biofuels landscape: comparative assessment of socio-environmental sustainability of lignocellulosic feedstocks. Cellul. Chem. Technol., 507–519.

Dincer, I., Rayash, A.A., 2020. Energy Sustainability. Elsevier, ISBN: 978-0-12-819556-7.

Egidi, G., Cividino, S., Paris, E., Palma, A., Salvati, L., Cudlin, P., 2021. The impact of multiple drivers of land sensitivity to desertification in a Mediterranean country. Environ. Impact Assess. Rev. 89, 106594.

EPA, 2021. https://www.epa.gov/sustainable-management-food/sustainable-management-food-basics. (Accessed 10 May 2021).

Eswaran, H., Lal, R., Reich, P.F., 2001. Land degradation: an overview. Responses to Land Degradation. In: Proc. 2nd. International Conference on Land Degradation and Desertification, Khon Kaen, Thailand. Oxford Press, New Delhi.

European Commission, 2015. FUTURE BRIEF: sustainable aquaculture. Science for Environment Policy (11).

Everard, M., Mcinnes, R., 2013. Systemic solutions for multi-benefit water and environmental management. Sci. Total Environ. 461–462C, 170–179.

Forest Europe, 2021. The Way Forward: Four Major Challenges to and Opportunities for Forest Sector Policy in Europe. https://www.foresteurope.org/docs/SoEF/reports/Future_Challenges_and_Opportunities_for_Forest_Policies_in_Europe_State_of_Europes_Forests_2011.pdf. (Accessed 15 May 2021).
Gontard, N., Sonesson, U., Birkved, M., Majone, M., Bolzonella, D., Celli, A., Angellier-Coussy, H., Jang, G.-W., Verniquet, A., Broeze, J., Schaer, B., Batista, A.P., Sebok, A., 2018. A research challenge vision regarding management of agricultural waste in a circular bio-based economy. Crit. Rev. Environ. Sci. Technol. 48 (6), 614–654.
Green Alliance, 2021. https://greenallianceblog.org.uk/2019/07/16/systemic-solutions-are-the-only-way-to-avoid-future-environmental-problems/. (Accessed 11 May 2021).
Jawed, K., Yazdani, S.S., Koffas, M.A.G., 2019. Advances in the development and application of microbial consortia for metabolic engineering. Metab. Eng. Commun. 9, e00095.
Joint Report AFD, EC, GIZ, 2017. Opportunities and Challenges for Aquaculture in Developing Countries.
Koukios, E., Karaoglanoglou, L., Papadaki, S., Koullas, D., Kourakos, N., 2020. New educational models to train engineers and executives on eco friendly technologies, products and sustainability policies; reference module in materials science and materials engineering. In: Encyclopedia of Renewable and Sustainable Materials. vol. 1, pp. 567–574.
Levine, A., 2012. Energy Efficiency in Agriculture: A Review of the Role of the Federal Government and State and Private Entities. An Alliance to Save Energy White Paper.
Nair, S., Johnson, J., Wang, C., 2013. Efficiency of irrigation water use: a review from the perspectives of multiple disciplines. Agron. J. 105 (2).
Panagos, P., Borrelli, P., Poesen, J., Ballabio, C., Lugato, E., Meusburger, K., Montanarella, L., Alewell, C., 2015. The new assessment of soil loss by water erosion in Europe. Environ. Sci. Pol. 54, 438–447.
Paull, J., 2014. Opportunities and challenges for organic food and agriculture: China and Australia. In: Mascitelli, B., O'Mahoney, B. (Eds.), Good Food for All: Developing Knowledge Relationships Between China and Australia. Connor Court Publishing, Ballarat, pp. 50–80 (Chapter 3).
POLITICO, 2008. https://www.politico.eu/article/the-challenge-of-sustainable-fish-farming/. (Accessed 5 May 2021).
Siddiqui, P.A., Moghadasian, M.H., 2020. Nutraceuticals and nutrition supplements: challenges and opportunities. Nutrients 12 (6), 1593.
Smith, L.G., Williams, A.G., Pearce, B.D., 2014. The energy efficiency of organic agriculture: a review. Renewable Agric. Food Syst. 30 (03), 280–301.
Usubiaga, A., Butnar, I., Schepelmann, P., 2018. Wasting food, wasting resources: potential environmental savings through food waste reductions. J. Ind. Ecol. 22 (3), 574–584. Special Issue: the Global Multi Regional Input Output Database "EXIOBASE".
Wysokiński, M., Domagała, J., Gromada, A., Golonko, M., Trębska, P., 2020. Economic and energy efficiency of agriculture. Agric. Econ. 66 (8), 355–364.

CHAPTER 3

Universal access to vaccines in post-COVID bioeconomy: Redesign for variants of concern and patent protection

Cristina Possas[a], Adelaide Maria de Souza Antunes[b,c], Alessandra Moreira de Oliveira[b], Suzanne de Oliveira Rodrigues Schumacher[b], and Akira Homma[a]
[a]Bio-Manguinhos, Oswaldo Cruz Foundation, Rio de Janeiro, Brazil
[b]School of Chemistry, Federal University of Rio de Janeiro, Rio de Janeiro, Brazil
[c]National Institute of Industrial Property, Rio de Janeiro, Brazil

1. Introduction

There is consensus in the scientific community that the most important instrument to control virus spread in human populations is the development of herd immunity, from infection or from vaccines. Vaccines are considered the most important strategy to achieve herd immunity, since natural herd immunity via infection, letting people become infected by an emerging virus, such as COVID-19, with very little scientific knowledge on the sequelae and long-term pathologies, should never be a policy (Possas et al., 2021a). Natural herd immunity just refers to an equilibrium between virus and host, a default mechanism for achieving interruption of viral transmission (James, 2020). Therefore, universal access to vaccines is the only way to achieve extensive global immunization and reach the required 75% populational threshold for herd immunity. Constraints such as "vaccine hesitancy" and other operational barriers in health services limiting access to vaccines, particularly affecting the poorest populations in developing countries, indicate that it will be impossible to reach globally this 75% immunity threshold without providing universal access to vaccines. There is no scientific evidence supporting that the spread of SARS-CoV-2 might stop naturally before at least 75% of the population has become immune, in opposition to previous mathematical modeling studies indicating that this threshold for herd immunity from natural infection could be much lower, from 10% to 20% (Aguas et al., 2020). Moreover, developing effective new COVID-19 vaccines and providing

Agricultural Bioeconomy
https://doi.org/10.1016/B978-0-323-90569-5.00015-9

universal access to them is not sufficient, since the duration of immune protection from these vaccines is not clear yet. An important scientific controversy has emerged on "waning immunity": there is a possibility that we may never reach herd immunity from SARS-CoV-2 infections or vaccines since, differently from many other viruses, the immunity would be temporary, lasting from 6 months to 1 year and then fading, with a possible risk of reinfection. Additional studies will be necessary to determine the extent of immune protection and to examine the possibility of chronic persistence of SARS-CoV-2 viral infection, which might require, such as Influenza, annual immunization. A recent study (Seow et al., 2020) provides evidence indicating that transient neutralizing antibody (nAb) response is a feature shared by both a SARS-CoV-2 infection that causes low disease severity and the circulating seasonal coronaviruses that are associated with common colds. Moreover, the emergence of new and concerning recent SARS-CoV-2 variants (United Kingdom B.1.1.7, South-Africa B.1.351, Brazil P1, India B.1.617) will probably require from scientists a redesign of approved vaccines. Finally, other issues such as the limited global access to multipatented vaccines (Possas et al., 2015; WHO, 2021a) and the sharp decline in the last decade in global vaccine coverage, related to "vaccine hesitancy," vaccine refusal and operational issues in immunization programs will probably contribute as well to reduce the impact of the new SARS-CoV-2 vaccines (Possas et al., 2021a,b). This scenario of very low COVID-19 vaccine coverage, threatening global economy and the future of mankind, supports the urgency of global and local policies providing universal access to vaccines, aiming the 75% global vaccine immunity threshold. From this perspective, we discuss SARS-CoV-2 variants and the novel technologies related to nucleic acid COVID-19 vaccines, based on a global search of applied patents and examining BRICS' countries applications. The chapter is structured as follows:

- Universal access to vaccines and pandemic preparedness: conceptual framework
- COVID-19 vaccines: global distribution and related variants
- SARS-CoV-2 variants and genomic vaccines: overview
- BRICS' vaccines: development and production capacity
- DNA and RNA vaccines in clinical trials: patent landscape

2. Universal access to vaccines and pandemic preparedness: Conceptual framework

The extraordinarily accelerated development of COVID-19 vaccines has been followed by successful and extensive immunization, with few adverse effects. Notwithstanding, their global distribution has so far favored developed nations with 75% of vaccines concentrated in 10 high-income countries (WHO, 2021a). This inequitable scenario is both unethical and inefficient from the global pandemic preparedness perspective, considering that the threshold for herd immunity from vaccines of 75% of the global population immunized. Universal access to vaccines, immunizing the more vulnerable populations in the world would contribute to maximize the number of lives saved and to pandemic control, slowing transmission and reducing the vulnerability to viral variants of concern. Low and middle-income countries started to receive in May 2021 their first supplies of vaccines and APIs from manufacturers through the COVAX Facility and other international collaborations. It has been estimated that maintaining this pace, without immediate action aiming universal access to vaccines, it may take years before their populations can be immunized (Organisation for Economic Co-operation and Development—OECD, 2021) From this perspective, an innovative global governance strategy should be urgently conceived by the G-7 and the G-20 countries to expand the COVID-19 vaccine production, providing rapid supply of the required vaccines to the more vulnerable populations in greater need, aiming not only immediate response to the COVID-19 pandemic but a new global preparedness model for possible future pandemics in the post-COVID era. This innovative global governance mechanism should urgently put in place strategies to accelerate immunization globally, strengthening vaccine distribution by COVAX and other mechanisms, supporting developing countries' manufacturers to scale up production to provide vaccine supplies to nations in greatest need. This novel governance strategy should be supported by sustainable legal patent flexibility and knowledge governance strategies to assure the feasibility of universal access to vaccines, sharing intellectual property and facilitating technology transfer, through negotiations in G-7 and G-20 countries and multilateral negotiations in the World Trade Organization (Fig. 1).

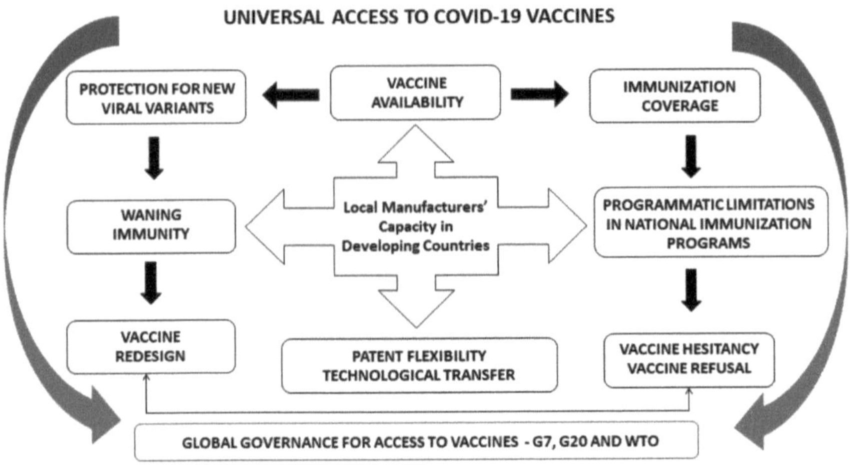

Fig. 1 Conceptual framework: universal access to COVID-19 vaccines. *(Sources: Elaborated by the authors.)*

3. COVID-19 vaccines: Global distribution and related variants

The emergence of new SARS-CoV-2 variants was expected, due to a known mutation mechanism by which every virus changes its genetic sequence. The term variant is actually the virus modified from the original virus (WHO, 2021b). In May 2021, The Centers for Disease Control and Prevention (CDC, 2021a) updated the SARS-CoV-2 variants classifications and definitions. This document identifies three classes of variants as defined by the US government interagency group: 1. Variant of High Consequence: No SARS-CoV-2 currently; 2. Variant of Interest: Variants accounting for the increasing number of cases and with specific genetic markers that can affect the immune response scape, diagnostics, transmission and therapeutics and 3. Variant of Concern (VOC): These ones are responsible for more severe disease, increasing transmissibility, hospitalizations and deaths, the reason why they will be discussed in this chapter. These VOC require new diagnostic and treatment strategies and can reduce the antibodies' neutralization generated by the vaccines (CDC, 2021a; WHO, 2021b). Despite of this constraint, COVID-19 vaccines have been designed to provide extensive immune protection against SARS-CoV-2 virus and for this reason it is expected that vaccines effectiveness against viral mutations and new variants will not be significantly reduced. Notwithstanding this broad immune protection from vaccines, the need for vaccine redesign is a possibility: if a vaccine promotes a satisfactory immune response against any variant, it will be possible to make important changes in vaccines' composition (WHO, 2021c). These SARS-CoV-2 variants require strictly monitoring vaccine efficacy, vaccine redesign and updates against future mutations, a challenge that could be facilitated by the flexibility of the breakthrough mRNA-based vaccine technology (Xie et al., 2021). From this perspective, according to Gupta (2021), new information is needed regarding the protection generated by vaccines against SARS-CoV-2 in the VOCs' scenario, such as those on the role of T-cell mediated immunity and how they act in the selection of these variants. From the WHO document (WHO, 2021d), four variants[a] of concern (VOCs) were chosen:

- B.1.1.7 lineage (Alpha variant first detected in the United Kingdom): Passos and Santos (2021) confirmed that the mutation N501Y, P681H and other mutations in the variants' strain (B.1.1.7) (receptor-binding domain of the Spike protein) promote a stronger interaction with ACE2 receptor. This variant is associated with higher transmissibility with rapid growth in the United Kingdom and all over the world. Based on this molecular finding, the high SARS-CoV-2's spread and infectivity of this lineage is better understood (CDC, 2021a; Pango Lineages, 2021a).
- B.1.351 lineage (Beta variant, first detected in South Africa): The SARS-CoV-2 variant was first identified in the Republic of South Africa. In the United States, B.1.351 has been detected in at least nine states (National Institutes of Health, 2021). The spike mutations of interest are: K417N, E484K, N501Y, D614G, A701V (European Centre for Disease Prevention and Control—ECDC, 2021). It has been estimated that when compared with previous variants, B.1.531 can spread 50% more rapidly. Moreover, it presents, face to natural infection/vaccine, moderate reduction of immune response (Virginia Departament of Health—VDH, 2021).
- P1 lineage (Gamma variant, first detected in Brazil—city of Manaus and Japan). This variant is associated with a higher severity and transmissibility of the disease. The spike mutations of interest for it were K417T, E484K,

[a] For the present work, it will be used the Pango Lineage for the Lineages of Concern (SARS-CoV-2) (Pango Lineages, 2021b). However, it is important to highlight that the World Health Organization has named the SARS-CoV-2 variants using the letters of the Greek alphabet (for non-scientific audiences), i.e., for lineage B.1.1.7 (Alpha), B.1.351 (Beta), P1 (Gamma) and B.1.617.2 (Delta). The later will not be used along this text (WHO, 2021d).

N501Y, D614G, H655Y (CDC, 2021a; ECDC, 2021; Funk et al., 2021).
- B.1.617.2 lineage (Delta variant, first detected in India): L452R, T478K, D614G, P681R are the spike mutations of interest for this variant (ECDC, 2021). It presents higher transmissibility and reduced neutralization, which can impact vaccine immune responses (WHO, 2021e).

4. SARS-CoV-2 variants and genomic vaccines: Overview

A search [b] for scientific articles was carried out from a specialized PubMed database and complemented with a second search[c] developed at Cortellis Drug Discovery Intelligence (CDDI) (Clarivate Analytics) database to retrieve journal articles, clinical trials and press releases from companies. In addition, a supplementary search strategy[d] was used for Sinovac's vaccine, i.e., CoronaVac, because no documents were retrieved based on the previous strategies. From the search strategies above described, it was possible to identify the studies related to launched, pre-registered or registered vaccines and VOC's neutralization data, despite the vaccine's characteristics, shown in Fig. 2 and Table 1.

The Fig. 2 shows the VOCs according to the location of emergence (where was first detected).

Besides the fact that these variants can cause severe diseases, a great challenge is to determine if these mutations make the current vaccines, which were designed for the earlier versions of COVID-19 virus, less/no effective. Moreover, it appeared that most of the vaccines (recombinant vaccines) or

[b] The search was carried out on the Pubmed Database (from National Center for Biotechnology Information, U.S. National Library of Medicine), using the following terms in title's document: "*variant, COVID-19 (or SARS-CoV-2), vaccine.*" The search was carried out on May 5, 2021. From Pubmed Database, 12 scientific articles were recovered (Three articles were excluded: 2. For evidence of vaccine escape and it was not about a VOC).

[c] The strategy search used the term "*vaccines*" for product category associated to "*severe acute respiratory syndrome coronavirus 2 (SARS-CoV-2) infection (COVID-19)*" (CDDI—Clarivate Analytics) for condition in Drugs and Biologics' advanced search. 280 registers were recovered. From each, literature and clinical trials' data related to vaccine's efficacy against VOCs were retrieved. The search was carried out in May 26, 2021.

[d] The search was carried out in the server medRxiv (founded by Cold Spring Harbor Laboratory, Yale University, and BMJ) where it is possible to find unpublished, but complete manuscripts. It was possible to include preprints from bioRxiv. For a specific strategy advanced search, the terms CoronaVac and variant were used in the title. From this search, two documents were retrieved.

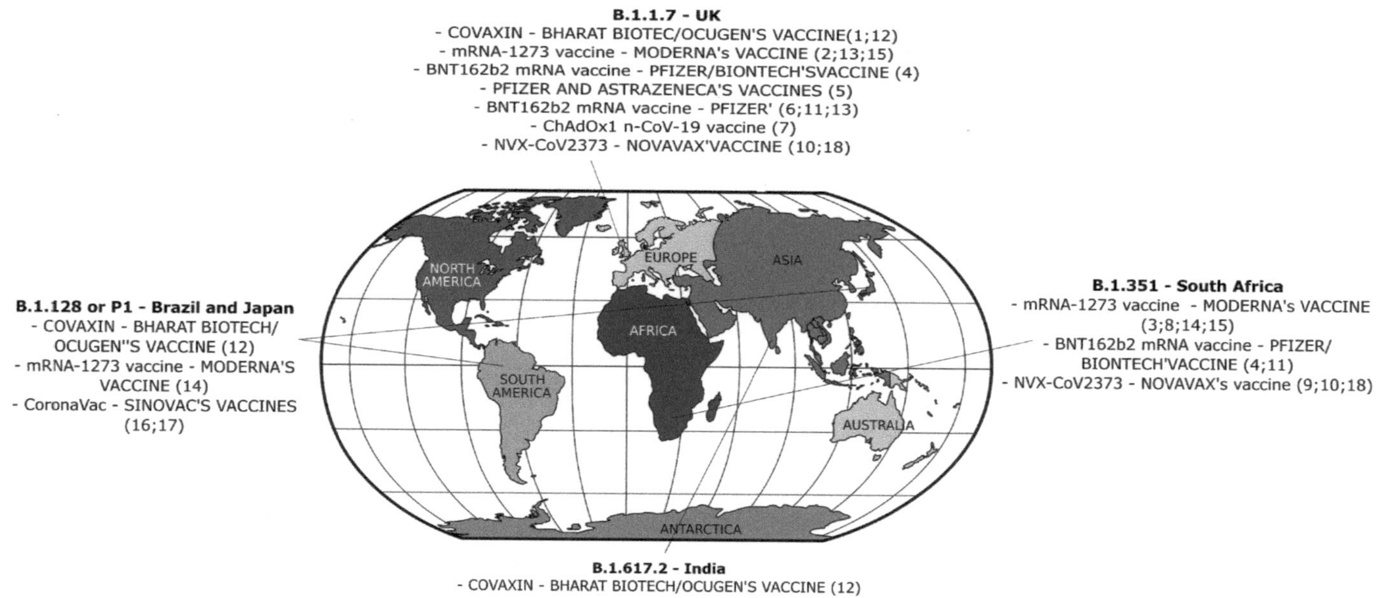

Fig. 2 Variants of concern according to location of emergence and related vaccines. *(Sources: Elaborated by the authors from CDDI (Cortelis Drug Discovery Inteligence—CDDI from Clarivate Analytics), n.d.; CDC, 2021a; Roberts, 2021; Sapkal et al., 2021 (1); Edara et al., 2021a (2); Edara et al., 2021b (3); Tada et al., 2021 (4); Supasa et al., 2021 (5); Sansone et al., 2021 (6); Emary et al., 2021 (7); Edara et al., 2021c (8); Shinde et al., 2021 (9); Novavax, 2021a (10); Xie et al., 2021 (11); Ocugen, 2021 (12); Garcia-Beltran et al., 2021 (13); Moderna, 2021a (14); Moderna, 2021b (15); Hitchings et al., 2021 (16); Ranzani et al., 2021 (17); Novavax, 2021b (18).)*

Table 1 Launched, pre-registered and registered vaccines according to CDDI Database—The main COVID-19 Vaccine's characteristics and their immunogenicity variants's data.

Vaccine's code name	Vaccine's brand name	Organizations	Highest phase	Vaccine's description	Mechanism of action	Immunogenicity variants' data[a]
(1) Gam-COVID-Vac	Sputnik V	Gamaleya Research Institute of Epidemiology and Microbiology (Originator) Res Inst Epidemiol Microbiol Gamalei (Gamaleya Research Institute of Epidemiology and Microbiology)	Launched 2020	*"Heterologous prime-boost SARS-CoV-2 (Covid-19 coronavirus) vaccine consisting of recombinant adenovirus type 26 (Ad26) vector (with deleted E1/E3 sites) and recombinant adenovirus type 5 (rAd5) vector (deleted E1/E3 sites), both encoding SARS-CoV-2 spike (S) glycoprotein"*	*"Spike Glycoprotein (S) (SARS-CoV-2; COVID-19 Virus)-Directed Immunity Inducers"*	No data
(2) Sputnik Light	–	Gamaleya Research Institute of Epidemiology and Microbiology (Originator)	Registered 2021	*"Recombinant adenovirus type 26 (Ad26) vector (with deleted E1/E3 sites) encoding SARS-CoV-2 spike (S) glycoprotein"*	*"Spike Glycoprotein (S) (SARS-CoV-2; COVID-19 Virus)-Directed Immunity Inducers"*	No data
(3) BBV-152	Covaxin	Bharat BioNtech (Originator) Ocugen	Launched 2021	*"Whole-virion beta-propiolactone-inactivated SARS-CoV-2 (NIV-2020-770 strain) containing the Asp614Gly mutation at the spike protein; formulated with Algel-IMDG (imidazoquinoline class molecule (TLR7 and TLR8 agonist) adsorbed onto Algel)"*	*"Human Coronavirus (SARS-CoV-2; COVID-19 Virus) Proteins-Directed Immunity Inducers"*	– B.1.1.7 (UK variant) – B.1.617 (Indian double mutant variant) – B.1.128.2 or P1 (Brazilian variant: with E484K mutation) (1)

(4) CoronaVac PiCoVacc	–	Sinovac (Originator) Instituto Butantan	Launched 2020	*"Human SARS-CoV-2 (Covid-19 coronavirus) inactivated vaccine; expressed in Vero cells"*	*"Human Coronavirus (SARS-CoV-2; COVID-19 Virus) Proteins-Directed Immunity Inducers"*	No data
(5) Ad26 SARS-CoV-2 Ad26-COVS1 Ad26.COV2.S Ad26COVS1 JNJ-78436735 VAC-31518	–	Janssen (Originator) Janssen Pharmaceutical (JP) (Originator)	Launched 2021	*"Human SARS-CoV-2 (Covid-19 coronavirus) vaccine consisting of replication-incompetent adenovirus vector serotype 26 (Ad26) encoding full-length and stabilized SARS-CoV-2 spike protein; produced based on AdVac(R) technology and expressed in human PER.C6® cell line"*	*"Spike Glycoprotein (S) (SARS-CoV-2; COVID-19 Virus)-Directed Immunity Inducers"*	Data were excluded[b]
(6) BNT-162b2 PF-07302048 RBP-020.2	Comirnaty (Generic name: Tozinameran)	BioNTech (Originator) Pfizer Shanghai Fosun Pharmaceutical (Group)	Launched 2020	"Human SARS-CoV-2 (Covid-19 coronavirus) vaccine comprising messenger RNA (mRNA), 5′-capped, encoding a full-length, codon-optimized pre-fusion stabilized conformation variant (K986P and V987P) of the SARS-CoV-2 spike (S) glycoprotein, flanked by 5′ and 3′ untranslated regions and a 3′ poly(A) tail, and containing N1-methylpseudouridine instead of uridine"	*"Spike Glycoprotein (S) (SARS-CoV-2; COVID-19 Virus)-Directed Immunity Inducers"*	B.1.1.7, B.1.429, and B.1.351 (2;3)

Continued

Table 1 Launched, pre-registered and registered vaccines according to CDDI Database—The main COVID-19 Vaccine's characteristics and their immunogenicity variants's data—cont'd

Vaccine's code name	Vaccine's brand name	Organizations	Highest phase	Vaccine's description	Mechanism of action	Immunogenicity variants' data
(7) TAK-919 mRNA-1273	–	Moderna (Originator) National Institutes of Health (NIH) Takeda	Launched 2020	*The vaccine consists of "mRNA CX-024414 encoding S-2P antigen comprising SARS-CoV-2 Spike (S) glycoprotein with a transmembrane anchor and an intact S1-S2 cleavage site, harboring two consecutive proline substitutions at positions 986 and 987 of the S2 subunit; encapsulated in lipid nanoparticles composed of the proprietary ionizable lipid SM-102, cholesterol, 1,2-distearoyl-sn-glycero-3-phosphocholine (DSPC) and PEG2000 DMG"*	*"Spike Glycoprotein (S) (SARS-CoV-2; COVID-19 Virus)-Directed Immunity Inducers"*	B.1.351, B.1.1.1.7, and P1 (4;5)
(8)	Covishield Vaxzevria	Jenner Vaccine Foundation; University of Oxford (Originators) Oxford BioMedica AstraZeneca	Launched 2020	The vaccine consists of *"of a replication-deficient chimpanzee adenovirus Oxford 1(ChAdOx1) vector encoding the SARS-CoV-2 spike (S) gene"*	*"Spike Glycoprotein (S) (SARS-CoV-2; COVID-19 Virus)-Directed Immunity Inducers"*	B.1.1.7 (6)

(9) Ad5-nCoV	Convidecia	Beijing Institute of Biotechnology; CanSino Biologics (Originators) NPO Petrovax Pharm	Launched —2021	The vaccine consists of "vaccine consisting of a replication-defective adenovirus type 5 vector encoding SARS-CoV-2 spike (S) gene"	"*Spike Glycoprotein (S) (SARS-CoV-2; COVID-19 Virus)-Directed Immunity Inducers*"	No data
(10) Covovax NVX-CoV-2373 NVX-CoV2373 TAK-019	–	Novavax (Originator) SK Bioscience Takeda	Pre-registered	The vaccine comprises a "*recombinant mutant full-length SARS-CoV-2 spike (S) glycoprotein harboring R682Q, R683Q and R685Q mutations in the S1/S2 cleavage site and stabilized by introducing two proline substitutions at K986P and V987P; encapsulated in nanoparticles*"	"*Spike Glycoprotein (S) (SARS-CoV-2; COVID-19 Virus)-Directed Immunity Inducers*"	B.1.351 (7;8;9)
(11) BBIBP-CorV	–	China National Pharma Group (SINOPHARM), Chinese Academy of Medical Sciences, Chinese Center Disease Control Prevent, National Institutes Food Drug Control (Originators)	Launched 2020	The vaccine comprises "*comprising beta-propiolactone (BPL)-inactivated SARS-CoV-2 19nCoV-CDC-Tan-HB02 (HB02) strain; propagated in Vero cells*"	"*Human Coronavirus (SARS-CoV-2; COVID-19 Virus) Proteins-Directed Immunity Inducers*"	No data

Continued

Table 1 Launched, pre-registered and registered vaccines according to CDDI Database—The main COVID-19 Vaccine's characteristics and their immunogenicity variants's data—cont'd

Vaccine's code name	Vaccine's brand name	Organizations	Highest phase	Vaccine's description	Mechanism of action	Immunogenicity variants' data
(12) –	–	China National Pharma Group (SINOPHARM), Wuhan Institute of Biological Products (Originators)	Launched 2021	The vaccine comprises "*beta-propiolactone-inactivated SARS-CoV-2 WIV04 strain; propagated in Vero cells*"	"*Human Coronavirus (SARS-CoV-2; COVID-19 Virus) Proteins-Directed Immunity Inducers*"	No data

The search was carried out in May 2021.
[a]Here all the variants' classifications from the studies were included.
[b]Excluded data: "Ad26.COV2.S protects Syrian hamsters against G614 spikevariant SARS-CoV-2 and does not enhance respiratory disease."
Sources: elaborated by the authors from CDDI (Clarivate Analytics): Ocugen, 2021 (1) (press release); Xie et al., 2021 (2); Garcia-Beltran et al., 2021 (3); Moderna, 2021a (4) (press release); Moderna, 2021b (5) (press release); Emary et al., 2021 (6); Novavax, 2021a (7) (press release); Novavax, 2021b (8) (press release); Shinde et al., 2021 (9).

those which target the original D614G ancestral sequence of the virus spike (single epitope) might fail to generate an immune response capable of neutralizing these new variants (Roberts, 2021; Sapkal et al., 2021). Adjustments to some spike sequences in vaccines can determine the impact of emerging COVID-19 lineages with specific mutations (Chen et al., 2021).

About the variant B.1.1.7, Sapkal et al.' data (2021) provided evidence indicating that the indigenous inactivated BBV152/COVAXIN vaccine works against the B.1.1.7 lineage. A messenger RNA vaccine (mRNA 1273) also neutralizes COVID-19 B.1.17 variant by a live virus neutralization assay against a panel of COVID-19's variants and showed protective immune response against COVID-19 B.1351 variant (Edara et al., 2021a,b). Tada et al. (2021) demonstrated the neutralized activity of BNT162b2 mRNA in vaccinated individuals. While Supasa et al. (2021) demonstrated a modest reduction (for Pfizer and AstraZeneca vaccine) of neutralization titers for the B.1.1.7, but with no evidence of vaccine evasion. Moreover the vaccine (BNT162b6) also confirmed protective effect against the VOC (B.1.1.7) (Sansone et al., 2021). Concerning the VOC B.1.1.7, it was observed that the vaccine ChAdOx1 n-C0V-19 provided a reduced neutralization, despite the vaccine efficacy against this variant (Emary et al., 2021). For the three mutant virus (English and South African variants), neutralization geometric mean titers of vaccinated human sera (two BNT162b2 doses) indicate small effects on neutralization (Xie et al., 2021). According to press release (Pfizer, 2021), the immunogenicity and safety of the BNT162b2 vaccine third dose from Pfizer and BioNTech will be evaluated as a booster against circulating SARS-CoV-2 variants. A study conducted by Indian Council of Medical Research suggested that COVAXIN (Bharat Biotech's vaccine) neutralizes Indian double mutant variant (B.1.617), Brazilian variant (B1.128.2, with E484K mutation) and UK variant (B.1.1.7) which promotes a reduction of a mutant virus escape (Ocugen, 2021). According to Garcia-Beltran et al. (2021), among people who received two full doses of BNT162b2 or mRNA-1273 vaccines, strains as B.1.1.7 (United Kingdom) and B.1.429 are still potently neutralized, while other circulating variants can escape. For B.1.351 data by human immune sera, the live virus neutralization and the antibody binding was compared against the emerging B.1.351 variant and circulating COVID-19 B1 variant. The study showed that the mRNA 1273 (from Moderna which encodes the full length stabilized the protein from the virus' spike) can induce immune response to neutralize the South African variant (B.1.351) promoting protective humoral immunity (Edara et al., 2021b,c; Tada et al., 2021). The phase 2 study conducted by the biotechnology company Moderna (2021a) demonstrated that a single 50 μg dose of the vaccine, i.e., mRNA-1273, can give a booster to individuals who were previously vaccinated. Initial data demonstrated that for SARS-CoV-2 and its VOCs, P1 (GAMMA) and B.1.351, there is a an increase of the neutralizing antibody titers. The results from the in vitro neutralization studies of sera were conducted for the Moderna COVID-19 vaccine. For this vaccine, sera from vaccinated individuals were tested and it was observated the production of neutralizing titer against B.1.1.7 and B.1.351 and all others key emerging variants (Moderna, 2021b). A study showed data about the neutralizing activity from Pfizer BioNtech BNT162b2 (lipid nanoparticle-formulated) mRNA vaccine against the B.1.351 and B.1.1.7 spike proteins (Tada et al., 2021). In the moment of dominant circulation of the B.1.351, NOVAVAX's vaccine with recombinant nanoparticle technology (NVX-CoV2373 vaccine) induced cross-protection during the pandemic and against (B.1.1.7) (Shinde et al., 2021; Novavax, 2021a,b). Furthermore, a test negative control study was conducted in Manaus, Brazil (where 75% of genotyped SARS-CoV was P.1) with 53,176 healthcare workers. The results for this study were suggestive, but inconclusive: one vaccine dose promotes the effectiveness against the infection's symptoms (P.1 transmission), but with a low estimated effectiveness with the two doses (Hitchings et al., 2021). Another test-negative case control study was conducted for the elderly population about the effectiveness of the Sinovac's vaccine in São Paulo, Brazil during circulation of P.1 variant. Coronavac was 42% effective, but it was not observed a significant protection until complete dose regimen. Moreover it is important to highlight that the decline of vaccine's effectiveness with the increasing age (Ranzani et al., 2021), due to immunosenescence, sustained by multifaceted immune remodeling of the innate and adaptive branches of the human immune system, yet to be adequately understood (Connors et al., 2021). Unfortunately, current vaccine strategies and platforms do not sufficiently protect older individuals (Goodwin et al., 2006).

In the same direction, the analysis of the studies provided in Fig. 1, our search also provided results for Table 1: vaccines launched, pre-registered or registered. This table also presents the following information: vaccine's code and brand names, organizations, highest phase of development, vaccine's description and mechanism of action and the.

5. BRICS' vaccines: Development and production capacity

The flexibility of patent protection for vaccines has become a major geopolitical issue in the World Trade Organization and other international fora such as G7, G-20 and the BRICS (Brazil, Russia, India, China and South Africa). This issue has gained momentum and amplified during the COVID-19 pandemics, with the BRICS' countries accounting for a large part of the production of vaccines and their active pharmaceutical ingredients (API), particularly China, India and Russia. This extraordinary capacity becomes evident when we consider that from the 12 leading vaccines approved worldwide, 6 of them have been developed and produced in the BRICS, with China accounting for four of them (Cansino Ad5-nCoV; Sinopharm BBIBP-CorV; Sinovac-Coronavac and Sinopharm-Wuhan); while Russia for two of them (Gamaleya Research Institute-Sputnik V and Vector Institute EpiVacCorona) and India one (Bahrat Biotech-Covaxin). China has accelerated its capacity for vaccine development and production in the last 20 years and plays a central role in the innovation and production of vaccines for infectious diseases and global vaccine patent deposits. India is home to the Serum Institute, the world's largest vaccine production manufacturer and other private manufacturers. Russia has a large scientific and technological capacity and significant capacity for vaccine development and production in the Gamaleya Research Institute and other manufacturers. Brazil has also important local production capacity and potential for vaccine development. In Brazil Bio-Manguinhos/Fiocruz and the Butantan Institute, the two major manufacturers, play a crucial role supplying domestic vaccine demands in a country with continental dimensions as well as for other Latin American countries. Despite its large production capacity, Brazil has a significant dependence on imports of vaccine API, with reduced local manufacturers' capacity for development and production of vaccine inputs, which has declined from 50% 20 years ago to just 5% now. This scenario of dependence of imports will require from the Brazilian government a redefinition of scientific, technological development and production vaccine policies and strategies. This will require a novel approach to incentives to technological innovation and to patent deposits, which are very low in the country. As for South Africa the local capacity is very limited (four manufacturers with just "fill and finish" capacity), reflecting the constraints of the African continent, which has imported for decades 99% of its vaccines. In spite of these local constraints, South Africa is recognized for its public health and scientific leadership in the continent and can play an important role in contributing to accelerate the region's strategy. It is expected thus that these strategies for vaccine development in the BRICS' countries will significantly contribute to accelerate their local innovation, development and production strategies and will certainly be an stimulus to accelerate as well their vaccine patent deposits.

6. DNA and RNA vaccines in clinical trials: Patent landscape

Nucleic acids vaccines are leading the vaccines tests in Clinical Trial Protocols—CTP (32%). When considered the distribution of vaccines candidates in development against COVID-19, the participation of these innovative platforms should be highlighted: 12% of these CTP corresponds to the use of genetic engineering of plasmid containing DNA. While RNA-based vaccines (saRNA and mRNA) corresponds to 20% of the CTP, which is related to encoding the Spike protein (antigen) (Rego et al., 2020). This extraordinary vaccine development and manufacturing capacity becomes evident when we consider, as mentioned above, that 6 of the 12 leading COVID-19 vaccines have been developed and produced in the BRICS, with China accounting for three of them, while India and Russia for one each. As for the nucleic acids vaccines' production, DNA or RNA are genetically engineered for the direct production of an antigen, which is easy to design. The DNA vaccines are highly safe with "*high titer neutralizing antibodies*" as an important characteristic. In addition, the RNA vaccines have a high degree of adaptability and can induce a strong immune response. The messenger RNA (mRNA) vaccine has emerged as a novel technology for COVID-19 vaccines, but has been studied for decades and due to this previous technological development can be done with the strategic use of readily available materials (CDC, 2021b; Chung et al., 2021). During the last decade, mRNA vaccine platform has progressed in technological innovation. Mainly those against COVID-19, which are the first licensed for distribution and were developed in a unprecedented high speed (Huang et al., 2021). The development of these mRNA vaccines is faster than the traditional ones. These vaccine are well known and have been studied before for other diseases, such as cytomegalovirus, rabies, flu and Zika. The process of production of mRNA vaccines can be scaled up and standardized, and considering COVID-19 pandemic scenario, scientists developed a mRNA with instructions for cells to build the virus' spike protein into the vaccine (CDC, 2021b).

6.1 DNA vaccines in clinical trials[e]: INO-4800; GX-19; ZyCoV-D and AGO31

These vaccines stand out for the relevance of their technologies for the prevention of SARS-CoV-2 (Ferraz, et al., 2020)[f]. According these authors, it was possible to retrieve the patent documents which were applied in any BRICS' countries. Fig. 3 shows the scenario with the total retrieved documents related to the DNA vaccines and the related applications in BRICS' countries. The number of documents for each of these BRICS' countries is shown in Table 2.

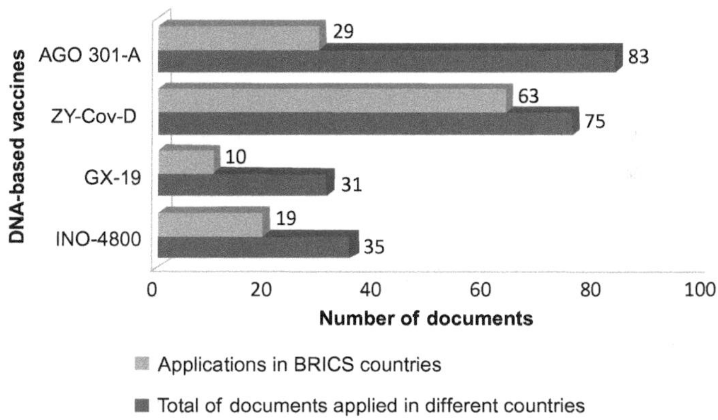

Fig. 3 Total applied patent documents related to DNA vaccines in different countries versus the number of applications in BRICS's countries. Note: For data acquisition, it was considered whether or not the document was applied in at least one of BRICS' countries. *(Source: Elaborated by the authors based on Ferraz, L.G.W., d'Urso de Souza Mendes, C., Von der Weid, I., Lourenço, A.L., Chedid, N.G.B., Carestiato, T., Brazilian National Institute of Industrial Property, 2020. Vacinas Baseadas em DNA Para a Prevenção da COVID-19: Mecanismo de ação, Ensaios Clínicos e Pedidos de Patentes. Observatório de Tecnologias Associadas à COVID-19. Ministério da Economia, Brazil. Retrieved from: https://www.gov.br/inpi/pt-br/servicos/patentes/tecnologias-para-covid-19/Arquivos%20Textos/Estudo8Vacinas.pdf.)*

[e] The study from Brazilian National Institute of Industrial Property—INPI recovered documents related to different vaccines technologies: DNA, RNA, viral vectors, attenuated viral and protein subunit vaccines (Ferraz et al., 2020). For the present work, it was recovered the documents related to the nucleic acid vaccines.

[f] The INPI developed a study (Ferraz et al., 2020) where one of the aspects was to carry out the analysis of patent applications related to DNA vaccines at a more advanced clinical stage in the current scenario of the COVID-19 pandemic. The vaccines chosen for that study were INO-4800; GX-19; ZyCoV-D (nCov Vaccine) and AGO31.

Table 2 Applied documents in BRICS' countries concerning to DNA-based technology.

DNA-based technology			Number of applied documents for BRICS's countries[a]				
Vaccine	**Developer Consortium**	**Total of applied patents considering each partner**	**Brazil**	**Russia Federation**	**India**	**China**	**South Africa**
INO-4800	Inovio Pharmaceuticals (United States)	35	6	7	0	19	1
GX-19	Genexine Inc. (Republic of South Korea) (partnership with Korean companies: Binex, International Vaccine Institute, Korea Advanced Institute of Science and Technology—KAIST, Pohang University of Science and Technology—POSTECH)	31	4	4	0	10	0
Zy-Cov-D	Zydus Cadila Healthcare Ltd. (India)	75	41	3	40	39	28
AGO 301-A	AnGes (Japan) Osaka University (Japan) Takara Bio (Japan)	83	1	0	5	27	0

[a]The document can be applied in more than one BRICS' country.
Source: elaborated by the authors based on Ferraz, L.G.W., d'Urso de Souza Mendes, C., Von der Weid, I., Lourenço, A.L., Chedid, N.G.B., Carestiato, T., Brazilian National Institute of Industrial Property, 2020. Vacinas Baseadas em DNA Para a Prevenção da COVID-19: Mecanismo de ação, Ensaios Clínicos e Pedidos de Patentes. Observatório de Tecnologias Associadas à COVID-19. Ministério da Economia, Brazil. Retrieved from: https://www.gov.br/inpi/pt-br/servicos/patentes/tecnologias-para-covid-19/Arquivos%20Textos/Estudo8Vacinas.pdf.

Based on this search strategy, we analyzed in Table 2 the documents applied in these countries.

According to the documents' analysis, the following information was provided about DNA related technologies.

INO-4800 Vaccine: Most of the documents from these BRICS' countries were applied in China. This DNA vaccine is a candidate for COVID-19, with good results of safety, tolerability and immunogenicity in all studied age groups. Here there were retrieved the documents related to DNA technique of one of the members of the consortium: The vaccine is administered by intradermal injection. Moreover, it is followed by electroporation, which is a technology used to develop DNA medicines/delivery mechanism (Inovio, 2021). Regarding the technologies related to some participant in the INO-4800 vaccine consortium, it was observed: Several documents were applied in China on this technique: An electroporation device used to deliver the antigen which can be used in DNA vaccine (WO Patent No. 2013066427A1—INOVIO PHARMACEUTICAL INC) (Broderick et al., 2013); a tissue electroporation device which allows the substances taking up the drug (e.g.) (WO Patent No. 2007095140A2—INOVIO PHARMACEUTICAL INC) (Kjeken et al., 2007); an "*electroporation device of needle-free jet injection system*" to introduce the drug into the tissues (WO Patent No. 2017117273A1—INOVIO PHARMACEUTICAL INC) (Mccoy et al., 2017). An oral electroporation device with a syringe which stores the pharmaceutical formulation for delivery was also claimed (WO Patent No. 2011137221A1—INOVIO PHARMACEUTICAL INC) (Kemmerrer et al., 2011). Moreover, for the introduction of implant agents, such as DNA into cells, an electroporation apparatus was developed (WO Patent No. 2000023563A1—INOVIO PHARMACEUTICAL INC) (Nanda et al., 2000). Moreover, an object of protection with application in China is the use of an expression vector to encode a protein, which can be utilized to prepare pharmaceutical compositions with polypeptide chains aiming to produce antibodies in a subject (e.g.) (WO Patent No. 2003059952A1—INOVIO PHARMACEUTICAL INC) (Bogen et al., 2003). As for other applications in Brazil, Russia and China, it can also be highlighted an electroporation device for patients treatment (BR Patent No. 112018013312A2—INOVIO PHARMACEUTICAL INC) (Campillo-Agusti et al., 2018) and a handset for electroporation use (BR Patent No. 112018013242A2—INOVIO PHARMACEUTICAL INC) (Stadelmann, 2018).

GX-19 Vaccine: It consists in a DNA vaccine which expresses SARS-CoV-2 protein antigen by Genexine, Inc. (National Institutes of Health, 2020). Concerning the retrieved documents related to DNA technique of one of the members of the consortium: the claimed compositions applied in China were about a new fusion protein which can be used to prevent/treat immune diseases (WO Patent No. 2015178746A1—GENEXINE INC, POHANG UNIVERSITY OF SCIENCE AND TECHNOLOGY) (Sung et al., 2015) or for other diseases, such as HPV-triggered cancer/cervical cancer (WO Patent No. 2012020871A1—BIOD CO LTD, GENEXINE INC; WO Patent No. 2007119896A1—GENEXINE INC, POHANG UNIVERSITY OF SCIENCE AND TECHNOLOGY) (Sung et al., 2007, 2012). With regard to the applications in Brazil, Russia and China by the company Genexine, it is important to highlight the compositions to treat growth hormone deficiency. One of them comprises a carrier and a recombinant human growth hormone (BR Patent No. 112018016817A2—TEVA PHARMACEUTICAL INDUSTRIES LTD., GENEXINE INC) (Kim et al., 2018). Another composition claimed was about a composition of polynucleotides which encodes fusion protein and it is used to treat cervical cancer (BR Patent No. 112017002809A2—GENEXINE INC) (Jin et al., 2017). Other documents which claimed immunoglobulin fusion proteins were also applied in these countries (HK Patent No. 1219490A1—GENEXINE INC, POHANG UNIVERSITY OF SCIENCE AND TECHNOLOGY; HK Patent No. 1219284A1—POHANG UNIVERSITY OF SCIENCE AND TECHNOLOGY | GENEXINE INC) (Sung, 2017; Sung and Yang, 2017).

ZyCoV-D Vaccine: It is a plasmid-based plasmid DNA[g] vaccine from Zydus Cadila that works against SARS-CoV-2's membrane protein. It is also the second indigenously-developed one, after Covaxin (from Bharat Biotech) against COVID-19 virus (Chandna, 2021). From the patent documents related to DNA technique of one of the members of the consortium, which has been also applied in BRICS' countries, two documents applied in the Russia Federation were recovered. One of them is about a protein which is expressed by cancer cells used to treat cancer (WO Patent No. 2013005164A2—CADILA HEALTHCARE LTD) (Desai et al., 2013). And a vaccine use to the treatment of cancer. This one comprises cancer cells to express the antigen (WO Patent No. 2012117323A1—CADILA HEALTHCARE LTD) (Darji et al., 2012). A retrieved document applied in Russia and China had as object of protection a "*new Mycobacterium w strain*" to treat desmocollin-3 expressing cancer (e.g., "*squamous type of*

[g] It is used in genetic engineering and consists in a circular, small and extrachromosomal bacterial DNA (Chandna, 2021).

non-small cell lung cancer") (WO Patent No. 2012095789A1—CADILA HEALTHCARE LTD) (Khamar and Modi, 2012). Indeed, for the applications in Brazil, China and South Africa, combined vaccines were claimed: a measles-malaria vaccine, a recombinant measles vaccine expressing malaria antigens (HK Patent No. 1221635A1—CADILA HEALTHCARE LTD; BR Patent No. PI1007584A2—CADILA HEALTHCARE LTD) (Billeter et al., 2016, 2017). Concerning applied documents in Brazil, China, India and South Africa, it is necessary to mention a process to produce antibodies with "glycans and/or charge species variants" (BR Patent No. 112015032800A2—CADILA HEALTHCARE LTD) (Bandyopadhyay et al., 2017). About the documents which were applied in India by CADILA HEALTHCARE LTD, the results were about a process of a subunit-based vaccine to prevent paratyphoid and typhoid fevers and Salmonella fevers (IN Patent No. 244897B) (Patel and Sinha, 2010) or a another process to isolate and purify polyribosyl ribitaol—the "*capsular polysaccharide of Haemophilus influenzae type B*" (IN Patent No. 257275B) (Dholakia et al., 2013). The development of new compounds were also claimed: to treat/prevent diseases as obesity and diabetes, substituted phenyl compounds (G-protein coupled receptor agonists) (IN Patent No. 201401495I3) (Desai and Srivastava, 2016). Or to treat hyperlipidemia or atherosclerosis, substituted heterocyclic derivatives ("*G-protein coupled receptor-40 agonists*") were object of protection (IN Patent No. 201502025I3) (Desai, 2016). Other heterocyclic compounds used in pharmaceutic compositions were objects of protection to treat arthritis or cancer (IN Patent No. 201621026238A) (Agarwal et al., 2018) or (IN Patent No. 201621026240A) (Agarwal et al., 2019).

AGO301 Vaccine: The DNA-based vaccine was developed by the Osaka University/Anges. The mechanism of action is the codification by the plasmid of the pathogen's protein in the cell. The gene transfer technology was developed by Takara Bio (Ferraz et al., 2020). From 83 documents related to DNA technique of one of the members of the consortium, 29 were applied at least in one of the BRICS' countries. The majority of patent fillings was in China (26 documents), with only one patent filled both in Brazil and China. Concerning Chinese applications, there are a variety of objects of protection: hepatitis virus polypeptides were claimed to treat and diagnose of hepatitis infections (EP Patent No. 933426A1 —OSAKA UNIVERSITY) (Fuke et al., 1999). To improve amplification systems and DNA synthesis, thermostable polypeptide factors, which promote the DNA polymerase activity, were claimed (WO Patent No. 1999000506A1—TAKARA HOLDINGS INC.) (Uemori et al., 1999). For disease treatment (e.g.) a new gene transfer method into cells (using adenovirus vector) with a high efficiency was claimed (WO Patent No. 2001032899A1—TAKARA HOLDINGS INC.) (Ueno et al., 2001). Moreover, a method to transfer a mutation into a nucleic acid useful in knockout organism' production and gene therapy was object of protection (WO Patent No. 2004007715A1—TAKARA BIO INC) (Cao et al., 2004). Furthermore, a new method to produce at low temperatures, target proteins was claimed (WO Patent No. 2005113768A1—TAKARA BIO INC) (Shodai et al., 2005). The document which was applied in China and Brazil was about a method to prepare plasmid/products of expression of one plasmid. Concerning this method, viral proteins were produced in quantity that can be used in vaccines and diagnostics (e.g.) (BR Patent No.199007566A—OSAKA UNIVERSITY) (Nakata et al., 1992). The applications by OSAKA UNIVERSITY and ANGES MG INC in China and India were about an agent (expression vector) used in injectable formulation to treat/improve diseases, such as hyperlipidemia/hypertension (WO Patent No. 2012141280A1) (Koriyama et al., 2012) or to treat/prevent cancer, another agent, also an expression vector, was claimed (WO Patent No. 2014034735A1) (Kyutoku et al., 2014).

6.2 RNA vaccines in clinical trials[h]: LNP-mRNA-1273, mRNA BNT-162b2, CVnCoV, ARCT-021, LNP-nCoVsaRNA and ARCoV

In the same way as DNA vaccines, RNA-based vaccines stand out for the relevance of their technologies for the prevention of SARS-CoV-2 (Von der Weid et al., 2020[i]) According these authors, it was possible to retrieve the patent documents which were applied in any BRICS' countries. Fig. 4 shows the scenario with the total of retrieved documents related to the RNA vaccines and the related applications in BRICS. The number of documents for each BRICS is shown in Table 3.

Based on this strategy search, we analyzed the documents which were applied in the BRICS countries (Table 3).

[h] The INPI study recovered documents related to different vaccine technologies: DNA, RNA, viral vectors, attenuated viral and protein subunit vaccines (Ferraz et al., 2020). For the present chapter, we recovered the documents related to the nucleic acid vaccines.

[i] The INPI developed a study (Von der Weid et al., 2020) where one of the aspects was to carry out the analysis of patent applications related to RNA vaccines at a more advanced clinical stage in the current scenario of the COVID-19 pandemic. The vaccines chosen for that study were **LNP-mRNA-1273, mRNA BNT-162b2, CVnCoV, ARCT-021, LNP-nCoVsaRNA and ARCoV.**

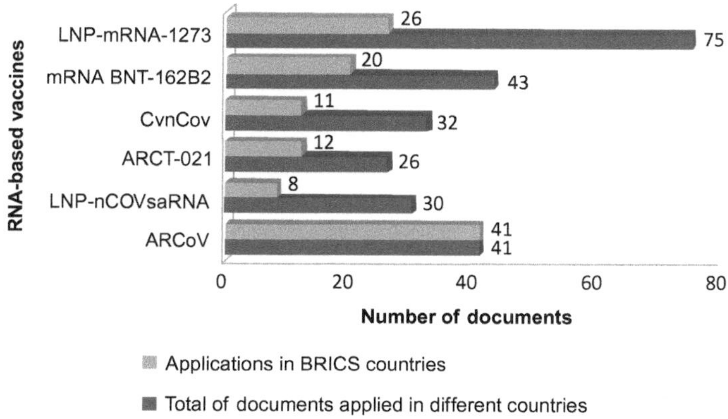

Fig. 4 Total of patent applied documents related to RNA vaccines and the number of applications in BRICS's countries. Note: For data acquisition, it was considered whether or not the document was applied in at least one of BRICS' countries. *(Source: elaborated by the authors based on Von der Weid, I., d'Urso de Souza Mendes, C., Ferraz, L.G., Chedid, N.G.B, Lourenço, A.L., Carestiato, T., Brazilian National Institute of Industrial Property, 2020. Panorama dos Documentos de Patente Relacionados às Vacinas de RNA em Testes Clínicos Para a Prevenção da COVID-19. Observatório de Tecnologias Associadas à COVID-19. Ministério da Economia, Brazil. Retrieved from: https://www.gov.br/inpi/pt-br/servicos/patentes/tecnologias-para-covid-19/Arquivos%20Textos/Estudo9vacinasmRNAdez2020.pdf.)*

From the scenario of patent applications containing relevant technologies related to RNA vaccines, the deposits rescued from the study of Von der Weid et al. (2020) were related to RNA vaccines for viral infections, in particular RNA viruses and respiratory viruses. They were also related to platforms that use RNA for gene therapy specifically for the development of antiviral vaccines. RNA and nanoliposomal modification methods have also been included (where applicable).

According to the documents' analysis, the following information was provided about RNA related technologies:

LNP-mRNA-1273 Vaccine: The novel lipid nanoparticle (LPN)-encapsulated messenger RNA-vaccine can encode SARS-CoV-2 pike protein (National Institutes of Health, 2019). From the patent documents related to RNA technique of one of the members of the consortium, modified polynucleotides were object of protection for diverse uses in documents applied in Brazil, Russia Federation, China and South Africa: to produce secreted proteins to the treatment of diseases such as lupus (e.g.) (US Patent No. 10385106B2—MODERNA THERAPEUTICS INC) (De Fougerolles and Guild, 2019a) or to produce nuclear proteins. It can

Table 3 Applied documents in BRICS' countries concerning to RNA-based technology.

RNA-based technology			Number of applied documents for BRICS's countries[a]				
Vaccine	**Developer Consortium**	**Total of applied patents considering each partner**	**Brazil**	**Russia Federation**	**India**	**China**	**South Africa**
LNP-mRNA-1273	Moderna/NIAID (United States)	74	22	21	6	26	11
mRNA BNT-162b2	BioNTech (Germany) Pfizer (United States)	43	13	9	11	18	7
CvnCoV	Curevac (Germany)	32	8	7	7	10	0
ARCT-021	Arcturus (United States) Duke-NUS (Singapore)	26	1	0	3	12	1
LNP-nCoVsaRNA	Imperial College London (United Kingdom)	30	2	0	4	5	2
ARCoV	Walvax Biotechnology/Academy of Military Medical Sciences (China)	41	0	0	0	41	0

[a]The document can be applied in more than one the BRICS' countries.
Source: elaborated by the authors based on Von der Weid, I., d'Urso de Souza Mendes, C., Ferraz, L.G., Chedid, N.G.B, Lourenço, A.L., Carestiato, T., Brazilian National Institute of Industrial Property, 2020. Panorama dos Documentos de Patente Relacionados às Vacinas de RNA em Testes Clínicos Para a Prevenção da COVID-19. Observatório de Tecnologias Associadas à COVID-19. Ministério da Economia, Brazil. Retrieved from: https://www.gov.br/inpi/pt-br/servicos/patentes/tecnologias-para-covid-19/Arquivos%20Textos/Estudo9vacinasmRNAdez2020.pdf.

be also used in the polypeptide of interest as antibodies, biologics, vaccines. It also can be used to the treatment of diseases such as diabetes (e.g.) (US Patent No. 20170028085A1—MODERNA THERAPEUTICS INC; EP Patent No. 3501550A1—MODERNA THERAPEUTICS INC) (Bancel et al., 2017, 2019). The modified polynucleotides *can encode "aryl hydrocarbon receptor nuclear translocator*" (US Patent No. 9050297B2—MODERNA THERAPEUTICS INC) (Chakraborty et al., 2015). Moreover it can be used to treat infectious diseases such as giardiasis and influenza (e.g.) (US Patent No. 10501512B2—MODERNA THERAPEUTICS INC; US Patent No. 9254311B2—MODERNA THERAPEUTICS INC) (Bancel et al., 2016; De Fougerolles and Guild, 2019b). Other object of protections applied in these countries was a terminally "*synthetic isolated terminally optimized messenger RNA*" which is used to reduce the antigen-mediated immune response (US Patent No. 9597380B2—MODERNA THERAPEUTICS INC) (Afeyan et al., 2017). Indeed a messenger RNA in a lipid nanoparticle can be used to treat acute diseases as stroke and sepsis (US Patent No. 20190290742A1—MODERNA THERAPEUTICS INC) (Afeyan et al., 2019). Documents applied in Brazil, Russia Federation and China are all of them about vaccine's development: "*Human cytomegalovirus RNA vaccines*" (US Patent No. 10064935B2 —MODERNA THERAPEUTICS INC) (Ciaramella et al., 2018); a "*Betacoronavirus mRNA vaccine*" (US Patent No. 10702600B1—MODERNA THERAPEUTICS INC) (Ciaramella and Himansu, 2020); "*combination PIV3/ hMPV RNA vaccines*" (US Patent No. 10272150B2—MODERNA THERAPEUTICS INC) (Ciaramella and Himansu, 2019a); "*Chikungunya virus RNA vaccines*" (US Patent No. 10238731B2—MODERNA THERAPEUTICS INC) (Ciaramella et al., 2019); sexually transmitted disease RNAvaccine (US Patent No. 10493143B2—MODERNA THERAPEUTICS INC) (Ciaramella and Himansu, 2019b). For Brazil, Russia Federation China and India, the applications were about influenza virus RNA vaccines (BR Patent No. 112018008078A2—MODERNA THERAPEUTICS INC; BR Patent No. 112016024644A2—MODERNA THERAPEUTICS INC) (Bouchon et al., 2017; Huang et al., 2018); a Herpes simplex virus RNA vaccine (BR Patent No. 112018008090A2—MODERNA THERAPEUTICS INC) (Bett et al., 2018); or respiratory syncytial virus RNA vaccine (BR Patent No. 112018008102A2—MODERNA THERAPEUTICS INC) (Espeseth et al., 2018).

mRNA BNT-162b2 Vaccine: The formulation is a RNA-lipid nanoparticle of modRNA (nucleoside modified mRNA) which encodes a P2 mutant spike protein of SARS-CoV-2 (WHO, 2020). From the applied patent documents related to RNA technique of one of the members of the consortium, about those which were applied in China, we recovered: a method of production of a liposome colloid where are included a method to prepare RNA lipoplex particles (WO Patent No. 2019077053A1—BIONTECH AG) (Bates et al., 2019). A pharmaceutical composition claimed to prevent infectious disease and cancer which comprises RNA to encode an antigen (EP Patent No. 3427723A1—BIONTECH AG; UNIV MAINZ TRON TRANSLATIONALE ONKOLOGIE) (Diken et al., 2019b) or RNA formulation used for immunotherapy (LT Patent No. 2830593T—BIONTECH AG; TRON TRANSLATIONALE ONKOLOGIE AN DER UNIV DER JOHANNES GUTENBERG UNIV MAINZ GEMEINNUETZIGE GMB) (Diken et al., 2019c). Applied documents in Brazil, Russia Federation and China referred to an aqueous formulation for pharmaceutical composition to ameliorate, treat or prevent diseases (BR Patent No. 112017005821A2—BIONTECH AG; WO Patent No. 2016045732A1—BIONTECH AG) (Haas and Borquez, 2016, 2017). With regard to the applications in Brazil, Russia Federation, China and South Africa, the documents claimed for a system of RNA construction and RNA replicon to produce protein in a subject (BR Patent No. 112018068391A2—BIONTECH AG; UNIV MAINZ TRON TRANSLATIONALE ONKOLOGIE) (Beissert et al., 2019a). Moreover, a system has been claimed used for RNA trans-replication with RNA construct and RNA replicon (WO Patent No. 2017162265A1—BIONTECH AG; UNIV MAINZ TRON TRANSLATIONALE ONKOLOGIE) (Beissert et al., 2017) and a method to provide single-stranded RNA (BR Patent No. 112018069417A2—BIONTECH AG) (Baiersdörfer and Karikó, 2019). Brazilian and Chinese documents claimed pharmaceutical compositions comprising single stranded RNA (e.g.) to introduce RNA into cell to prevent cancer and infectious diseases (e.g.) (BR Patent No. 112018077163A2—BIONTECH AG; UNIV MAINZ TRON TRANSLATIONALE ONKOLOGIE) (Diken et al., 2019a) and used in vaccine compositions and therapy (WO Patent No. 2018010815A1—BIONTECH AG; UNIV MAINZ TRON TRANSLATIONALE ONKOLOGIE) (Diken et al., 2018). Moreover, to the application in Brazil, Russia Federation, China, South Africa and India, the document claimed for RNA replicon which comprises "*a 5′ replication recognition sequence*," "*replicated by the functional alphavirus non-structural protein*" (BR Patent No. 112018068381A2—BIONTECH AG; UNIV MAINZ TRON TRANSLATIONALE ONKOLOGIE) (Beissert et al., 2019b). For applications in India, RNA's modification (US Patent No. 9476055B2) (Holtkamp et al., 2016) and recombinant vaccines (US Patent No. 8178653B2) (Kreiter et al., 2012) were claimed.

CvnCoV Vaccine: The mRNA-based vaccine from CureVac as a vaccine candidate against COVID-19 comprises a mRNA which was not chemically modified. Its development has begun in January 2020. The formulation is lipid nanoparticle-based and the mRNA is able to encodes the SARS-CoV-2's "*prefusion stabilized full-length spike protein*" (CureVac, 2021). From the applied patent documents related to RNA technique of one of the members of the consortium which were applied only in China, it can be considered a combination vaccine. Here the vaccine has at least two antigens. Both of them at least comprises RNA which is able to encode (with antigenic function): derivatives or variants of proteins, proteins or fragments (US Patent No. 10588959B2—CUREVAC AG) (Kallen et al., 2020). Indeed a method to produce RNA (different molecule species) for compositions was claimed. This method is performed in vitro in single reaction vessel: "*simultaneous RNA in vitro transcription of mixture of DNA molecule species*" (BR Patent No. 112019008481A2—ACUITAS THERAPEUTICS INC; CUREVAC AG) (Baumhof et al., 2020). For documents with applications in countries of interest as Brazil, Russia Federation and China, the applications referred to a composition of mRNA complexed with free mRNA and cationic compounds (adjuvant components) used in the treatment of tumors/cancers (e.g.) (US Patent No. 20170202957A1—CUREVAC AG) (Fotin-Mleczek and Voss, 2017). Furthermore, a new nucleic acid sequence was claimed to treat infections diseases as HIV, dengue and rabies (e.g.) (US Patent No. 10166283B2—CUREVAC AG; BR Patent No. 112014016361A2—CUREVAC AG) (Probst et al., 2018, 2019). For applications in China, Russia, Brazil and India, a vaccine against respiratory syncytial virus infections with a new mRNA sequence was object of protection (BR Patent No. 112016003361A2—CUREVAC AG) (Kramps et al., 2017). Indeed, a pharmaceutical composition containing RNA was claimed to prevent/treat cancer/tumor (BR Patent No. 112017018368A2—CUREVAC AG) (Baumhof et al., 2018) and an immunolatory composition to treat diseases as cancer or HIV (BR Patent No. PI0915844A2—CUREVAC AG) (Fotin-Mleczek and Voss, 2015). Besides a new nucleic acid with many activities as immunostimulant, virucide and antimicrobial, for example (US Patent No. 10441653B2—CUREVAC AG) (Hoerr et al., 2019), was applied in Russia Federation and China. The unique application in Brazil and India (among BRICS) was about a vaccine against Zika which comprises an artificial nucleic acid (BR Patent No. 112018016755A2—CUREVAC AG; SANOFI PASTEUR) (Girerd-Chambaz et al., 2018).

ARCT-021 Vaccine: Almost all documents were applied only in China which the focus of protection is the development of new compounds to increase RNA delivery (e.g.). About them, it can be described a cationic lipid compound used for the treatment of diseases such as inflammatory diseases and cancer. It is also used as material for excipients, additive, coating and bioengineering applications. From the applied patent documents related to RNA technique of one of the members of the consortium, it is important to highlight that these lipid nanoparticles can increase RNA's lifetime in the circulation (also applied in India) (AU Patent No. 2017379059B2—ARCTURUS THERAPEUTICS INC) (Chivukula et al., 2020a). Or a new carbonyl derivative used in pharmaceutical composition (with a mRNA to encode the interest protein) to treat cancer/inflammatory diseases (e.g.) which is in a lipid nanoparticle (US Patent No. 20200109113A1—ARCTURUS THERAPEUTICS INC) (Chivukula et al., 2020b). Moreover, new substituted ester compounds to treat renal, inflammatory and cardiovascular diseases (e.g.) are used to deliver the nucleic acid to the target cells. The compound can be used in a pharmaceutical composition in a lipid nanoparticle which encapsulates mRNA (US Patent No. 20190388562A1—ARCTURUS THERAPEUTICS INC) (Chivukula et al., 2019).

Or a new substituted oxycarbonyl compounds to deliver therapeutic and biologically active molecules (e.g., DNA, eDNA, mRNA, immunoglobulin proteins) applied in China and India to the treatment of immune or inflammatory diseases (e.g.) (IL Patent No. 245638A—ARCTURUS THERAPEUTICS INC) (Arcturus Therapeutics INC, 2020). Furthermore other ionizable cationic lipid compound to introduce the nucleic acid into the cell to the treatment of diseases (e.g., genetic abnormality) were also claimed (AU Patent No. 2015350561B2—ARCTURUS THERAPEUTICS INC) (Chivukula and Payne, 2020).

LNP-nCoVsaRNA Vaccine: The lipid nanoparticles (LNP) are used to formulate RNA vaccine with self-amplifying RNA (encapsulated saRNA), which is a relevant platform to the development of RNA-based vaccines (McKay et al., 2020). The applications related to the current platform were analyzed. From the applied patent documents related to RNA technique of one of the members of the consortium, the documents applied in Brazil were about a new dengue vaccine. For this, the immunogenic fragment (envelope dimer epitope—EDE—from the dengue virus) is encoded by the nucleic acid (BR Patent No. 112017001340A2—IMPERIAL INNOVATIONS LTD) (Barba-Spaeth et al., 2017). More one time, a nucleic acid encodes EDE to prevent disease caused by one or more flaviviruses, as dengue disease was applied in Brazil and India: (BR Patent No. 112018075396A2—IMPERIAL INNOVATIONS LTD; INSTITUT PASTEUR) (Rouvinski et al., 2019) was claimed The object of protection of the documents applied in China, India and South Africa were about a method to produce a vaccinal virus (from Reoviridae family) strain. To introduce the mutation into the virus, the cell's transfection with "*viral single stranded RNA*" is done for further cell culture (US Patent No. 8673315B2—UNIVERSITY OF LONDON) (Boyce and Roy, 2014). Vaccinal viral strain/isolated viral single stranded RNA (ssRNA) was claimed to vaccinate non-human animal against African horse sickness virus (e.g.) (US Patent No. 10688176B2—UNIVERSITY OF LONDON) (Roy, 2020).

Finally, a Chinese application was about nucleic acid constructs to produce (e.g.) retroviral vectors, such as lentiviral vector (US Patent No. 20190055568A1—UNIVERSITY COLLEGE LONDON) (Mekkaoui and Pulé, 2019).

ARCoV Vaccine: The LNP-encapsulated m-RNA vaccine is a thermostable vaccine which targets the SARS-CoV-2's receptor binding domain (Zhang et al., 2020). From the applications related to RNA technique of one of the members of the consortium, all the recovered documents were not applied outside Chinese territory. About the objects of protection, key findings emerged, such as an anti-coronavirus vaccine of recombinant coronavirus ("*RNA as genetic material and complementary DNA obtained by reverse transcription of the RNA*") (CN Patent No. 103316337B—ACADEMY OF MILITARY MEDICAL SCIENCES) (Chang et al., 2014) and about inventions related to functional/structural proteins in coronavirus (SARS) that can be used as immunogens (CN Patent No. 1566144A) (Fuchu et al., 2005). To prepare SARS vaccine/medicine and to treat SARS (e.g.), it was used the reovirus genome and the coding protein of the virus (CN Patent No. 1565628A—ACADEMY OF MILITARY MEDICAL SCIENCES) (Hong and Qing, 2005). What's more, a new kind of coronavirus which can produce a specific reaction to diagnostics, was claimed (CN Patent No. 1176210C—ACADEMY OF MILITARY MEDICAL SCIENCES) (Qingyu, 2004). Antibodies against SARS-CoV were also claimed to treat and prevent SARS-CoV-2, such as "*one kind of chicken anti-SARS-CoV IgY antibody egg*" (CN Patent No. 1268643C—ACADEMY OF MILITARY MEDICAL SCIENCES) (Xiliang, 2006) or "*an anti-SARS coronavirus F(ab')2 antibody fragment*" (CN Patent No. 1300172C—ACADEMY OF MILITARY MEDICAL SCIENCES) (Xiliang, 2007). Regard to compositions to treat and diagnose SARS, one comprises SARS polypeptides. Polynucleotides were also object of protection, which were used to encode not only these peptides, but also antigen cells, specific T-cells and antibodies (CN Patent No. 1566342B—ACADEMY OF MILITARY MEDICAL SCIENCES) (Dong et al., 2010). About RNA technology, it was also important to consider an applied document about small RNA molecules ("*small single-stranded or double-stranded RNA molecules*") used to prevent and cure SARS (CN Patent No. 1194003C—ACADEMY OF MILITARY MEDICAL SCIENCES) (Guang et al., 2005).

7. Discussion

We provided information on variants of concern according to location of emergence and related vaccines. The main challenge here is to determine if these mutations make the current vaccines, which were designed for the earlier versions of COVID-19 virus, less/no effective. There is a possibility that most of the vaccines (recombinant vaccines) or that which target the original D614G ancestral sequence of the virus spike (single epitope) might fail to generate an immune response capable of neutralizing these new variants. We also made a patent analysis of nucleic acid COVID-19 vaccines (DNA and RNA-based) at a more advanced clinical stage, which provided relevant technological information on deposits in BRICS countries, related to the development of compounds, formulations, delivery systems and novel technologies against diseases other than COVID-19, such as Zika, malaria, measles, tuberculosis and hepatitis. The patents for RNA and DNA technologies from deposits in the BRICS revealed that China is a leading country in terms of deposits given its great potential for the development of these new technologies. Considering the deposits in the BRICS of technologies related to DNA vaccines, we observed, from the total documents applied in different countries a variation of 32% to 84% of applications in at least one of the BRICS countries. For RNA vaccine technology, there was a variation of 27% to 100%, due to Chinese developers (Walvax Biotechnology/Academy of Military Medical Sciences) with priority deposits in China. It is important to consider that many of the analyzed patents will still be filed (those with WO registration). It also should be noted that considering the 18-month secrecy period, many documents related to these technologies could not be retrieved.

8. Conclusion

From a viral evolution perspective, universal access to fast, effective and widespread vaccination is the only way to respond to the COVID-19 pandemic, slowing down viral mutations and minimizing exposure of the more vulnerable populations in developing countries to concerning new COVID-19 viral variants. Immediate and universal global immunization is thus key to prevent new COVID-19 strains from evolving into resistance to vaccines, avoiding future aggravation of the pandemic. In our study we describe four new SARS-CoV-2 viral variants of concern (VOCs) which have emerged and globally disseminated in few months. These VOCs have important implications for vaccine development, since it is not clear yet how the immune system of vaccinated people will respond to contact with these variants: B.1.1.7, first detected in the United Kingdom; B.1.351 (first detected in South Africa); P1 (first detected in Brazil) and B.1.617.2 (first detected in India). Although additional research is needed, these variants of concern might be associated with important changes in morbidity and mortality profiles of populations and consequently a redesign of approved vaccines might be needed. Concerns on the protection provided by current vaccines and on the risk of immune escape have been a global issue in vaccine preparedness for the pandemic. As current vaccines have been designed for extensive immune response, it is expected that vaccines against one variant will protect against similar variants but not always against highly divergent variants, requiring redesign of vaccines and multivalent vaccines (Darby and Hiscox, 2021). Developing countries have been much slower to vaccinate their populations, since they have been excluded from access to most COVID-19 vaccines, concentrated in just 10 developed countries (WHO, 2021a). A global governance strategy for immediate universal immunization should be urgently put in place by the G7 and G-20 countries and multilateral negotiations in WTO, supporting developing countries vaccine manufacturers (DCVMN), particularly in the BRICS countries with significant vaccine manufacturing capacity, led by China, as indicated in our patent study. This new global governance strategy, incorporating patent flexibility and incentives for innovation, will be key not only to control the COVID-19 pandemic but to build sustainable global vaccine preparedness for possible future pandemics.

References

Afeyan, N.B., Bancel, S., Chakraborty, T., Hoge, S.G., Roy, A., de Fougerolles, A., 2019. US Patent No. 20190290742A1. United States Patent and Trademark Office.

Agarwal, S., Bahekar, R., Desai, R., 2018. IN Patent No. 201621026238A. Intellectual Property India.

Agarwal, S., Bahekar, R., Desai, R., 2019. IN Patent No. 201621026240A. Intellectual Property India.

Afeyan, N.B., Bancel, S., Chakraborty, T., Hoge, S.G., Roy, A., de Fougerolles, A., 2017. US Patent No. 9597380B2. United States Patent and Trademark Office.

Aguas, R., Corder, R.M., King, J.G., Gonçalves, G., Ferreira, M.U., Gomes, G.M., 2020. Herd immunity thresholds for SARS-CoV-2 estimated from unfolding epidemics. medRxiv. https://doi.org/10.1101/2020.07.23.20160762.

Arcturus Therapeutics INC, 2020. Patent No. 245638A. Israel Patent Office.

Baiersdörfer, M., Karikó, K., 2019. BR Patent No. 112018069417A2. Brazilian National Institute of Industrial Property.

Bancel, S., Chakraborty, T., De Fougerolles, A., Ejebe, K., Elbashir, S.M., Ellsworth, J.L., Guild, J., Hatala, P., John, M., Roy, A., Schrum, J.P., Whoriskey, S., Wood, K.M., 2016. US Patent No. 9254311B2. United States Patent and Trademark Office.

Bancel, S., Chakraborty, T., De Fougerolles, A., Ejebe, K., Elbashir, S.M., Ellsworth, J.L., Guild, J., Hatala, P., John, M., Roy, A., Schrum, J.P., Weiss, R., Whoriskey, S., Wood, K.M., 2017. US Patent No. 20170028085A1. United States Patent and Trademark Office.

Bancel, S., Chakraborty, T., De Fougerolles, A., Ejebe, K., Elbashir, S.M., Ellsworth, J.L., Guild, J., Hatala, P., John, M., Roy, A., Schrum, J.P., Whoriskey, S., Wood, K.M., 2019. EP Patent No. 3501550A1. European Patent Office.

Bandyopadhyay, S., Mendiratta, S.K., Patel, S., 2017. BR Patent No. 112015032800A2. Brazilian National Institute of Industrial Property.

Barba-Spaeth, G., Duquerroy, S., Guardado-Calvo, P., Mongkolsapaya, J., Rey, F.A., Rouvinski, A., Screaton, G., Vaney, M.C., 2017. BR Patent No. 112017001340A2. Brazilian National Institute of Industrial Property.

Bates, F., Borquez, I.H.E., Haas, H., Hiller, T.M., & Hörner, S. (2019). WO Patent No. 2019077053A1. World Intellectual Property Organisation.Bates, F., Borquez, I.H.E., Haas, H., Hiller, T.M., Hörner, S., 2019 WO Patent No. 2019077053A1. World Intellectual Property Organization.

Baumhof, P., Fotin-Mleczek, M., Heidenreich, R., Kallen, K.J., Kowalczyk, A., Probst, J., 2018. BR Patent No. 112017018368A2. Brazilian National Institute of Industrial Property.

Baumhof, P., Fotin-Mleczek, M., Heidenreich, R., Hope, M.J., Jasny, E., Lazzaro, S., Lin, P.J.C., Lutz, J., Mui, B., Petsch, B., Rauch, S., Schwendt, K.E., Tam, Y., 2020. BR Patent No. 112019008481A2. Brazilian National Institute of Industrial Property.

Beissert, T., Perkovic, M., & Sahin, U. (2017). WO Patent No. 2017162265A1. World Intellectual Property OrganisationBeissert, T., Perkovic, M., Sahin, U., 2017. WO Patent No. 2017162265A1. World Intellectual Property Organization.

Beissert, T., Perkovic, M., Sahin, U., 2019a. BR Patent No. 112018068391A2. Brazilian National Institute of Industrial Property.
Beissert, T., Perkovic, M., Sahin, U., 2019b. BR Patent No. 112018068381A2. Brazilian National Institute of Industrial Property.
Bett, A.J., Casimiro, D.R., Ciaramella, G., John, S., 2018. BR Patent No. 112018008090A2. Brazilian National Institute of Industrial Property.
Billeter, M., Fazio, A., Gianino, V., Glueck, R., 2016. BR Patent No. PI1007584A2. Brazilian National Institute of Industrial Property.
Billeter, M., Fazio, A., Gianino, V., Glueck, R., 2017. HK Patent No. 1221635A1. Intellectual Property Department, The Government of the Hong Kong Special Administrative Region.
Bogen, B., Mathiesen, I., & Tjelle, T.E. (2003). WO Patent No. 2003059952A1. World Intellectual Property Organisation.Bogen, B., Mathiesen, I., Tjelle, T.E., 2003. WO Patent No. 2003059952A1. World Intellectual Property Organization.
Bouchon, A., Huang, E.Y.C., Ciaramella, G., 2017. BR Patent No. 112016024644A2. Brazilian National Institute of Industrial Property.
Boyce, M., Roy, P., 2014. US Patent No. 8673315B2. United States Patent and Trademark Office.
Broderick, K., Kemmerrer, S., & Mccoy, J. (2013). WO Patent No. 2013066427A1. World Intellectual Property Organisation.Broderick, K., Kemmerrer, S., Mccoy, J., 2013. WO Patent No. 2013066427A1. World Intellectual Property Organization.
Campillo-Agusti, A., Ho, E.E., Lovell, N., Masterson, S., Stadelmann, B., Kemmerrer, S., 2018. BR Patent No. 112018013312A2. Brazilian National Institute of Industrial Property.
Cao, C., Sagawa, H., & Kato, I. (2004). WO Patent No. 2004007715A1. World Intellectual Property Organisation.Cao, C., Sagawa, H., Kato, I., 2004. WO Patent No. 2004007715A1. World Intellectual Property Organization.
Centers for Disease Control and Prevention—CDC, 2021a. SARS-CoV-2 Variant Classifications and Definitions. Retrieved from: https://www.cdc.gov/coronavirus/2019-ncov/cases-updates/variant-surveillance/variant-info.html.
Centers for Disease Control and Prevention—CDC, 2021b. Understanding mRNA COVID-19 Vaccines. Retrieved from: https://www.cdc.gov/coronavirus/2019-ncov/vaccines/different-vaccines/mrna.html.
Chandna, H., 2021. ZyCoV-D, the Made-in-India COVID Vaccine by Zydus Cadila, Likely to Hit Markets in June. The Print. Retrieved from: https://theprint.in/health/zycov-d-the-made-in-india-covid-vaccine-by-zydus-cadila-likely-to-hit-markets-in-june/657272/.
Chakraborty, T., De Fougerolles, A., Weiss, R., 2015. US Patent No. 9050297B2. United States Patent and Trademark Office.
Chang, G., Hu, Y., Kang, X., Li, J., Lin, L., Liu, H., Luo, Y., Sun, W., Wu, X., Yang, Y., Zhang, Y., Zhu, Q., 2014. CN Patent No. 103316337B. China National Intellectual Property Administration.
Chen, R.E., Zhang, X., Case, J.B., Winkler, E.S., Liu, Y., Van Blargan, L.A., Diamond, M.-S., 2021. Resistance of SARS-CoV-2 variants to neutralization by monoclonal and serum-derived polyclonal antibodies. Nat. Med. 27, 717–726. https://doi.org/10.1038/s41591-021-01294-w.
Chivukula, P., Payne, J.E., 2020. AU Patent No. 2015350561B2. Australian Patent Office.
Chivukula, P., Karmali, P., Payne, J.E., Tanis, S.P., 2019. US Patent No. 20190388562A1. United States Patent and Trademark Office.
Chivukula, P., Karmali, P., Payne, J.E., Tanis, S.P., 2020a. AU Patent No. 2017379059B2. Australian Patent Office.
Chivukula, P., Karmali, P., Payne, J.E., Tanis, S.P., 2020b. US Patent No. 20200109113A1. United States Patent and Trademark Office.

Chung, J.Y., Thone, M.N., Kwon, Y.J., 2021. COVID-19 vaccines: the status and perspectives in delivery points of view. Adv. Drug Deliv. Syst. 170, 1–25. https://doi.org/10.1016/j.addr.2020.12.011.

Ciaramella, G., Himansu, S., 2019a. US Patent No. 10272150B2. United States Patent and Trademark Office.

Ciaramella, G., Himansu, S., 2019b. US Patent No. 10493143B2. United States Patent and Trademark Office.

Ciaramella, G., Himansu, S., 2020. US Patent No. 10702600B1. United States Patent and Trademark Office.

Ciaramella, G., John, S., Mousavi, K., 2018. US Patent No. 10064935B2. United States Patent and Trademark Office.

Ciaramella, G., Elbashir, S.M., Himansu, S., Huang, E.Y.C., Zaks, T., 2019. US Patent No. 10238731B2. United States Patent and Trademark Office.

Connors, J., Bell, M.R., Marcy, J., Kutzler, M., Haddad, E.K., 2021. The impact of immuno-aging on SARS-CoV-2 vaccine development. Geroscience 43, 31–51. https://doi.org/10.1007/s11357-021-00323-3.

Cortelis Drug Discovery Inteligence—CDDI. Clarivate Analytics. Retrieved from: https://access.cortellis.com/login?app=drugdiscovery&referrer=%2Fdrugdiscovery%2Fhome.

CureVac, 2021. Celonic and Curevac Announce Agreement to Manufacture Over 100 Million Doses of CureVac's COVID-19 Vaccine Candidate, CVnCoV. Retrieved from: https://www.curevac.com/en/2021/03/30/celonic-and-curevac-announce-agreement-to-manufacture-over-100-million-doses-of-curevacs-covid-19-vaccine-candidate-cvncov/.

Darby, A.C., Hiscox, J.A., 2021. COVID-19: variants and vaccination. We have the tools to track variants and adapt vaccines as required. BMJ 372 (771), 1–2. https://doi.org/10.1136/bmj.n771. 33757984.

Darji, A.D., Desai, N.M., Khamar, B.M., Modi, I.A., & Shukla, C.P. (2012). WO Patent No. 2012117323A1. World Intellectual Property Organisation.Darji, A.D., Desai, N.-M., Khamar, B.M., Modi, I.A., Shukla, C.P., 2012. WO Patent No. 2012117323A1. World Intellectual Property Organization.

De Fougerolles, A., Guild, J., 2019a. US Patent No. 10385106B2. United States Patent and Trademark Office.

De Fougerolles, A., Guild, J., 2019b. US Patent No. 10501512B2. United States Patent and Trademark Office.

Desai, R., 2016. IN Patent No. 201502025I3. Intellectual Property India.

Desai, R.C., Srivastava, B.K., 2016. IN Patent No. 201401495I3. Intellectual Property India.

Desai, N.M., Khamar, B.M., Modi, I.A., Shukla, C.P., & Thakkar, V.S. (2013). WO Patent No. 2013005164A2. World Intellectual Property Organisation.Desai, N.M., Khamar, B.M., Modi, I.A., Shukla, C.P., Thakkar, V.S., 2013. WO Patent No. 2013005164A2. World Intellectual Property Organization.

Dholakia, A., Hakawat, S.S., Patel, P.R., Sinha, A., 2013. IN Patent No. 257275B. Intellectual Property India.

Diken, M., Erbar, S., Haas, H., Hörner, S., Kreiter, S., Herrero, J.M., Reuter, K., Sahin, U., Schlegel, A., Vogel, A., & Zucker, D. (2018). WO Patent No. 2018010815A1. World Intellectual Property Organisation.Diken, M., Erbar, S., Haas, H., Hörner, S., Kreiter, S., Herrero, J.M., Reuter, K., Sahin, U., Schlegel, A., Vogel, A., Zucker, D., 2018. WO Patent No. 2018010815A1. World Intellectual Property Organization.

Diken, M., Erbar, S., Haas, H., Hörner, S., Kreiter, S., Herrero, J.M., Reuter, K., Sahin, U., Schlegel, A., Vogel, A., Zucker, D., 2019a. BR Patent No. 112018077163A2. Brazilian National Institute of Industrial Property.

Diken, M., Fritz, D., Haas, H., Kranz, L.M., Kreiter, S., Meng, M., Reuter, K., Sahin, U., 2019b. EP Patent No. 3427723A1. European Patent Office.

Diken, M., Fritz, D., Haas, H., Kranz, L.M., Kreiter, S., Meng, M., Reuter, K., Sahin, U., 2019c. LT Patent No. 2830593T. State Patent Bureau of the Republic of Lithuania.

Dong, Z., Fu, A., Li, Q., Liu, G., Sun, M., 2010. CN Patent No. 1566342B. China National Intellectual Property Administration.

Edara, V.V., Floyd, K., Lai, L., Gardner, M., Hudson, W., Piantadosi, A., Suthar, M.S., 2021a. Infection and mRNA-1273 antibodies neutralize SARS-CoV-2 UK variant. medRvix. https://doi.org/10.1101/2021.02.02.21250799.

Edara, V.V., Norwood, C., Floyd, K., Lai, L., Davis-Gardner, M.E., Hudson, W.H., Suthar, M.S., 2021b. Reduced binding and neutralization of infection- and vaccine-induced antibodies to the B.1.351 (South Africa) SARS-CoV-2 variant. bioRxiv. https://doi.org/10.1101/2021.02.20.432046.

Edara, V.V., Norwood, C., Floyd, K., Lai, L., Davis-Gardner, M.E., Hudson, W.H., Suthar, M.S., 2021c. Infection- and vaccine-induced antibody binding and neutralization of the B.1.351 SARS-CoV-2 variant. Cell Host Microbe 29, 516–521. https://doi.org/10.1016/j.chom.2021.03.009.

Emary, K.R.W., Golubchik, T., Aley, P.K., Ariani, C.V., Angus, B., Bibi, S., Ramasamy, M.N., 2021. Efficacy of ChAdOx1 nCoV-19 (AZD1222) vaccine against SARS-CoV-2 variant of concern 20201201 (B.1.1.7): an exploratory analysis of a randomised controlled trial. Lancet 10 (397), 1351–1362. https://doi.org/10.1016/S0140-6736(21)00628-0.

Espeseth, A., Bett, A.J., Wang, D., Ciaramella, G., Bahl, K., 2018. BR Patent No. 112018008102A2. Brazilian National Institute of Industrial Property.

European Centre for Disease Prevention and Control—ECDC, 2021. SARS-CoV-2 Variants of Concern as of 3 June 2021. Retrieved from: https://www.ecdc.europa.eu/en/covid-19/variants-concern.

Ferraz, L.G.W., d'Urso de Souza Mendes, C., Von der Weid, I., Lourenço, A.L., Chedid, N.G.B., Carestiato, T., Brazilian National Institute of Industrial Property, 2020. Vacinas Baseadas em DNA Para a Prevenção da COVID-19: Mecanismo de ação, Ensaios Clínicos e Pedidos de Patentes. Observatório de Tecnologias Associadas à COVID-19, Ministério da Economia, Brazil. Retrieved from: https://www.gov.br/inpi/pt-br/servicos/patentes/tecnologias-para-covid-19/Arquivos%20Textos/Estudo8Vacinas.pdf.

Fotin-Mleczek, M., Voss, S., 2015. BR Patent No. PI0915844A2. Brazilian National Institute of Industrial Property.

Fotin-Mleczek, M., Voss, S., 2017. US Patent No. 20170202957A1. United States Patent and Trademark Office.

Fuchu, H., Qingyu, Z., Xiaohong, Q., 2005. CN Patent No. 1566144A. China National Intellectual Property Administration.

Fuke, I., Mori, C., Okayama, H., Takamizawa, A., Yoshida, I., 1999. EP Patent No. 933426A1. European Patent Office.

Funk, T., Pharris, A., Spiteri, G., Bundle, N., Melidou, A., Carr, M., Adlhoch, C., 2021. Characteristics of SARS-CoV-2 variants of concern B.1.1.7, B.1.351 or P.1: data from seven EU/EEA countries, weeks 38/2020 to 10/2021. Euro Surveill. 26 (16). https://doi.org/10.2807/1560-7917.ES.2021.26.16.2100348.

Garcia-Beltran, W.F., Lam, E.C., Denis, K.S., Nitido, A.D., Garcia, Z.H., Hauser, B.M., Balazs, A.B., 2021. Multiple SARS-CoV-2 variants escape neutralization by vaccine-induced humoral immunity. Cell 184, 2372–2383. https://doi.org/10.1016/j.cell.2021.03.013.

Girerd-Chambaz, Y., Jasny, E., Petsch, B., 2018. BR Patent No. 112018016755A2. Brazilian National Institute of Industrial Property.

Goodwin, K., Viboud, C., Simonsen, L., 2006. Antibody response to influenza vaccination in the elderly: a quantitative review. Vaccine 24 (8), 1159–1169. https://doi.org/10.1016/j.vaccine.2005.08.105.

Guang, Y., Jie, L., Ningsheng, S., 2005. CN Patent No. 1194003C. China National Intellectual Property Administration.
Gupta, R.K., 2021. Will SARS-CoV-2 variants of concern affect the promise of vaccines. Nat. Rev. Immunol. 21, 340–341. https://doi.org/10.1038/s41577-021-00556-5.
Haas, H., & Borquez, I.H.E. (2016). WO Patent No. 2016045732A1. World Intellectual Property Organisation.Haas, H., Borquez, I.H.E., 2016. WO Patent No. 2016045732A1. World Intellectual Property Organization.
Haas, H., Borquez, I.H.E., 2017. BR Patent No. 112017005821A2. Brazilian National Institute of Industrial Property.
Hitchings, M.D.T., Ranzani, O.T., Torres, M.S.S., de Oliveira, S.B., Almiron, M., Said, R., Siqueira, A.M., 2021. Effectiveness of Corona Vac in the setting of high SARS-CoV-2 P.1 variant transmission in Brazil: a test-negative case-control study. medRxiv. https://doi.org/10.1101/2021.04.07.21255081.
Hoerr, I., Ketterer, T., Probst, J., Scheel, B., 2019. US Patent No. 10441653B2. United States Patent and Trademark Office.
Holtkamp, S., Kreiter, S., Sahin, U., Tuereci, O., 2016. US Patent No. 9476055B2. United States Patent and Trademark Office.
Hong, Z., Qing, D., 2005. CN Patent No. 1565628A. China National Intellectual Property Administration.
Huang, E.Y.C., Ciaramella, G., Flynn, J.A., Babaoglu, K., Zhang, L., 2018. BR Patent No. 112018008078A2. Brazilian National Institute of Industrial Property.
Huang, Q., Zeng, J., Yan, J., 2021. COVID-19 mRNA vaccines. J. Genet. Genomics, 1–8. https://doi.org/10.1016/j.jgg.2021.02.006.
Inovio, 2021. Inovio Annouces Positive Data From Phase 2 Segment of Clinical Trial Evaluating INO-4800, Its COVID-19 DNA Vaccine. Retrieved from: https://ir.inovio.com/news-releases/news-releases-details/2021/INOVIO-Announces-Positive-Data-from-Phase-2-Segment-of-Clinical-Trial-Evaluating-INO-4800-its-COVID-19-DNA-Vaccine/default.aspx.
James, J.J., 2020. Yin and Yang and Herd immunity. Disaster Med. Public Health Prep. 14 (6), e12–e14. https://doi.org/10.1017/dmp.2020.229.
Jin, H.T., Lim, H.S., Suh, Y.S., Nam, E.J., 2017. BR Patent No. 112017002809A2. Brazilian National Institute of Industrial Property.
Kallen, K.J., Kramps, T., Schnee, M., Voss, D., 2020. US Patent No. 10588959B2. United States Patent and Trademark Office.
Kemmerrer, S.V., Mccoy, J., & Broderick, K. (2011). WO Patent No. 2011137221A1. World Intellectual Property Organisation.Kemmerrer, S.V., Mccoy, J., Broderick, K., 2011. WO Patent No. 2011137221A1. World Intellectual Property Organization.
Khamar, B.M., & Modi, I.A. (2012). WO Patent No. 2012095789A1. World Intellectual Property Organisation.Khamar, B.M., Modi, I.A., 2012. WO Patent No. 2012095789A1. World Intellectual Property Organization.
Kim, T.K., Ahn, Y.J., Woo, J.W., Cha, J.E., Lee, J.Y.J., Jang, W.I., 2018. BR Patent No. 112018016817A2. Brazilian National Institute of Industrial Property.
Kjeken, R., Mathiesen, I., Tjelle, T.E., & Mchugh, G. (2007). Patent No. 2007095140A2. World Intellectual Property Organisation.Kjeken, R., Mathiesen, I., Tjelle, T.E., Mchugh, G., 2007. Patent No. 2007095140A2. World Intellectual Property Organization.
Koriyama, H., Morishita, R., Nakagami, F., Nakagami, H., & Yoshida, N. (2012). WO Patent No. 2012141280A1. World Intellectual Property Organisation.Koriyama, H., Morishita, R., Nakagami, F., Nakagami, H., Yoshida, N., 2012. WO Patent No. 2012141280A1. World Intellectual Property Organization.
Kramps, T., Petsch, B., Schnee, M., Voss, D., 2017. BR Patent No. 112016003361A2. Brazilian National Institute of Industrial Property.

Kreiter, S., Sahin, U., Tuereci, O., 2012. US Patent No. 8178653B2. United States Patent and Trademark Office.

Kyutoku, M., Morishita, R., Nakagami, H., & Tomioka, H. (2014). WO Patent No. 2014034735A1. World Intellectual Property Organisation.Kyutoku, M., Morishita, R., Nakagami, H., Tomioka, H., 2014. WO Patent No. 2014034735A1. World Intellectual Property Organization.

Mccoy, J., Broderick, K., & Kemmerer, S. (2017). WO Patent No. 2017117273A1. World Intellectual Property OrganisationMccoy, J., Broderick, K., Kemmerer, S., 2017. WO Patent No. 2017117273A1. World Intellectual Property Organization.

McKay, P.F., Hu, K., Blakney, A.K., Samnuan, K., Bouton, C.R., Rogers, P., Shattock, R.-J., 2020. Self amplifying RNA SARS-CoV-2 lipid nanoparticle vaccine induces equivalente preclincal antibodies titer and viral neutralization to recovered COVID-19 patients. BioRxiv. https://doi.org/10.1101/2020.04.22.055608.

Mekkaoui, L., Pulé, M., 2019. US Patent No. 20190055568A1. United States Patent and Trademark Office.

Moderna, 2021a. Moderna Announces Positive Initial Booster Data Againt SARS-CoV-2 Variants of Concern. Retrieved from: https://investors.modernatx.com/news-releases/news-release-details/moderna-announces-positive-initial-booster-data-against-sars-cov/#:~:text=Statements%20%26%20Perspectives-,Moderna%20Announces%20Positive%20Initial%20Booster%20Data%20Against,CoV%2D2%20Variants%20of%20Concern&text=CAMBRIDGE%2C%20Mass.&text=A%20booster%20dose%20of%20mRNA%2D1273.351%2C%20the%20Company's%20strain%2D,booster%20dose%20of%20mRNA%2D1273.

Moderna, 2021b. Moderna COVID-19 Vaccine Retains Neutralizing Activity Against Emerging Variants First Identified in the U.K. and The Republic of South Africa. Retrieved from: https://investors.modernatx.com/news-releases/news-release-details/moderna-covid-19-vaccine-retains-neutralizing-activity-against/#:~:text=Statements%20%26%20Perspectives-,Moderna%20COVID%2D19%20Vaccine%20Retains%20Neutralizing%20Activity%20Against%20Emerging%20Variants,the%20Republic%20of%20South%20Africa&text=1.351%2C%20first%20identified%20in%20the,1.1.

Nakata, A., Saito, A., Shinagawa, H., Tanaka, N., 1992. BR Patent No. 199007566A. Brazilian National Institute of Industrial Property.

Nanda, G.S., Laverdiere, R., & Hofmann, G.A. (2000). WO Patent No. 2000023563A1. World Intellectual Property Organisation.Nanda, G.S., Laverdiere, R., Hofmann, G.A., 2000. WO Patent No. 2000023563A1. World Intellectual Property Organization.

National Institutes of Health—NIH, 2019. Clinical Trials.gov. Safety and Immunogenicity Study of2019-nCoV Vaccine (mRNA-1273) for Prophylaxis of SARS-CoV-2 Infection (COVID-19). Retrieved from: https://clinicaltrials.gov/ct2/show/NCT04283461.

National Institutes of Health—NIH, 2020. Clinical Trials.gov. Safety and Immunogenicity Study of GX-19, a COVID-19 Preventive DNA Vaccine in Health Adults. Retrieved from: https://clinicaltrials.gov/ct2/show/study/NCT04445389.

National Institutes of Health—NIH, 2021. Clinical Trials.gov. NIH Clinical Trial Evaluating Moderna COVID-19 Variant Vaccine Begins. Retrieved from: https://www.nih.gov/news-events/news-releases/nih-clinical-trial-evaluating-moderna-covid-19-variant-vaccine-begins.

Novavax, 2021a. Novavax Announces Positive Data From Three Complementary Studies of COVID-19 Beta (B.1351) Variant Strain Vaccine. Retrieved from: https://ir.novavax.com/pressreleases?ae45096c_field_nir_news_category%5Bvalue%5D%5B%5D=4041&ae45096c_year%5Bvalue%5D=2021&op=Apply&ae45096c_widget_id=ae45096c&form_build_id=formKV_yv_GhPtrsSdMa2pE9jxPciOYpb25eeXb2qYTLxQk&form_id=widget_form_base#inner-content.

Novavax, 2021b. Novavax Confirms High Levels of Efficacy Against Original and Variant COVID-19 Strains in United Kingdom and Souh Africa trials. Retrieved from: https://ir.novavax.com/news-releases/news-release-details/novavax-confirms-high-levels-efficacy-against-original-and-0.

Ocugen, 2021. Ocugen Announces Studies Showing COVAXIN Potentially Effective Against Three Key Variants of SARS-CoV-2. Retrieved from: https://www.globenewswire.com/en/news-release/2021/05/03/2221269/0/en/Ocugen-announces-studies-showing-COVAXIN-potentially-effective-against-three-key-variants-of-SARS-CoV-2.html#:~:text=Ocugen%20is%20developing%20COVAXIN%2C%20Bharat,19%20disease%20(including%20hospitalization.

Organisation for Economic Co-operation and Development—OECD (2021). OECD Policy Responses to Coronavirus (COVID-19). Access to COVID-19 Vaccines: Global Approaches in a Global Crisis. Retrieved from: https://www.oecd.org/coronavirus/policy-responses/access-to-covid-19-vaccines-global-approaches-in-a-global-crisis-c6a18370/.Organisation for Economic Co-operation and Development—OECD, 2021. OECD Policy Responses to Coronavirus (COVID-19). Access to COVID-19 Vaccines: Global Approaches in a Global Crisis. Retrieved from: https://www.oecd.org/coronavirus/policy-responses/access-to-covid-19-vaccines-global-approaches-in-a-global-crisis-c6a18370/.

Pango Lineages, 2021a. Global Report Investigating Novel Coronavirus Haplotypes. Retrieved from: https://cov-lineages.org/global_report.html.

Pango Lineages, 2021b. Lineages Description List. Retrieved from: https://cov-lineages.org/lineage_description_list.html.

Passos, G.A., Santos, J.C., 2021. The high infectivity of SARS-CoV-2 B.1.1.7 is associated with increased interaction force between Spike-ACE2 caused by the viral N501Y mutation. bioRvix. https://doi.org/10.1101/2020.12.29.424708.

Patel, P.R., Sinha, A., 2010. IN Patent No. 244897B. Intellectual Property India.

Pfizer, 2021. Pfizer and BioNTech Initiate a Study as Part of a Broad Development Plan to Evaluate COVID-19 Booster and New Vaccine Variants. Retrieved from: https://www.pfizer.com/news/press-release/press-release-detail/pfizer-and-biontech-initiate-study-part-broad-development.

Possas, C., Antunes, A.M.S., Mendes, F.M.L., Schumacher, S.O.R., Martins, R.M., Homma, A., 2015. Access to new technologies in multipatented vaccines: challenges for Brazil. Nat. Biotechnol. 33 (6), 599–603. https://doi.org/10.1038/nbt.3244.

Possas, C., Marques Jr., E.T.A., Risi Jr., J.B., Homma, A., 2021a. COVID-19 and future Disease X in circular economy transition: redesigning pandemic preparedness to prevent a global disaster. Circ. Econ. Sustain. https://doi.org/10.1007/s43615-021-00060-x.

Possas, C., Antunes, A.M.S., de Oliveira, A.M., Mendes, C.D.S., Ramos, M.P., Schumacher, S.-O.R., Homma, A., 2021b. Vaccine innovation for pandemic preparedness: patent landscape, global sustainability and circular bioeconomy in post-COVID-19 era. Circ. Econ. Sustain. doi. https://doi.org/10.1007/s43615-021-00051-y.

Probst, J., Schlake, T., Thess, A., 2018. BR Patent No. 112014016361A2. Brazilian National Institute of Industrial Property.

Probst, J., Schlake, T., Thess, A., 2019. US Patent No. 10166283B2. United States Patent and Trademark Office.

Qingyu, Q.Z., 2004. CN Patent No. 1176210C. China National Intellectual Property Administration.

Ranzani, O.T., Hitchings, M., Nieto, M.D., D'Agostini, T.L., de Paula, R.C., de Paula, O.-F.P., Croda, J., 2021. Effectiveness of the CoronaVac vaccine in the elderly population during a P.1 variant-associated epidemic of COVID-19 in Brazil: a test-negative case control study. MedRxiv. https://doi.org/10.1101/2021.05.19.21257472.

Rego, G.N.A., Nucci, M.P., Alves, A.H., Oliveira, F.A., Marti, L.C., Nucci, L.P., Gamarra, L.F., 2020. Current clinical trials protocols and the global effort for immunization against SARS-CoV-2. Vaccines 8 (474), 1–44. https://doi.org/10.3390/vaccines8030474.

Roberts, M., 2021. What are the India, Brazil, South Africa and UK variants? BBC News. Retrieved from: https://www.bbc.com/news/health-55659820.

Rouvinski, A., Rey, F., Screaton, G., Barba-Spaeth, G., Mongkolsapaya, J., Vaney, M.C., 2019. BR Patent No. 112018075396A2. Brazilian National Institute of Industrial Property.

Roy, P., 2020. US Patent No. 10688176B2. United States Patent and Trademark Office.

Sansone, E., Tiraboschi, M., Sala, E., Albini, E., Lombardo, M., Castelli, F., De Palma, G., 2021. Effectiveness of BNT162b2 vaccine against the B.1.1.7 variant of SARS-CoV-2 among healthcare workers in Brescia, Italy. J. Infect. 7. https://doi.org/10.1016/j.jinf.2021.04.038.

Sapkal, G.N., Yadav, P.D., Ella, R., Deshpande, G.R., Sahay, R.R., Gupta, N., Bhargava, B., 2021. Inactivated COVID-19 vaccine BBV152/COVAXIN effectively neutralzes recently emerged B.1.1.7 variant of SARS-CoV-2. J. Travel Med. 27. https://doi.org/10.1093/jtm/taab051.

Seow, J., Grahan, C., Merrick, B., Acors, S., Steel, K.J.A., Hemmings, O., Doores, K.J., 2020. Longitudinal evaluation and decline of antibody responses in SARS-CoV-2 infection. medRvix. https://doi.org/10.1101/2020.07.09.20148429.

Shinde, V., Bhikha, S., Hoosain, Z., Archary, M., Bhorat, Q., Fairlie, L., Madhi, S.A., 2021. Efficacy of NVX-CoV2373 COVID-19 vaccine against the B.1351 variant. N. Engl. J. Med. 384 (20), 1898–1909. https://doi.org/10.1056/NEJMoa2103055.

Sung, Y.C., 2017. HK Patent No. 1219490A1. Intellectual Property Department, The Government of the Hong Kong Special Administrative Region.

Sung, Y.C., Yang, S., 2017. HK Patent No. 1219284A1. Intellectual Property Department, The Government of the Hong Kong Special Administrative Region.

Sung, Y.C., Jin, H.T. Seo, S.H., Park, S., & Youn, J.I. (2007). WO Patent No. 2007119896A1. World Intellectual Property Organisation.Sung, Y.C., Jin, H.T., Seo, S.H., Park, S., Youn, J.I., 2007. WO Patent No. 2007119896A1. World Intellectual Property Organization.

Sung, Y.C., Seo, S.H., & Suh, Y.S. (2012). WO Patent No. 2012020871A1. World Intellectual Property Organisation.Sung, Y.C., Seo, S.H., Suh, Y.S., 2012. WO Patent No. 2012020871A1. World Intellectual Property Organization.

Sung, Y.C., Lee, J.Y., Song, M.Y., Lim, H.S., & Lee, B.H. (2015). WO Patent No. 2015178746A1. World Intellectual Property Organisation.Sung, Y.C., Lee, J.Y., Song, M.Y., Lim, H.S., Lee, B.H., 2015. WO Patent No. 2015178746A1. World Intellectual Property Organization.

Stadelmann, B., 2018. BR Patent No. 112018013242A2. Brazilian National Institute of Industrial Property.

Shodai, T., Kobori, H., Sagara, T., Endo, H., Takakura, H., Tomono, J., Sagawa, H., Mukai, H., & Kato, I., (2005). WO Patent No. 2005113768A1. World Intellectual Property Organisation.Shodai, T., Kobori, H., Sagara, T., Endo, H., Takakura, H., Tomono, J., Sagawa, H., Mukai, H., Kato, I., 2005. WO Patent No. 2005113768A1. World Intellectual Property Organization.

Supasa, P., Zhou, D., Dejnirattisai, W., Liu, C., Mentzer, A.J., Ginn, H.M., Screaton, G.R., 2021. Reduced neutralization of SARS-CoV-2 B.1.1.7 variant by convalescente and vaccine sera. Cell 184, 2201–2211. https://doi.org/10.1016/j.cell.2021.02.033.

Tada, T., Dcosta, B.M., Samanovic-Golden, M., Herati, R.S., Cornelius, A., Mulligan, M.J., Landau, N.R., 2021. Neutralization of viruses with European, South African, and United States SARS-CoV-2 variant spike proteins by convalescente sera and BNT162b2 mRNA vaccine-elicited antibodies. BioRxiv. https://doi.org/10.1101/2021.02.05.430003.

Uemori, T., Sato, Y., Fujita, T., Miyake, K., Mukai, H., Asada K., & Kato I., (1999). WO Patent No. 1999000506A1. World Intellectual Property Organisation.Uemori, T., Sato, Y., Fujita, T., Miyake, K., Mukai, H., Asada, K., Kato, I., 1999. WO Patent No. 1999000506A1. World Intellectual Property Organization.

Ueno, T., Matsumura, H., Tanaka, K., Iwasaki, T., Ueno, M., Fujinaga, K., Asada, K., & Kato, I., (2001). WO Patent No. 2001032899A1. World Intellectual Property Organisation.Ueno, T., Matsumura, H., Tanaka, K., Iwasaki, T., Ueno, M., Fujinaga, K., Asada, K., Kato, I., 2001. WO Patent No. 2001032899A1. World Intellectual Property Organization.

Virginia Department of Health—VDH, 2021. Variants of the Virus That Causes COVID-19. Retrieved from: https://www.vdh.virginia.gov/coronavirus/variants/.

Von der Weid, I., d'Urso de Souza Mendes, C., Ferraz, L.G., Chedid, N.G.B., Lourenço, A.L., Carestiato, T., Brazilian National Institute of Industrial Property, 2020. Panorama dos Documentos de Patente Relacionados às Vacinas de RNA em Testes Clínicos Para a Prevenção da COVID-19. Observatório de Tecnologias Associadas à COVID-19, Ministério da Economia, Brazil. Retrieved from: https://www.gov.br/inpi/pt-br/servicos/patentes/tecnologias-para-covid-19/Arquivos%20Textos/Estudo9vacinasmRNAdez2020.pdf.

World Health Organization—WHO, 2020. mRNA Vaccines Against COVID-19: Pfizer-BioNTech COVID-19 Vaccine BNT162b2. Retrieved from: https://apps.who.int/iris/bitstream/handle/10665/338096/WHO-2019-nCoV-vaccines-SAGE_evaluation-BNT162b2-2020.1-eng.pdf?sequence=1&isAllowed=y.

World Health Organization—WHO, 2021a. Declaration of WHO Director General to Parliamentary Assembly of the Council of Europe (PACE) Covid-19 Vaccines Must be a 'Global Public Good, Available to Everyone, Everywhere'. https://pace.coe.int/en/news/8184/covid-19-vaccines-must-be-a-global-public-good-available-to-everyone-everywhere. (Accessed 5 April 2021).

World Health Organization—WHO, 2021b. An Update on SARS-CoV-2 Virus Mutations & Variants. The Latest on the COVID-19 Global Situation & The Emergence of News Mutations & Variants. Retrieved from: https://www.who.int/docs/default-source/coronaviruse/risk-comms-updates/update47-sars-cov-2-variants.pdf?sfvrsn=f2180835_4.

World Health Organization—WHO, 2021c. The Effects of Virus Variants on COVID-19 Vaccines. Retrieved from: https://www.who.int/news-room/feature-stories/detail/the-effects-of-virus-variants-on-covid-19-vaccines.

World Health Organization—WHO, 2021d. Tracking SARS-CoV-2 Variants. Retrieved from: https://www.who.int/en/activities/tracking-SARS-CoV-2-variants/.

World Health Organization—WHO, 2021e. News Updates. Latest News From WHO on COVID-19 and Other Breaking Health Stories. Retrieved from: https://www.who.int/news-room/news-updates.

Xie, X., Liu, Y., Liu, J., Zhang, X., Zou, J., Fontes-Garfias, C.R., Shi, P.Y., 2021. Neutralization of SARS-CoV-2 spike 69/70 deletion, E484K and N501Y variants by BNT162b2 vaccine-elicited sera. Nat. Med. 27, 620–621. https://doi.org/10.1038/s41591-021-01270-4.

Xiliang, D.W., 2006. CN Patent No. 1268643C. China National Intellectual Property Administration.

Xiliang, D.W., 2007. CN Patent No. 1300172C. China National Intellectual Property Administration.

Zhang, N.-N., Li, X.F., Deng, Y.Q., Zhao, H., Huang, Y.J., Yang, G., Qin, C.F., 2020. A thermostable mRNA vaccine against COVID-19. Cell 182, 1271–1283. https://doi.org/10.1016/j.cell.2020.07.024.

CHAPTER 4

Peptide production by molecular farming with antiviral effects

Masoumeh Kordi[a], Zeynab Borzouyi[b], Robab Salami[a], and Behnam Asgari Lajayer[c]
[a]Department of Cell & Molecular Biology, Faculty of Life Sciences & Biotechnology, Shahid Beheshti University, Tehran, Iran
[b]Agriculture Department, Islamic Azad University of Sabzevar, Sabzevar, Khorasan Razavi, Iran
[c]Department of Soil Science, Faculty of Agriculture, University of Tabriz, Tabriz, Iran

1. Introduction

The high demand for recombinant proteins used in the pharmaceutical industry has created the opportunity to employ different expression hosts for veterinary and human applications. The pharmaceutical industry recombinantly produces the desired protein with prokaryotic and eukaryotic expression systems such as *Escherichia coli*, yeast, insects, mammals, and plants (Shanmugaraj et al., 2020). All production systems have their own advantages and disadvantages. In the bacterial expression system, for example, we have rapid crop production and high yield, while the yeast system is able to apply post-translational modifications to the functional activity of recombinant proteins (Vieira Gomes et al., 2018). Most approved recombinant biopharmaceuticals are produced in mammalian cell lines (Lalonde and Durocher, 2017). In order to compete, new expression systems must have unique advantages to be able to overcome existing limitations. Plant systems have several advantages over conventional expression platforms. Advances in plant molecular production methods in the last decade have made them an attractive production system (Shanmugaraj and Ramalingam, 2014; Burnett and Burnett, 2020). Antimicrobial peptides are interact with a very wide range of human, plant and animal pathogenic microorganisms (Thevissen et al., 2007). Small peptides, called peptide antibiotics, are a new generation of agents for plant protection and the treatment of human and animal microbial diseases (Bowdish et al., 2005). Some cationic peptides have antiviral activity. The three-dimensional structure of the peptide and the interaction of the peptide with the microbial membrane through hydrophobic and electrostatic forces determine the mode of action of the peptide. Although most

Agricultural Bioeconomy
https://doi.org/10.1016/B978-0-323-90569-5.00016-0

antimicrobial peptides have an amphipathic structure, their secondary structures are between different classes (e.g., α-helices, β-sheets, disordered loops, and extended structures) (Holaskova et al., 2015).

2. Molecular farming for production drugs and vaccines

Molecular farming was first performed to produce medicinal proteins in tobacco and potatoes in 1990 (Sijmons et al., 1990). Fifteen years later, the first medicinal protein produced in the transgenic plant entered the market (Twyman et al., 2003). Several plant-derived proteins suitable for the treatment of human diseases such as gastriclipase for the treatment of cystic fibrosis and antibodies for the prevention of dental caries and the treatment of non-Hodgkins lymphoma are close to commercialization (Ma et al., 2003). Recently, several livestock-derived vaccines in plants are being developed (Aguirreburualde et al., 2016). With the advent of molecular agriculture, advances have been made in gene transfer methods, control of gene expression, protein accumulation, and alteration of the structural and functional properties of peptides (Leite et al., 2019). One of the advantages of using plants in protein production is that Manipulated plants have unparalleled potential for large-scale protein production. Large-scale production capacity of manipulated plants is one of the few solutions to overcome the commercial and ethical challenges to the production of drugs and vaccines. The cost of producing vaccines and protein therapies in plants is low, the low cost of production is the main factor in choosing the production method in developing countries. Another advantage of the plant system is the ability of plants to produce a variety of recombinant molecules (Laere et al., 2016). Plants are able to synthesize small peptides, polypeptides, and complex proteins that microbial systems do not. Plants are able to make post-translational modifications to the recombinant protein produced (Ma et al., 2005; Leite et al., 2019). When choosing the right plant host, the health risks of contamination with microbial toxins and human pathogens are minimized (Holaskova et al., 2015). Fig. 1 summarizes the steps of identification, production of antiviral peptide in the plant to its commercialization.

3. Antimicrobial peptide and antiviral effects

Antimicrobial peptides have common characteristics due to their different sources such as short length (10–50 amino acids), amphipathic and positive charge. Peptides show different properties based on their structural

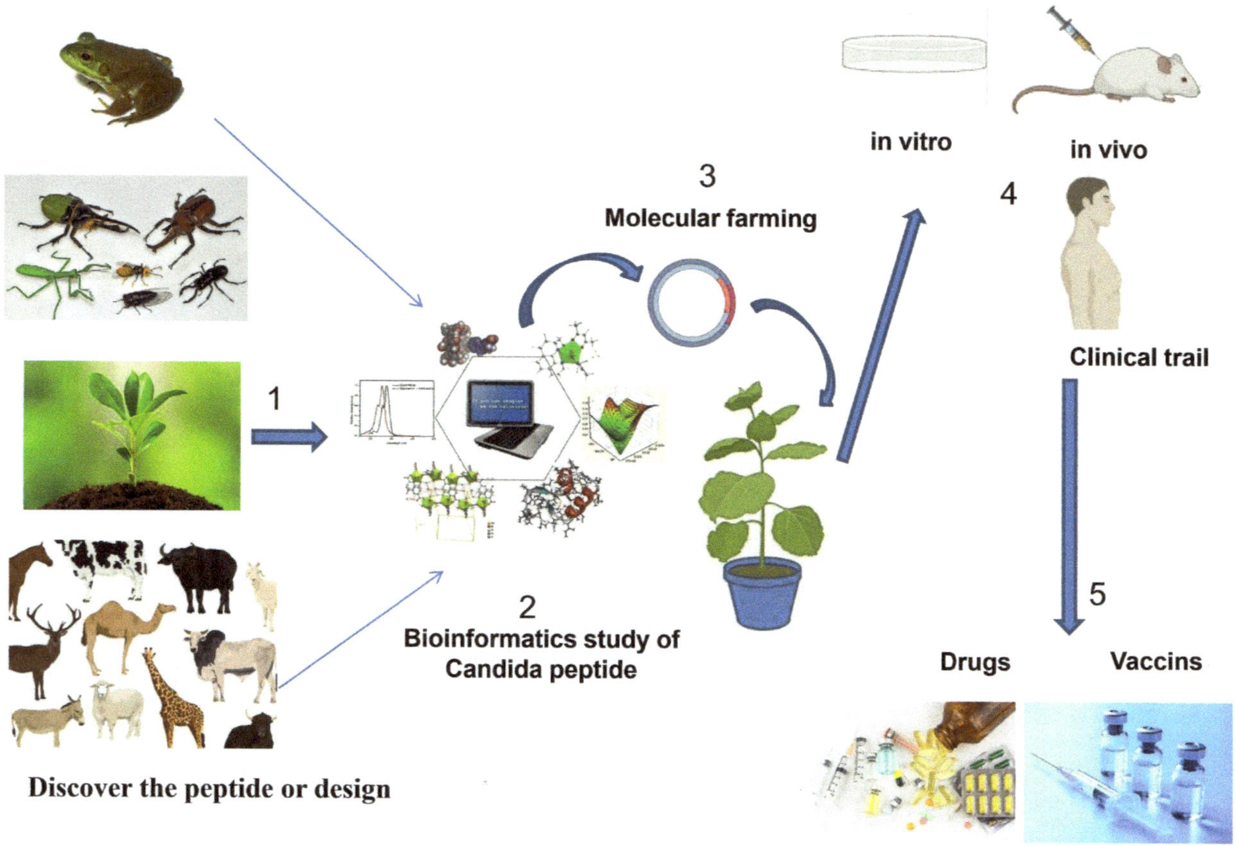

Fig. 1 Steps to identify and use antiviral peptide. (1) Identification of antiviral peptides from natural sources. (2) Design of in silico antivirus peptide and screening of peptides with antiviral effects in computer. (3) Production of antiviral peptide by genetic engineering in plants. (4) Perform in vitro and in vivo tests and perform clinical trial. (5) Commercialization and consumption of peptides as drugs and vaccines against the virus.

properties (Barlow et al., 2014). Biologically active peptides are called Cryptides or Crypteins, which have a wide range of biological activities (Pimenta and Lebrun, 2007). Antiviral peptides are obtained through various computational methods (bioinformatics), natural resources and biological resources. There are many peptide databases in which peptides have been identified as having antiviral properties (Qureshi et al., 2013). The first approved peptide drug is a peptide against the HIV virus (Teissier et al., 2011). After computer design of peptide drugs with software, the desired peptides are modeled in computer space and screened in the face of the target molecule in silico space. After virtual docking of the peptides in the presence of the virus surface protein, the superior peptides are selected for experimental studies. Repeat in vivo/in vitro tests are required to validate the peptides (Nevola and Giralt, 2015). For example, a number of peptides were designed against Zika virus proteins and the designed peptides were screened insilico. After docking, the selected peptides were tested by ELISA, and the peptide with the best performance was identified for zika (Mascini et al., 2019). Antimicrobial peptides are present in various organisms such as amphibians, plants, mammals, arthropods, bacteria, fungi, viruses or protozoa as the immune system (Ahmed et al., 2019). Recently, antimicrobial peptides have been developed as an antiviral therapy. Antiviral peptides are naturally hydrophobic, amphipathic, and cationic positively charged. Cyclotides are a family of plant peptides that have been shown to have antiviral activity against human immunodeficiency virus (HIV), Dengue virus, and influenza virus (Boas et al., 2019). In vitro methods including phage display, yeast display, ribosome display, mRNA display and bacteria display are the most common peptide production technologies. These methods produce high quality and high affinity peptides to their targets (Ullman et al., 2011).

4. Peptide heterologous expression host systems

Recombinant proteins are expressed and produced as foreign proteins as drugs, vaccines or monoclonal antibodies in the Expressive host (Burnett and Burnett, 2020). Given the high market demand for high-value recombinant proteins as drugs, it provides an opportunity to develop different hosts for the expression of these proteins. Various prokaryotic (such as *Escherichia coli*) and eukaryotic expression systems such as yeast, insects, plant and mammals are used to produce these proteins (Schillberg et al., 2019). Each of the eukaryotic and prokaryotic systems has advantages and disadvantages, bacterial systems have rapid production and high performance, while yeast systems can make the necessary post-translational changes for recombinant protein activity (Schillberg et al., 2019). Plant expression system is a permissive and cost-effective system in the production of recombinant proteins for the production of vaccines, monoclonal antibodies and industrial enzymes (Shanmugaraj et al., 2020).

5. Peptide production in plants

Plants were first used as hosts for the expression of recombinant proteins in 1980 (Shanmugaraj et al., 2020), in 2012 the first recombinant protein in plants was commercialized for the treatment of Gaucher (Fox, 2012). Among the advantages of the plant expression system are easy cultivation, low cost, mass production and rapid recombinant protein, and post-translational modifications in the protein produced (Chen and Davis, 2016). Nicotiana plants are good hosts due to their high growth rate and easy genetic manipulation, and many recombinant proteins have been produced in tobacco so far (Burnett and Burnett, 2020). Many therapeutic proteins are in clinical trials and close to commercialization (Yao et al., 2015). To date, many Therapeutic proteins have been produced in plant by genetic manipulations and have been studied in vivo and in vitro. Many plant-derived medicinal and non-medicinal proteins are currently being clinically developed. The plant molecular farming research is developing rapidly (Shanmugaraj et al., 2020).

6. Purification of recombinant antimicrobial peptides

One of the challenges of producing recombinant protein is to purify the antimicrobial peptide produced (Holaskova et al., 2015). The method of peptide purification depends on the system used to produce the peptide heterogeneously. Antimicrobial peptides produced in plants must be recovered from inside the cell and not secreted into the environment (Haney et al., 2017). Non-secreted antimicrobial peptides are recovered by dialysis or chromatography (e.g., reverse phase HPLC, molecular exclusion chromatography, etc.) (Xu and Ng, 2015). There are many challenges in the clinical trial phase of antimicrobial peptides to the commercial stage. Antimicrobial peptides have poor therapeutic performance in vivo compared to in vitro (Costa et al., 2019). One of the limitations in the development of antimicrobial peptides is the protease digestion of peptides in the gastrointestinal tract (Lau and Dunn, 2018). Through chemical modification and genetic engineering, these defects in antimicrobial peptides can be eliminated and improved (De Oliveira et al., 2020).

7. Challenges of using peptides in treatment

Peptides are limited in drug production due to instability, short half-life, lower potency, inability to cross membrane barriers, and poor bioavailability due to protease degradation (Ali et al., 2013). By providing solutions and overcoming these challenges, peptides can be used effectively in treatment. For example, by binding the peptide to the polyethylene glycol polymer, we can increase the stability of the peptides by increasing the molecular weight (Chew et al., 2017). For efficient transfer of peptides, it can be encapsulated in nanoparticles and administered through prefilled syringes, auto injectors and biodegradable micro needles (Lee et al., 2019).

8. Peptide-based drugs

Recently, viruses have been the cause of many human diseases. Scientific advances have led to the widespread production and distribution of vaccines and antiviral drugs (Lee et al., 2019). Despite rapid advances in health care, infections caused by certain viruses lead to many complications. Viral epidemics have led scientists to constantly look for new molecules with antiviral potential. Candidate antiviral peptides are promising and have high potential for conversion to antiviral drugs, although peptides with low antiviral activity are rare (Ali et al., 2013; Lau and Dunn, 2018). There are more than 60 commercially available drugs on the market in the United States, Europe, Japan, and several Asian countries (Lau and Dunn, 2018). The number of peptide drugs undergoing clinical trials is increasing every year. Peptides can be obtained from various natural sources, Biological sources such as High-throughput screening and bioinformatics (Qureshi et al., 2013).

9. Conclusion

Given that most viral infections are still incurable and because the emergence of new viruses is a threat to the health of people around the world, it is important to identify and produce new compounds as drugs and vaccines against viral diseases. Meanwhile, antimicrobial peptides with antiviral effects are promising compounds. These small peptides can be produced in plants and used as drugs or vaccines for viral diseases.

References

Aguirreburualde, M.S.P., Petruccelli, S., Almonacid, F.B., Wigdorovitz, A., 2016. Plant-based vaccine for livestock: key points to unleash platform translation in developing countries. Curr. Mol. Biol. Rep. 2, 171–179.

Ahmed, A., Siman-Tov, G., Hall, G., Bhalla, N., Narayanan, A., 2019. Human antimicrobial peptides as therapeutics for viral infections. Viruses 11, 704.

Ali, R., Rani, R., Kumar, S., 2013. New peptide based therapeutic approaches. In: Ashraf, G.M., Sheikh, I.A. (Eds.), Advances in Protein Chemistry. OMICS Group eBooks, Jeddah.

Barlow, P.G., Findlay, E.G., Currie, S.M., Davidson, D.J., 2014. Antiviral potential of cathelicidins. Future Microbiol. 9 (1), 55–73.

Boas, L., Campos, M., Berlanda, R., Neves, N., Franco, O., 2019. Antiviral peptides as promising therapeutic drugs. Cell. Mol. Life Sci. 76 (18), 3525–3542.

Bowdish, D.M., Davidson, D.J., Lau, Y.E., Lee, K., Scott, M.G., Hancock, R.E., 2005. Impact of LL-37 on antiinfective immunity. J. Leukoc. Biol. 77, 451–459.

Burnett, M.J.B., Burnett, A.C., 2020. Therapeutic recombinant protein production in plants: challenges and opportunities. Plants People Planet. 2, 121–132.

Chen, Q., Davis, K.R., 2016. The potential of plants as a system for the development and production of human biologics. F1000Res. 5, F1000.

Chew, M.F., Poh, K.S., Poh, C.L., 2017. Peptides as therapeutic agents for dengue virus. Int. J. Med. Sci. 14, 1342–1359.

Costa, F., Teixeira, C., Gomes, P., et al., 2019. Clinical application of AMPs. In: Clinical Basic Immunodermatology, second ed, pp. 281–298.

De Oliveira, K.B.S., Leite, M.L., Rodrigues, G.R., Duque, H.M., Costa, R.A., Cunha, V.-A., Costa, L.S., Cunha, N.B., Franco, O.L., Dias, S.C., 2020. Strategies for recombinant production of antimicrobial peptides with pharmacological potential. Expert. Rev. Clin. Pharmacol., 1751–2441.

Fox, J.L., 2012. First plant-made biologic approved. Nat. Biotechnol. 30, 472.

Haney, E.F., Mansour, S.C., Hancock, R.E.W., 2017. Antimicrobial peptides. In: Hansen, P.R. (Ed.), Methods in Molecular Biology. Springer New York, New York, NY, pp. 3–22.

Holaskova, E., Galuszka, P., Frebort, I., et al., 2015. Antimicrobial peptide production and plant based expression systems for medical and agricultural biotechnology. Biotechnol. Adv. 33, 1005–1023.

Laere, E., Kiong Ling, A.P., Wong, Y.P., Koh, R.Y., Lila, M.A.M., Hussein, S., 2016. Plant-based vaccines: production and challenges. J. Bot., 4928637.

Lalonde, M.E., Durocher, Y., 2017. Therapeutic glycoprotein production in mammalian cells. J. Biotechnol. 251, 128–140.

Lau, J.L., Dunn, M.K., 2018. Therapeutic peptides: historical perspectives, current development trends, and future directions. Bioorg. Med. Chem. 26, 2700–2707.

Lee, A.C., Harris, J.L., Khanna, K.K., Hong, J.H., 2019. A comprehensive review on current advances in peptide drug development and design. Int. J. Mol. Sci. 20, 2383.

Leite, M., Sampaio, K., Costa, F., Franco, O., Dias, S., Cunha, N., 2019. Molecular farming of antimicrobial peptides: available platforms and strategies for improving protein biosynthesis using modified virus vectors. An. Acad. Bras. Cienc. 91, e20180124.

Ma, J.K., Drake, P.M., Christou, P., 2003. The production of recombinant pharmaceutical proteins in plants. Nat. Rev. Genet. 4, 794–805.

Ma, J., Barros, E., Bock, R., Christou, P., Dale, P., Dix, P., Fischer, R., Irwin, J., Mahoney, R., Pezzotti, M., Schillberg, S., Sparrow, P., Stoger, E., Twyman, R., 2005. Molecular farming for new drugs and vaccines. EMBO Rep. 6 (7), 593–599.

Mascini, M., Dikici, E., Robles Mañueco, M., Perez-Erviti, J.A., Deo, S.K., Compagnone, D., Wang, J., Pingarrón, J.M., Daunert, S., 2019. Computationally designed peptides for Zika virus detection: an incremental construction approach. Biomolecules 9, 498.

Nevola, L., Giralt, E., 2015. Modulating protein-protein interactions: the potential of peptides. Chem. Commun. (Camb.) 51, 3302–3315.

Pimenta, D.C., Lebrun, I., 2007. Cryptides: buried secrets in proteins. Peptides 28 (12), 2403–2410.

Qureshi, A., Thakur, N., Tandon, H., Kumar, M., 2013. AVPdb: a database of experimentally validated antiviral peptides targeting medically important viruses. Nucleic Acids Res. 42 (D1), D1147–D1153.

Schillberg, S., Raven, N., Spiegel, H., Rasche, S., Buntru, M., 2019. Critical analysis of the commercial potential of plants for the production of recombinant proteins. Front. Plant Sci. 10, 720.

Shanmugaraj, B., Ramalingam, S., 2014. Plant expression platform for the production of recombinant pharmaceutical proteins. Austin J. Biotechnol. Bioeng. 1, 4.

Shanmugaraj, B., Bulaon, C., Phoolcharoen, W., 2020. Plant molecular farming: a viable platform for recombinant biopharmaceutical production. Plants 9 (7), 842.

Sijmons, P.C., Dekker, B.M., Schrammeijer, B., Verwoerd, T.C., van den Elzen, P.J., Hoekema, A., 1990. Production of correctly processed human serum albumin in transgenic plants. Biotechnology (N. Y.) 8, 217–221.

Teissier, E., Penin, F., Pécheur, E.I., 2011. Targeting cell entry of enveloped viruses as an antiviral strategy. Molecules 16, 221–250.

Thevissen, K., Kristensen, H.H., Thomma, B.P., Cammue, B.P., Francois, I.E., 2007. Therapeutic potential of antifungal plant and insect defensins. Drug Discov. Today 12, 966–971.

Twyman, R.M., Stoger, E., Schillberg, S., Christou, P., Fischer, R., 2003. Molecular farming in plants: host systems and expression technology. Trends Biotechnol. 21, 570–578.

Ullman, C., Frigotto, L., Cooley, R., 2011. *In vitro* methods for peptide display and their applications. Brief. Funct. Genomics. 10 (3), 125–134.

Vieira Gomes, A.M., Souza Carmo, T., Silva Carvalho, L., Mendonca Bahia, F., Parachin, N.S., 2018. Comparison of yeasts as hosts for recombinant protein production. Microorganisms 6, 38.

Xu, C., Ng, D.T.W., 2015. Glycosylation-directed quality control of protein folding. Nat. Rev. Mol. Cell Biol. 16, 742–752.

Yao, J., Weng, Y., Dickey, A., Wang, K.Y., 2015. Plants as factories for human pharmaceuticals: applications and challenges. Int. J. Mol. Sci. 16, 28549–28565.

CHAPTER 5

An assessment of smallholder farmers' status in Umzinyathi and Harry Gwala districts of Kwazulu Natal Province, South Africa

Phokele Maponya[a] and Ikechi Kelechi Agbugba[b,c]
[a]Agricultural Research Council-Vegetable, Industrial and Medicinal Plants, Pretoria, Gauteng, South Africa
[b]Department of Agricultural & Applied Economics, Faculty of Agriculture, Rivers State University, Port Harcourt, Nigeria
[c]School of Agribusiness Management, Rome Business School (Nigeria Campus), Ikeja, Lagos State, Nigeria

1. Introduction

Globally, smallholder farmers produce about 70% of our food, but individually they are often faced with constraints and challenges, which results in increased poverty (WWF, 2019). Furthermore, according to Growth Commission (2008) agriculture remains the best opportunity for the estimated 1.5 to 2 billion people living in smallholder households to escape poverty. Studies show that income growth generated by agriculture is up to four times more effective in reducing poverty than growth in other sectors (Growth Commission, 2008). Smallholder farmers are constrained with bottlenecks that impede their capacity and potential to produce and market their produce. As a matter of fact, rural poverty has entrenched much of the Sub-Saharan Africa nations, and the urgency to transform the lives and concerns of smallholder farmers cannot be over emphasized. Indeed. majority of rural persons derive their livelihoods from agriculture. Hence, the urgent need for developing the sector. Efforts are being made by the South African government in launching land reform initiatives at the dawn of democracy on the assumption that bottlenecks related to land size and would be eliminated paving ways for a more inclusive agricultural economy (Mujuru and Obi, 2020).

In South Africa, sustainability of smallholder farmers is importance for government. In support of this stance, the South African Government

https://doi.org/10.1016/B978-0-323-90569-5.00011-1

introduced several programs in the agricultural sector with the aim of achieving a sustainable agricultural sector. Indeed, South Africa's National Development Plan (NDP) has assigned smallholder agriculture, to drive development in rural areas and to improve the livelihoods of at least 370,000 people, specifically in the separate-developed areas known as the former homelands (NPC, 2011). The ANC-led government has committed itself to expand the number of smallholder producers selling their produce from 200,000 to 250,000 by 2014, and to 500,000 smallholders toward 2020 (Aliber and Hall, 2012). On the other hand, the National Development Plan (NDP)—vision 2030, which builds on the New Growth Path, propose to create 11 million jobs by 2030, of which about 1 million jobs are expected to be created through agricultural development (Van Averbeke, 2013). This will result in economic growth, food security and poverty reduction.

The key rationale for the study was based on the goal of the government of South Africa focus on establishing agricultural markets in 44 districts of South Africa including Harry Gwala and UMzinyathi Districts. The Agricultural Research Council (ARC) was involved in an agricultural project in the districts in Kwazulu Natal Province where smallholder farmers were assessed, trained and mentored to implement sustainable production and develop markets. As an integral part of the study, the research was conducted with the overall aim of assessing the status of smallholder farmers in Harry Gwala and Umzinyathi Districts. The key objectives were: (1) To identify and describe the characteristics of selected agricultural projects in Harry Gwala and Umzinyathi Districts; (2) To determine the factors enabling smallholder farmers to participate in agricultural markets in the Harry Gwala and Umzinyathi Districts.

1.1 Theoretical background

According to FAO (2012), the well-being and sustainability of smallholder farmers features as priority from an international, continental, regional, as well as the national level. Sustainable agriculture can be described as "development that meets the needs of the present without compromising the ability of the future generations in meeting their needs." The world population is expected to exceed 9 billion by 2050 (Blignaut et al., 2014), and agriculture is expected to feed the growing population and the practice of sustainable agriculture by smallholder farmers is believed to contribute immensely to the food security. The World Farmers Organisation (WFA) has noted that providing farmers with knowledge, techniques and the right tool will make it possible for farmers around the world to address food security. It noted that urgent action must be taken by all governments around the world to ensure that smallholder farmers break the poverty and hunger by producing food in a sustainable way.

Furthermore, the report by Wolfenson (2013) on the United Nations Conference on Sustainable Development (UNCSD) held in Rio de Janeiro emphasized that smallholders are key actors in fostering sustainable development; the report encourages all countries to strengthen their support to smallholder farmers to guarantee food security. However, smallholder farmers face number of challenges in their quest to be sustainable. Examples of challenges were presented by FAO (2012), which includes (a) Labor availability; (b) Information; (c) Tenure security; (d) Social capita; (e) Risk and shocks; (f) Credit access and resource endowment and (g) Market access.

Similarly, FAO (2012) listed the following recommendations to all the country governments in ensuring that farmers are sustainable: (a) Allocation of agricultural resources; (b) Investment in agriculture; (c) Strengthening farmer's organizations; (d) Empowering women farmers; (e) Ensuring secure access to economic, natural productive resources; (f) Promoting market stability; (g) Promotion and sharing of research and adapted technologies.

Over the years, the government of South Africa has invested in agricultural projects by means of infrastructure and inputs support with the aim of linking smallholder farmers into commercial stream of agriculture including agro-processing industries. However, there are many factors influencing choice of smallholder's farmers to effectively engage into formal markets and industries. In fact, according to Jari and Fraser (2009) it was observed in South Africa that smallholder farmers find it difficult to participate in commercial markets due to a range of technical and institutional constraints. Factors such as poor infrastructure, lack of market transport, dearth of market information, insufficient expertise on grades and standards, inability to have contractual agreements and poor organizational support have led to the inefficient use of markets, hence, commercialization bottlenecks. This observation stresses technical, socio economic factors and institutional bottlenecks as responsible for low market participation of smallholder farmers.

According to ARC (2017) Kwazulu Natal Province good and reliable rainfall together with fertile soils, made its agricultural sector extremely productive and is known for its specialist capabilities across several types of farming. KwaZulu-Natal has a total of 6.5 million hectares of land for farming purposes, of which 82% is suitable for extensive livestock production, while 18% comprises arable land (KZNDARD, 2020). At the heart of such new concepts is agricultural growth and a commitment to ensuring the integrated development of sustainable rural enterprises capable of supporting further development within the province's agricultural sector. In fact, According to Moloi (2021) a R57.2 million support program is enabling smallholder farmers in KwaZulu-Natal (KZN) to buy the tractors, implements, fertilizer, seeds, and seedlings needed to make their farms commercially successful.

"The introduction of the Farmer Support Programme today, is a clear statement that we are determined and stand ready to provide a dedicated and efficient support package to all the agricultural value chains that are key drivers to the province's sustainable economic growth and transformation," (Moloi, 2021).

UMzinyathi District has extensive grasslands in the north supporting the primary agricultural sector based on cattle ranching for beef, smallholder sheep and mixed farming and maize cultivation (UDM, 2016). The district has high potential for growth in agricultural crop production (maize, soybean, and sugarcane). In rural areas, particularly at Umsinga and Nquthu, animals are not kept as farming commodities and are seldom slaughtered or sold. According to UDM (2016) there is a need for farmer support programs to improve stock and crop management. Product beneficiation is minimum as most products are exported to major markets for further processing and/or export.

There are a significant number of projects that have been initiated by the department in order to enhance agricultural development in the district of Umzinyathi. These projects are situated in the uMsinga Local Municipality. Extensive grasslands in the north supporting the primary agricultural sector based on cattle ranching for beef, small-scale sheep and mixed farming and maize cultivation. In the southern areas substantial forestry is prevalent. Sugar cane and smaller scale fruit farming such as avocado and kiwi fruit cultivation also occur. Within eNdumeni Local Municipality, Dundee has the main economic activities ranging from retail trade, tourism, and farming. To the east of the town of Tugela Ferry the town there are three agricultural co-operatives that have been formed and are located on land that ranges from good agricultural potential to very limited agricultural potential.

Harry Gwala district is richly endowed in natural resources and having some of the best soils in the province, abundant water, and good rainfall (HGDM, 2016). Hence it is no surprise that agriculture sector plays a decisively dominant role in Harry Gwala District. According HGDM (2016) agriculture contributed proportionately more employment in Harry Gwala District. Agriculture in the form of commercial farms and to large extent commercial plantations forms the bedrock of the economy of the district. The abundance of high-quality soils, high altitude, abundant water, climatic extremes make the area suitable for dairy farming and as a result a significant portion of dairy produce consumed in Kwazulu Natal is produced within the district. Maize is the most important grain crop in South Africa, and the second largest crop produced in the country after sugar cane, and it is grown in Harry Gwala District (KZNDARD, 2020).

More so, there is a high level of fruit production in many parts of Harry Gwala District Municipality, and great potential for improved returns if more farmers engage in processing of fruits. Case studies suggest that smallholder processing of fruits and vegetables has potential these improved returns will come as long as appropriate processing equipment, processing skills, packing materials, and marketing information are made available. Livestock farming has a very long history in the district. Kwa Sani Local Municipality alone contributes a substantial amount to sum-up the income of the key players in the agricultural sphere in the area. It is pertinent to note that the number of cattle's slaughtered has increased mainly due to increasing demand. The beef supply chain is also becoming increasingly and vertically integrated, with feedlots owning abattoirs. However, beef farming in Harry Gwala is hindered by the lack of an abattoir in the area. Also, the industry has faced some difficulty in the district environment (KZNDARD, 2020).

2. Methodology

2.1 Study area

Umzinyathi District is located in the North-Western part of KwaZulu-Natal Province. The area is comprised of four local municipalities which are Nquthu, Endumeni, Msinga and Umvoti. Geographically, the area covers approximately 858,965 ha (85 89.6 km^2) and has a total population of about 510,838 persons. The district is predominantly rural as it includes expansive rural settlements, extensive commercial farms, and few towns, which service the rural communities. Towns located within the district include Dundee (the main economic hub and administrative center located within Endumeni), Greytown, Kranskop, Tugela Ferry, Pomeroy, Keate's Drift and Nquthu. Main Economic Sectors: Community, social and personal services (27.4%), agriculture, hunting, forestry, and fishing (22.6%), wholesale and retail trade (15.9%), manufacturing (14.8%), financial, insurance, real estate and business services (7.3%), electricity, construction (6.3%), transport, storage and communication (3.5%).

The Southern two-thirds of the district receives a high (summer) rainfall that is slightly less than that of Alfred Nzo and Sisonke (mostly 800–900 mm). The northern parts are drier with areas receiving less than 700 mm. The summer period is fairly cool (Tmax in January 27–30 °C, except in valley situations). On the other hand, winters are mild in the seaboard south (Tmin in July of 7–8 °C) and cool to cold inland (Tmin in July of 3–4 °C). Frost may be expected in the north during the last week of May. Climatically, the area is thus, suited to rainfed arable agriculture where slopes and soils thrive. There is an abundance of moderately good to good soils, particularly south of Greytown. The dominant soils are moderately deep or deep, well-drained, sandy clay loams of clay loams with favorable physical characteristics. There is, however, a high probability of the soils having a lime requirement that has to be determined by on-site sampling and analysis.

The Harry Gwala (previously Sisonke) District Municipality (HGDM) is located in the Southwestern region of the KwaZulu-Natal Province with land size of approximately 1,112,800 ha (10,547 km^2) in geographical extent. Majority of the Ingonyama land is concentrated centrally within the Ingwe municipal area. The district comprises of five local municipalities and one District Management Area on the Western side of the municipal area. The local municipalities within the subject district are: KwaSani, Ingwe, Greater Kokstad, Umzimkhulu and uBuhlebezwe. Harry Gwala District Municipality (previously Sisonke District Municipality) is situated south of KwaZulu-Natal. The seat of Harry Gwala is Ixopo. The municipality is surrounded by uMgungundlovu to the North-East, Ugu to the South-East, OR Tambo to the South, Alfred Nzo to the South-West, the Kingdom of Lesotho to the North-West and uThukela to the North. It is also bordered by the Drakensberg Mountains, which form a 200 km-long World Heritage Site. Economically, the most prominent employment sectors are agriculture, construction, and small-scale manufacturing. Importantly, the district is typical of a progressive farming methods/systems. The unspoilt natural environment has high ecotourism and adventure-tourism potential. Cities/Towns: Creighton, Himeville, Ixopo, Kokstad, Underberg, uMzimkhulu. Main Economic Sectors: Community services and agriculture (60%).

Furthermore, the Harry Gwala district receives a median annual rainfall ranging mostly between 800 and more than 1000 mm, all things remaining equal. Climatically the area is well-suited to rainfed arable agriculture where slopes and soils permit. A strong summer peak in rainfall is evident. Summer rains commence in September–October with the wettest months as January and February; and the driest, May to August. However, despite some dissection of the landscape, the dominant soils are deep or moderately deep, well drained loams or clay loams with good physical properties. There is a high probability of the soils having a lime requirement that must be determined by on-site sampling and analysis. Harry Gwala District offers the best opportunities for the rainfed production of (cold tolerant) fruits and vegetables crops.

2.2 Study design

Quantitative and qualitative research designs were employed as copies of structured questionnaire written in English were developed to elicit information for the study. More so, the study incorporated group discussions, stakeholders' discussion, as well as field observations. As part of standard protocol for conducting the study, meetings were held with all stakeholders in the UMzinyathi and Harry Gwala districts namely: (1) Local municipalities, (2) Department of Agriculture, Forestry and Fisheries (DAFF), (3) Department of Rural Development and Land Reform (DRDLR), (4) Local economic agencies and (5) Local farmers. The goal of the meetings was intended to introduce and further explain the study aims and DRDLR approved the study as all participants agreed voluntarily to participate.

2.3 Sampling procedure and analytical technique

Purposive sampling technique was employed in selecting eighty-nine (89) smallholder farmers, in order to cover their uniformity and homogenous characteristics such as infrastructure requirements, skills availability, production challenges, agricultural training needs, among other factors. As indicated in Fig. 1, the smallholder farmers visited in the UMzinyathi and Harry Gwala districts were prioritized based on the agriculture potential of the area such as project type, numbers of projects around an area, size of the land, chances of extending production, water availability, commitment of members to their projects, internal conflicts and working material and infrastructure.

2.4 Correlation analysis among variables

Correlation is a bivariate analysis that measures the strengths of association between two variables and the direction of the relationship. In terms of the strength of relationship, the value of the correlation coefficient varies between $+1$ and -1. When the value of the correlation coefficient lies around ± 1, then it is said to be a perfect degree of association between

Fig. 1 Kwazulu Natal Province orientation map including Umzinyathi and Harry Gwala Districts.

the two variables. When the correlation coefficient value tends toward 0, the relationship between the two variables will become weaker. The direction of the relationship is simply the + (indicating a positive relationship between the variables) or—sign (indicating a negative relationship between the variables). Usually, in statistics, four types of correlation measured includes: Pearson correlation, Kendall rank correlation, Spearman correlation, and Point-Biserial correlation. In this example, Pearson correlation was used and the following variables were found to be significant: Type of market, Size of land and Extension services.

3. Results and discussion

As shown in Fig. 2, smallholder farmers were visited in the following five local municipalities: Umsinga (33), Greater Kokstad (5), UMzimkhulu (17), Ingwe (16) and Ubuhlebezwe (18). The smallholder farmers were prioritized based on their agricultural potential. Fig. 3 indicated the beneficiaries as follows: Umsinga (385), Greater Kokstad (525), UMzimkhulu (619), Ingwe (143) and Ubuhlebezwe (210).

The age distribution of the respondents indicated that the majority (37%) fell within the age group of 18–35. Furthermore, as indicated in Table 1, the rest of the age categories were 36–45; 46–50 and >61 had 26%, 19% and

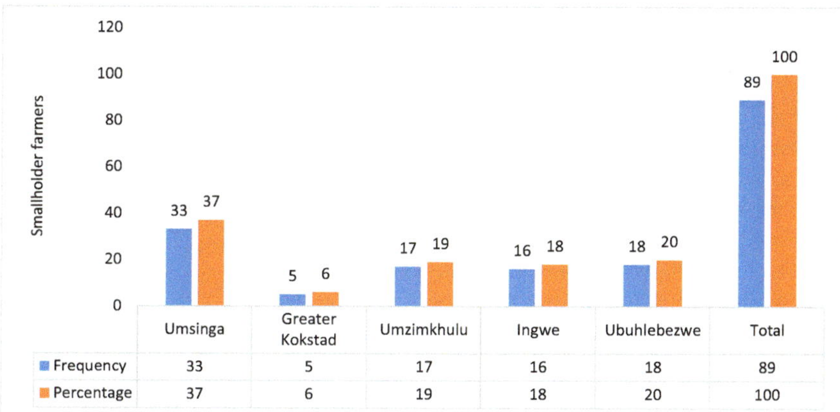

Fig. 2 Smallholder farmers visited per local municipality.

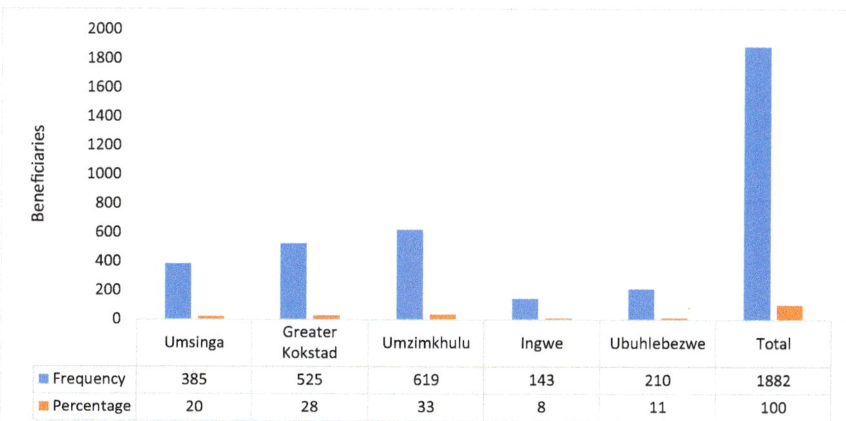

Fig. 3 Diagram showing a number of beneficiaries per local municipality.

Table 1 Distribution of smallholder farmers according to their socio-economic characteristics.

	Respondents	% Farmers socio economic characteristics
Gender		
Female	40	45
Male	49	55
Total	89	100
Age categories		
18–35	33	37
36–45	23	26
46–60	17	19
>61	16	18
Total	89	100
Level of education		
Primary education incomplete	14	15
Primary education completed	6	7
Secondary education incomplete	16	18
Secondary education completed	40	45
Tertiary education completed	6	7
Tertiary education incomplete	7	8
Total	89	100
Employment status		
Fulltime farmer	89	100
Total	89	100

Source: Field Survey, 2021.

18% respectively. In terms of education, 15% had incomplete primary education, 7% had completed primary education, 18% had incomplete secondary education, 45% of smallholder farmers had completed secondary education and 7% had tertiary education completed while 8% had incomplete tertiary education. The results in Table 1 indicated that the gender composition of females at 45% and males at 55%. As indicated in Table 1, all (100%) the farmers were farming full time and is line with assessing their sampling criteria, targeting fulltime farmers.

Results in Fig. 4 shows respondents' farming experience. About (39%) of respondents had 1–5 years in farming, 26% had 6–10 years, 19% had 11–20 years, 10% had 21–49 years and 6% had >50 years. According to Maponya (2021), farming experience, performance, socio-institutional

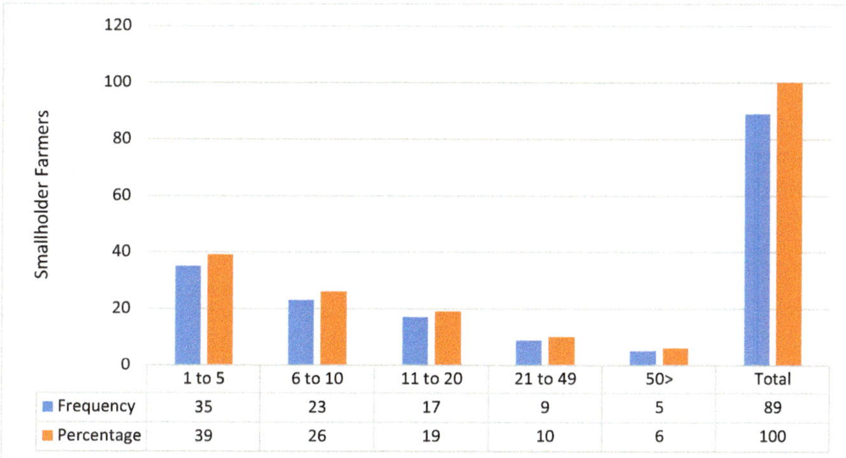

Fig. 4 Distribution of smallholder farmers according to their farming experience.

variables, and the stakeholder interaction plays an important role in the establishment and strengthening of local institutions that have capacity to break the farmers' capital constraint to invest in agriculture. Generally, these farmers had good farming experience thereby indicating an understanding of the rudiments of crop production. Their extensive experience makes them efficient in their decision-making processes and their willingness in taking risks is associated with the practice of improved technologies.

Results on land acquisition as indicated in Fig. 5 majority (73%) of projects operated on permission to occupy (PTO); thus, indicating that the role of traditional leaders in allocating land to farmers are as follows: own finance

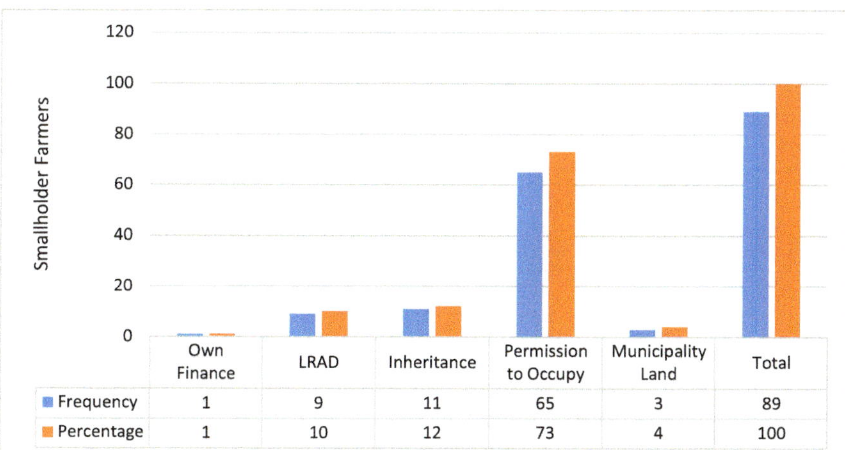

Fig. 5 Distribution of smallholder farmers according to their land acquisition.

(1%), land redistribution for agricultural development (LRAD) (10%), inheritance (12%), municipality land (4%). The challenge with the PTO approach is that whenever the headman or local chief feels that they do not want a particular farmer, the local chief can take that piece of land anytime; and as a result of this, majority of the farmers find it difficult to invest in infrastructure due to the fact that there no security on such land. Hence, land ownership is a challenge.

In general, agriculture requires large quantities of good quality water for use in various production processes. This supports the findings from FAO (2012) that irrigation uses up to 70% of all fresh water appropriated for human use. Furthermore, smallholder farmers used different water sources for irrigation, with rivers being the most utilized (67%), followed by boreholes (11%), rainwater (8%), municipal water (7%) and dams (7%), as the case may be (Table 2).

Table 3 indicated that most smallholder farmers cultivated vegetables and grain (66%), vegetables, citrus, deciduous (7%) and vegetables (27%). This will ensure increased production and consumption of fruit and vegetables, creating a complete and viable agro-value chain that will expand community-driven agricultural production and processing, and enhance the role of horticultural crops in addressing the immense malnutrition problem of vulnerable groups in Kwazulu Natal Province. In their study on

Table 2 Distribution of smallholder farmers according to their source of water.

Water source	Farmers	% of water sources
Dam	6	7
River	60	67
Borehole	10	11
Municipal water	6	7
Rainwater	7	8
Total	**89**	**100**

Table 3 Distribution of smallholder farmers according to their crops planted.

Crops	Farmers	% of crops planted
Vegetables	24	27
Grain, vegetables	59	66
Vegetables, citrus, deciduous	6	7
Total	**89**	**100**

Source: Field Survey, 2021.

Smallholder maize farmers and implication for extension service delivery, Agbugba et al. (2020) made a similar observation.

Table 4 showed that smallholder farmers employed different types of agricultural supplies inputs with results indicating that majority (62%) of projects had access to seeds, seedlings, kraal manure, fertilizer, and pesticides, 23% had access to kraal manure, fertilizer, and pesticides. This finding aligns with that of Harvey (2012), who stated that efficient use of farm inputs included fertilizers, crop protection chemicals, seeds and fuel and is, more or less essential in ensuring profitability, driving productivity growth and improving environmental sustainability of farm businesses.

Table 5 showed that majority (94%) of the smallholder farmers had access to extension services compared to the 6% that did not have access to the extension service. The results further indicated the form of extension services as follows: formal extension (43%), farmer-to-farmer (1%), formal extension, farmer-to-farmer (9%) and formal extension, farmer-to-farmer, neighbors, and family support (47%). Similarly, Agbebi (2012) found a

Table 4 Distribution of smallholder farmers according to their supply inputs.

Inputs	Farmers	% of inputs used
Kraal manure, fertilizer, pesticides	34	80
Seeds, kraal manure, fertilizer, pesticides, seedlings	55	20
Total	**89**	**100**

Source: Field Survey, 2021.

Table 5 Distribution of smallholder farmers according to their access to extension services.

Extension services	Farmers	% of extension services
Access to extension service		
Yes	87	94
No	2	6
Form of extension service		
Formal extension	38	43
Farmer to farmer	1	1
Formal extension, farmer to farmers	7	9
Formal extension, farmer to farmer, neighbors and family support	43	47
Total	**89**	**100**

Source: Field Survey, 2021.

positive relationship between farmers' access to extension services and their profitability, with productivity at its highest point at the extreme ends of distribution of yield residuals.

Majority (50%) of the smallholder farmers had access to 1–5 ha of land, while only 11%, 22% and 7% of smallholder farmers had 6–10, 11–20 and 21–49 ha, respectively (as shown in Fig. 6). Quite a few number of smallholder farmers had land of >50 ha (10%). Mathye et al. (2000) emphasized that the size of the farm (land) used for production is positively related to market participation. When farmers have more land, their production will be higher.

According to Maponya et al. (2014) smallholder farmers struggle to penetrate the market, especially the formal market due to various challenges (constant supply, transportation, lack of information). Fig. 7 indicated that majority (93%) of the smallholder farmers were engaged with trading in informal markets while quite a number of smallholder farmers were engaged in formal market (7%).

As indicated in Table 6, there is a positive correlation among type of market, size of land and extension services. It was not surprising to realize a positive association among extension services, size of land training and type of market as a well-trained and advised farmer is empowered to make informed decisions and to identify market opportunities where they exist. Moreover, farmers with large land sizes could allocate their land for different enterprise

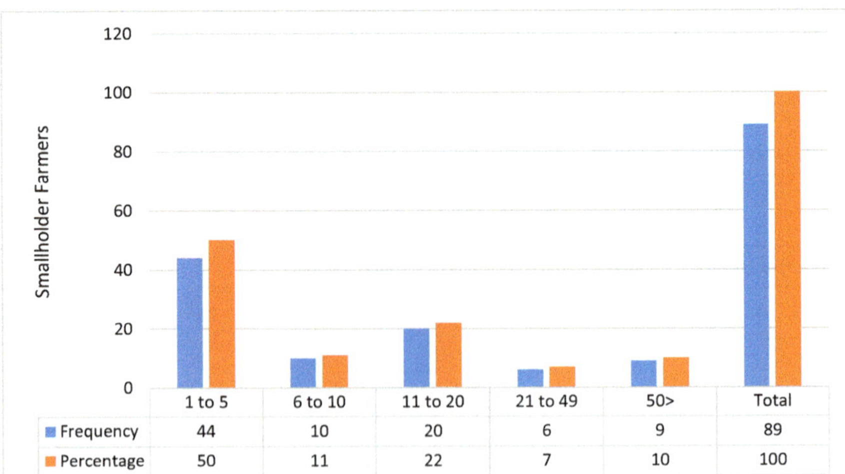

Fig. 6 Distribution of smallholder farmers according to their land sizes.

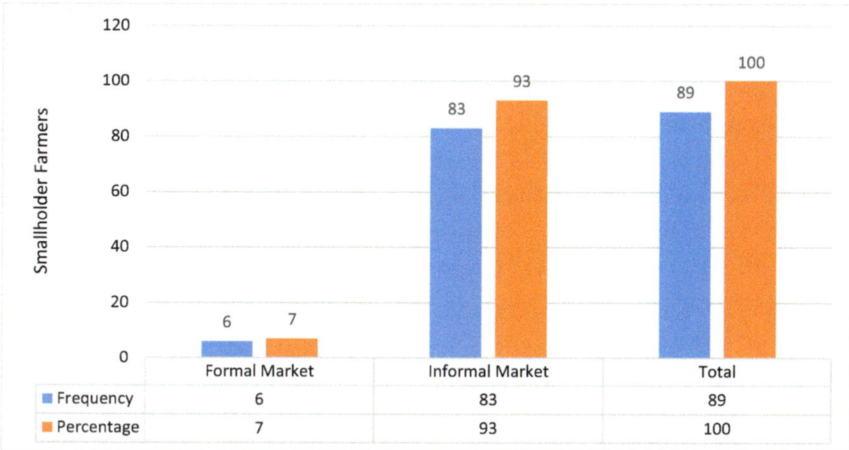

Fig. 7 Distribution of smallholder farmers according to their market access.

Table 6 Pearson correlation coefficients among variables.

Variable	TYM[a]	SOL[b]	EXS[c]
TYM[a]	1.00	0.050[d]	0.0129[e]
SOL[b]	0.050[d]	1.00	−0.67
EXS[c]	0.0129[e]	−0.67	1.00

[a]Type of market.
[b]Size of land.
[c]Extension services.
[d]5% significant level.
[e]1% significant level.
Source: Field Survey, 2021.

giving them better position to participate in the market. These results are in line with smallholder farmer surveys conducted by Maponya et al. (2014) in other South African districts.

4. Conclusion and recommendations

The chapter attempted to identify and analyze the socio-economic characteristics as well as the relationship among different variables in order to understand the status of smallholder farmers in Harry Gwala and Umzinyathi districts. The smallholder farmers in the Umzinyathi and Harry Gwala districts are facing multiple challenges exacerbated by socio-economic challenges; hence, the establishment of a R57.2 million support program was launched lately to enable smallholder farmers in KwaZulu-Natal (KZN) Province to buy tractors, implements, fertilizer, seeds and seedlings needed to make their farms commercially successful. In order words, it is important to understand the status of smallholder farmers so as to establish remedial actions that guide policy makers and other smallholder farmers supporting bodies. This could also help these supporting bodies understand the key factors responsible for the status of smallholder farmers in the region thereby enabling them to make better decision. Furthermore, there is need for government to recognize, reward as well as assist informal ventures belonging to smallholder farmers that have access in transforming informal markets into more formalized ventures. The study recommends that organizations overseeing the improvement of socio-economic conditions, policy, food security and rural development should use the results of the study as a guide for planning and decision-making activities for improved smallholder farmers' livelihoods.

Declaration statements

Funding (information that explains whether and by whom the research was supported)

Funding was allocated by the Department of Rural Development and Land Reform.

Conflicts of interest/competing interests (include appropriate disclosures)

No conflict of interests that are directly or indirectly related to the work submitted for publication.

Availability of data and material (data transparency)

Not applicable.

Code availability (software application or custom code)

Not applicable.

Authors' contributions (optional: Please review the submission guidelines from the journal whether statements are mandatory)

The author contributed to the study as a project manager and main author.

Additional declarations for articles in life science journals that report the results of studies involving humans and/or animals

Not applicable.

Ethics approval (include appropriate approvals or waivers)

Ethical approval was received from the funder (DRDLR) and my employer (ARC).

Consent to participate (include appropriate statements)

Consent to participate statement was included in the questionnaire for the respondents to read and sign before the interviews.

Consent for publication (include appropriate statements)

The author gives consent for publication.

References

Agbebi, F.O., 2012. Assessment of the impact of extension services on fish farming in Ekiti State, Nigeria. Asian J. Agric. Rural Dev. 2 (1), 62–68.

Agbugba, I.K., Christian, M., Obi, A., 2020. Economic analysis of smallholder maize farmers: implications for public extension services in Eastern Cape. S. Afr. J. Agric. Ext. 48 (2), 50–63.

Agricultural Research Council ISCW (ARC-ISCW), 2017. Pretoria, South Africa.

Aliber, M., Hall, R., 2012. Support for smallholder farmers in South Africa: challenges of scale and strategy. Dev. South. Afr. 29 (4), 548–562.

Blignaut, J.N., De Wit, M.P., Midgley, S., Drimie, S., Knot, J., Crookes, D., 2014. Sustainable farming as a viable option for enhanced food and nutritional security and a sustainable productive resource base. Inception report. Prepared for the Green Fund, Development Bank Southern Africa. Pretoria.

FAO, 2012. The State of Food and Agriculture—An FAO Perspective. Food and Agricultural Organisation, Rome.

Growth Commission, 2008. The Growth Report: Strategies for Sustained Growth and Inclusive Development. World Bank, Washington, DC.

Harry Gwala District Municipality (HGDM), 2016. The Agriculture Status in the Harry Gwala District, Kwazulu Natal Province, 2015/16 Integrated Development Plan.

Harvey, M., 2012. The efficiency with farm inputs—A recipe for productivity report. Rabobank, The Netherlands.

Jari, B., Fraser, G.C.G., 2009. An analysis of institutional and technical factors influencing agricultural marketing amongst smallholder farmers in the Kat River Valley, Eastern Cape Province, South Africa. Afr. J. Agric. Res. 4 (11), 1129–1137.

KZNDARD (Kwazulu Natal Department of Agriculture and Rural Development), 2020. Annual-Performance-Plan-2020_21.

Maponya, P., 2021. Opportunities and constraints faced by smallholder farmers in the Vhembe District, Limpopo Province in South Africa. Circ. Econ. Sustain., 1–14.

Maponya, P.I., Modise, D., Van Den Heever, E., Mahlangu, S., Baloyi, T., Maluleke, A., Carsten, J., Van Der Walt, M., Chauke, D., Mkhari, R., Sole, L., Duba, M., Malebana, J., Mphahlele, M., 2014. Establishment of fruits and vegetables markets: a case of Waterberg District in Limpopo Province. J. Agric. Sci. 6 (6), 1–12.

Mathye, M.M., Makhura, M.T., Kirsten, J.F., 2000. Transaction costs in the marketing of bananas: explaining market participation of smallholders in the Northern Province. In: Paper Presented at the 38th Annual AEASA Conference, Sun City, pp. 27–29.

Moloi, B.S., 2021. Support Programme Enabling KZN Small-Scale Farmers to Potential Commercial Farmers. Kwazulu Natal Department of Agriculture and Rural Development.

Mujuru, N.M., Obi, A., 2020. Effects of cultivated area on smallholder farm profits and food security in rural communities of the Eastern Cape Province of South Africa. Sustainability 12, 1–17.

NPC, 2011. Diagnostic Report. National Planning Commission, Pretoria, South Africa.

UMzinyathi District Municipality (UDM), 2016. The Agriculture Status in the UMzinyathi District, Kwazulu Natal Province, 2015/6 Integrated Development Plan.

Van Averbeke, W., 2013. Improving plot holder livelihood and scheme productivity on smallholder canal irrigation schemes in the Vhembe district municipality. Limpopo Province, South Africa.

Wolfenson, K.D.M., 2013. Coping with the Food and Agriculture Challenge: Smallholders' Agenda. Natural Resources Management and Environment Department Food and Agriculture Organization of the United Nations, Rome.

WWF (World Wild Fund), 2019. Integrated Annual Report 2019, South Africa.

CHAPTER 6

Using geospatial technologies to manage COVID-19

Karim Ennouri[a,d], Slim Smaoui[b], Olfa Ben Braiek[c], and Mohamed Ali Triki[a]
[a]Laboratory of Amelioration and Protection of Olive Genetic Resources, Olive Tree Institute, University of Sfax, Sfax, Tunisia
[b]Laboratory of Microbial, Enzymatic Biotechnology and Biomolecules (LBMEB), Center of Biotechnology of Sfax, University of Sfax, Sfax, Tunisia
[c]Laboratory of Transmissible Diseases and Biologically Active Substances (LR99ES27), Faculty of Pharmacy, University of Monastir, Monastir, Tunisia
[d]Digital Research Centre of Sfax, Technopark of Sfax, Sfax, Tunisia

1. Introduction

The COVID-19 epidemic caused instability in all sectors of the economy, with impacts felt by each sector in several ways (Hiscott et al., 2020). Government agencies are struggling to adapt to the evolving problem on a frequent basis, while most service-oriented businesses are becoming used to full-time remote jobs. Both developed and industrialized nations are coming to an impasse, with stringent lock-down measures being implement across countries and towns through the planet (Ahani and Nilashi, 2020). The space industry is witnessing the disaster effects, with launches abandoned, operating projects scaled down and private firms shutting down operations. Nevertheless, the downstream space market, comprising telecommunications, planet observations and exploration, is increasingly in demand, due to the useful informations facilitated by satellite pictures and location-based services that make it easy to detect the humanitarian problem from space (Masrur et al., 2020).

Geographical awareness is important to identify, understand and respond to epidemic of infectious diseases, particularly in the event of a pandemic such as Coronavirus disease. The origin of diseases, how they spread and the severity of control, prevention and monitoring efforts of Public health authorities, decision makers and managers require Geographical information to consider disease trends in real time, to classify at-risk communities and to prepare tailored strategies, such as evaluating available services or rising healthcare options (Zhou et al., 2020). In addition, good coordination with

Agricultural Bioeconomy
https://doi.org/10.1016/B978-0-323-90569-5.00014-7

other service organizations and people is required to ensure a cohesive response. Remote Sensing (RS) is defined as a technique capable of collecting information on the Earth's surface without any contact-allows continuous and frequent tracking of the Earth's environmental conditions. It furnishes massive quantities of geographical information, and offers potential of obtaining climatic and environmental details (Diffenbaugh et al., 2020). This knowledge, added with effective field studies, can be employed to recognize and chart possible disease habitats; forecast demographic trends, track quantitative and qualitative changes in the corresponding habitats (Tu et al., 2018); and prepare management systems to identify regions where the infection risk is greater or maybe lesser (Lan et al., 2021).

Moreover, space exploration technologies, particularly satellite observations, networking, mapping and positioning, offer far-reaching strategies to urgent problems facing mankind, varying from health, and more specifically the current COVID-19, to food safety, farming, energy, catastrophe risk mitigation (Djalante et al., 2020). The convergence of satellite imagery, established data and surface details and utilization of new information sources, in addition to analytics, tools and techniques will be helpful in obtaining appropriate, accurate and quality information. The collected data are essential for authorities, companies, societies and people to take informed action and make evidence-based actions (Rashid and Wang, 2021). In fact, specialists predict the expansion of aerospace industry during next years. Novel aerospace start-ups are transforming the marketplace, capitalizing on digitalization, the Cloud computing and contributing to the spill over effects of aerospace investment in research and innovation (Chamola et al., 2020). As a common factor, Remote Sensing has the opportunity to build this common operating image for numerous organizations. By means of this technology (such as geographical analysis, imaging and location analysis), the health authorities and government departments will map confirmed and active incidents, deaths and recoveries to assess where COVID-19 diseases have happened (Ding et al., 2020). Besides, time-based maps are able to demonstrate how pathogens propagate over time and when users may want to focus treatments. Coronavirus has an undue effect on some age groups, such as the seniors and people having chronic medical disorders (Franch-Pardo et al., 2020).

Over the last years, the supply of Satellite remote sensing data has risen exponentially mainly due to two indicators: technical developments contributing to the New Space economy and open data policies leading to the liberalization of spatial data (Zhu et al., 2019). On the one side, the rapid evolution of economic satellite technology has reduced entry barriers, especially due to reduced production methods and the rise of commercial off-the-shelf modules. This, in turn, is conceivable to result in increasingly output and access to data attributable to the improved regional spread and the regular re-examination times that these arrangements will achieve (Pettorelli et al., 2018). As well, more data collected also means the creation of better solutions with enhanced detection of change, such as imaging that changes can be more visible to both human eyes and efficient Artificial Intelligence algorithms (Maxwell et al., 2018), which is crucial to provide insight into the Coronavirus crisis, and thus drive the huge demand for space records.

This availability of data from numerous instruments in the Sentinel constellation provides the public with valuable insights into the use of satellites to track global developments, as for example to track global emissions decrease during COVID-19 crises (Sun et al., 2021). Another example consists to monitor border protection practices between countries, particularly during this crisis, when certain countries decide to close their frontiers (Pan and Zhang, 2020). Additionally, Satellite data may therefore play an important role in economic policy, especially with regard to environmental and climate policy as well as financial experts, to make information on manufacturing processes widely accessible to insurers and policymakers (Tassa, 2020).

The chapter is structured in three main lines:

- Overview of the major characteristics of sensors, which provide a basis for comparison of their use;
- The modern state of the use of remote sensing introduced in endemic regions;
- Information provided via satellites and type of used sensors.

2. Principal sensors, characteristics and applications

Remote Sensing tests the energy reflected or released in distinct wavelengths of the electromagnetic spectrum through instruments, typically on-board satellites. In the case of environmental surveillance, the visible infrared areas can be used and the microwave range is employed (Reddy, 2018). Remote Sensing sensors may be categorized as active or passive inactive systems, as radar, and radio detector, the sensor emits a spectral microwave radiation flux that communicates with objects on the Earth's surface and collects the reflected portion of the sensor (Xiao et al., 2018). The key benefits of radar are as follows: it operates day and night; the cloud layer does not preclude it; and it identifies a disparity between the specific structures and slopes (Sun et al., 2017).

Passive systems record the energy of the atmosphere (light and heat) reflected and released by the sensor. The TM (Thematic Mapper) sensor of the LANDSAT (Land Satellites) series and the HRV (High Resolution Visible) sensor of the SPOT (Satellite Pour l'Observation de la Terre) series are examples of passive sensor systems (Pimentel et al., 2016). The various kinds of resolution that are properties of each sensor describe the information quality of each acquired scene. Furthermore, the spectral resolution refers to the spectral range of each band. This indication is used to identify such targets. For instance, a high reflectance value in the near-infrared range and a low value in the visible red band are observed in the agricultural land (Fan et al., 2018). Additionally, spatial resolution is defined as the smallest potential area of land that can be personalized (Zhong et al., 2015), and temporal resolution means the time interval between two satellites traveling around the same position on planet (Houborg et al., 2015). The presence of the LANDSAT ETM+ sensor, is due to the huge mass of data produced throughout its active time, between 1999 and 2003 (Claverie et al., 2015).

Initially interacting with Remote Sensing images, the user interprets the image by specific elements such as tone, form, scale, shape, softness, shadow, and combinations. By this approach, it is possible to differentiate between damaged wood blinds from softwood such as towns and rural areas, urban or suburban areas, native areas, gravel, fields, meadows and pastures (Farmakis-Serebryakova and Hurni, 2020). In terms of choices of image, those with a range of space, spectral and temporal elements should be used for study purposes. In the case of images with specific resolution, the main task is to estimate the size of images. By identifying small farm areas, trees, houses, roof features and distances between residential buildings, a maximum of 5 m of resolution can be used (Mahabir et al., 2018). A 15-m resolution can detail forest areas or classify city barriers, while a resolution of 20 to 30 m can be applied to describe towns, streets and airports-planes, forests and agriculture, as well as land-use characteristics (Dou et al., 2014). Depending on the study area, a spatial resolution of 10 to 100 m is employed to detect flooded areas, and a resolution of 1000 m can be used to calculate vegetation and temperature (Zhang et al., 2020a,b). Additionally, the images provided by the LANDSAT and SPOT satellites provide comprehensive ranges, which are not accessible from the NOAA (National Oceanic and

Table 1 Remote sensing satellites worldwide.

System	Operator	Satellites	Type	Resolution (m)
EROS	ImageSat International	EROS A	Optical	1.5
		EROS B	Optical	0.7
		EROS C	Optical	0.7
Ikonos	GeoEye	IKONOS-2	Optical	0.8
OrbView	GeoEye	OrbView-2	Optical	1.000
GeoEye	GeoEye	GeoEye-1	Optical	0.41
QuickBird	DigitalGlobe	EarlyBird	Optical	3
		QuickBird-1	Optical	1
		QuickBird	Optical	0.6
Radarsat	MDA	Radarsat-1	Radar	8
		Radarsat-2	Radar	3
SPOT	Spot Image	SPOT2	Optical	10
		SPOT4	Optical	10
		SPOT5	Optical	2.5
WorldView	DigitalGlobe	WorldView-1	Optical	0.5
Disaster Monitoring Constellation	DMC International Imaging	AISAT-1 (Algeria)	Optical	32
		NigeriaSAT-1 (Nigeria)	Optical	32
		UK-DMC (United Kingdom)	Optical	32
		Beijing-1 (China)	Optical	4
TerraSar		Terrasar-X	Radar	1
RapidEye	RapidEye	RapidEye-1	Optical	6
		RapidEye-2	Optical	6
		RapidEye-3	Optical	6
		RapidEye-4	Optical	6
		RapidEye-5	Optical	6

Adapted from www.oneonta.edu/faculty/baumanpr/geosat2/RS-Introduction/RS-Introduction.html.

Atmospheric Administration) weather satellites (Table 1). However, the latter has a higher time resolution and can collect images from the same location at a frequency of 12h, while the time resolution of LANDSAT and SPOT is 16 days and 26 days respectively (Kganyago et al., 2018). Typically, pictures that cover a wider area are employed when performing research within a country or continent. In this scenario, it is possible to use NOAA

Table 2 Landsat 7 and 8 bands (https://landsat.gsfc.nasa.gov/landsat-8/landsat-8-overview/).

Satellite	Sensor(s)	Band (wavelength, μm)	Spatial resolution (m)
Landsat 7	Enhanced Thematic Mapper plus (ETM+)	1 (0.450–0.515)	30
		2 (0.525–0.605)	30
		3 (0.630–0.690)	30
		4 (0.775–0.900)	30
		5 (1.550–1.750)	30
		6 (10.40–12.50)	60
		7 (2.080–2.350)	30
		8 (0.520–0.900)	15
Landsat 8	Operational Land Imager (OLI)	1 (0.435–0.451)	30
		2 (0.452–0.512)	30
		3 (0.533–0.590)	30
		4 (0.636–0.673)	30
		5 (0.851–0.879)	30
		6 (1.566–1.651)	30
		7 (2.107–2.294)	30
		8 (0.503–0.676)	15
		9 (1.363–1.384)	30
	Thermal Infrared Sensor (TIRS)	10 (10.60–11.19)	100
		11 (11.50–12.51)	100

or TERRA satellite imagery (Guo et al., 2017). A large number of LANDSAT images is needed for the same region, which will entail additional costs (Table 2). IKONOS and QUICKBIRD photos with better spatial resolution may be of interest to study densely populated urban areas (Alphan and Çelik, 2016). Radar images can be useful in tropical areas with extreme cloudiness. Furthermore, active microwave sensors are especially useful for the monitoring of flooded areas. In addition to spatial and temporal resolution, the spectral spectrum that can better identify the target should also be notified. Flooded woods can be tracked using the SAR-L band of JERS-1 radar; and the panchromatic band of SPOT or IKONOS can be used for urban characteristics (De Alban et al., 2018). Ecotones are classified with better precision using spectrum bands between 0.4 and 1.3 μm, including visible and near-infrared bands, utilizing LANDSAT, SPOT, or IKONOS satellites, based on the desired specificity. As well, factors linked to endemics (as for example the impact of climate change on vegetation or population development and erosion) need a temporary follow-up. A mixture of sensors from various spacecraft, such as a combination of radar and optical pictures, can be applied in a complementary manner (Qadir and Mondal, 2020).

On other side, the image processing technology based on spectral response will generate new information. For example, bands via hybrid modeling are generated for the proportions of water, soil and vegetation. The activity of the NDVI (Normalized Difference Vegetation Index) also generates biomass information belts that indicate disease vectors and potential reservoir habitats (Gandhi et al., 2015; Riva et al., 2017). This operation transforms the spectral components represented in the RGB (red, green, blue) color space into a specific method that provides information on the strength, shade and saturation that can be applied for image fusion, allowing the usage of SPOT-PAN space while retaining the LANDSAT-TM spectral resolution (Sepuru and Dube, 2018).

TERRA, the Latin word for "Land," a territory monitoring satellite Launched by NASA in 1999, has tremendous potential for endemic research. This satellite has a range of sensors, including MODIS (Moderate Resolution Imaging Spectroradiometer) and ASTER (Advanced Spaceborne Thermal Emission and Reflection Radiometer). The former has 36 spectral bands of 250, 500, and 1000-m resolution and 2-day temporal resolution. The latter has three 15-m resolution bands in the 0.5–0.9 μm region, six 30-m resolution bands in the 1.6–2.5 μm region, and five 90-m resolution bands in the 8–12 μm region, with a 16-day time resolution these sensors are useful for measuring surface environmental conditions in different areas (Wu et al., 2015; Obata et al., 2017).

These instruments are able to measure environmental shift parameters with improved spectral, spatial and temporal resolution. The literature on the use of Remote Sensing in vector-borne disease research is growing, and new research opportunities are opening up with the advent of improved resolution sensors (Garni et al., 2014; Jamison et al., 2015). Most of information obtained by satellites is related to vegetation using NDVI, an indicator extracted from frequency band operations (Pang et al., 2017). For example, in the case of NOAA, using bands 1 and 2 from the AVHRR sensor, vegetation can be highlighted by the following operations: band 2/band 1 or (band 2 − band 1)/band 2 + band 1. For the same purpose, TM4 and TM3 frequency bands can also be used, and these operations are used based on LANDSAT. Compared with soil in Band 4 of the TM sensor and Band 2 of AVHRR, using these operations enhances the spectral response of vegetation (Yagoub et al., 2014).

3. COVID-19: Impact assessments with insights from satellite data

Financial systems are planning for a deep recession attributable to inter-industry instability triggered by the novel COVID-19. As countries in the world are lock-in, these policies are destructive to the economy, with several industries intended to be impacted by the International Shutdown Sectors (Bauer and Weber, 2020). Nevertheless, from an ecological point of view, the projected effect is largely favorable, considering the reduction of pollution caused by manufacturing impact. There are various ways to assess and understand the crisis impact, including aircraft surveillance, drones and information based on cellular network pings (Chamola et al., 2020). Satellite measurements provide a detailed, reliable and cost-effective approach to quantifying the effects of blockages on industry and environment (Filippini et al., 2020).

3.1 Industrial impacts

Several sectors are impacted by the COVID-19 crisis. Each field is defined with different proportions from space and thus exhibits various impact degrees. While the graphic picture comparison indicates only the effects of Coronavirus crisis, the exploitation of photos using data fusion and machine learning methods allows users to measure the crisis effect on different industries (Tuli et al., 2020). The resulting association explains clearly how satellite imaging, composed with other data sources, will provide information in order to measure the impacts of the latest COVID-19 epidemic on diverse industries (Arora et al., 2020).

3.2 Manufacturing: Monitoring changes in production

International production is strictly associated with current distribution networks for both small and strong industries on side and the needs of consumer goods on other side, then such display of images over various periods offers an approach for visual understanding and predict tendency. In this specific case, space exploration techniques and algorithms are considered as helpful and can quantify the number of artifacts on images and consequently quantify the level of effect on organizations and policy makers (Pal et al., 2021). Performing such an analysis at regional and national levels would provide more information on the crisis effects and industry of transporting goods, particularly on opportunity to examine a given chronological period and to predict objective patterns (Wang and Su, 2020).

3.3 Shipping: Monitoring changes in port and airport activities

In view of the decline in worldwide demand and specially the importance on critical supplies, there has been a pronounced decline in port activities. In reality, ports expect smaller container supplies as demand slows down and industries shut down across the world (Zhang et al., 2020a,b). Furthermore, the gradual closing of borders directly influenced the aviation industry across the world from January 2020 onwards. The blocking of steps implemented by more than 40 nations, which resulted in a drastic decrease in the number of operating flights worldwide, leading to threats airline partners (Serrano and Kazda, 2020). Leisure travel has stopped all over the country, and most of the flights have been landed at main airports. As a direct result of this decrease in the export and import of goods and corporate holiday travel, shipping and aviation activities around the world are expected to be largely limited. Besides, satellite data help to measure the effects that they would have on the shipping and aviation sectors and, more generally, on the international economic consequence as a whole (Sun et al., 2020).

4. Remote sensing as a tool to survey endemic diseases

Satellite measurements play a vital role in identifying the beneficial impact of lockout on the climate, due to the various types of satellites monitoring different indicators. For example, the Sentinel-5P satellite provides monitoring data on contaminants, including ozone, sulfur dioxide, nitrogen dioxide, carbon monoxide, formaldehyde and methane (Omrani et al., 2020). The information from these satellites play an increasingly role in helping both decision makers and populations to recognize the environmental effects of Coronavirus crisis (Vîrghileanu et al., 2020).

More specifically, Nitrogen dioxide is emitted by vehicle engines, power stations and other manufacturing operations and is supposed to intensify respiratory disorders such as asthma, which are also seeing a massive decrease in pollution since this time since a national lock was implemented (Ogen, 2020). Sentinel-5P data is important not only to explain the environmental effects of the COVID-19 crisis, but also to help policy makers identify crucial lessons that can support future decision making, particularly carbon and climate change strategies. As the industrial pollution is considered as the biggest pollution provider, the deductions from satellite data will allow decision makers to collect data that can enhance policy making in concordance with climate change in the following years (Solberg et al., 2021).

Millions of people have become contaminated with the epidemic of a quickly respiratory virus identified as Coronavirus since 2020. Satellite technology and space-based structures have played a major role in managing all population subdivisions and managing the transmission of contagious diseases during the fight against COVID-19 (Fig. 1). Spatial technologies, like Mapping and Geoinformation (GIS)-driven infrastructure for geographical medical informations, satellite imagery, telemedicine, satellite communications, artificial intelligence (AI)-driven systems and global positioning system service, play an essential role in diagnosis, confinement and care of patients, as well as accelerated creation of hospitals and thorough decontamination operations (Fong et al., 2020). In the meantime, space applications have made significant contributions to national online learning, immediate transport requirements for vital goods among regions, and the vast mobilization of healthcare professionals around the world, among other items. In consideration of the infectious existence of Coronavirus, it has been shown that the increasing use and development of space-based technology significantly enhances the efficacy of helping against disease expansion by eliminating clinical face-to-face communication (Dang et al., 2019; Fong et al., 2020). Under these exceptional conditions, the additional spread of space-based tools between nations and the

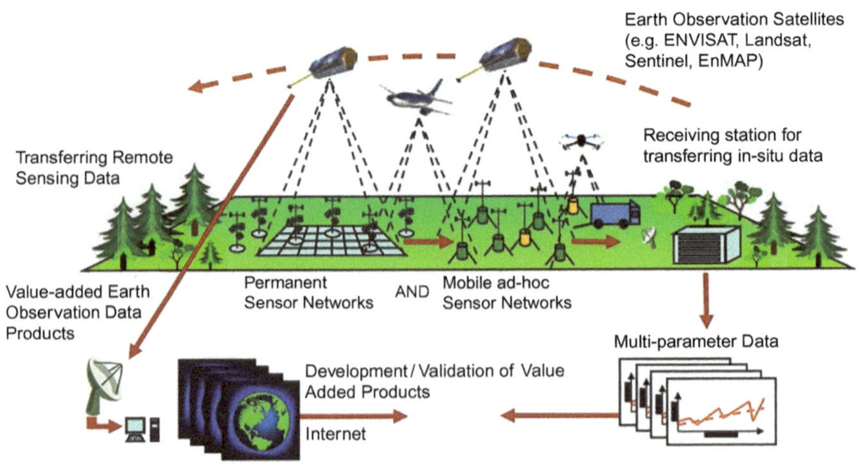

Fig. 1 Differences of monitoring and observation approaches (Lausch et al., 2018).

advancement of closer intercontinental coordination in employing civil aerospace science and high-tech could make a major participation to handling exhaustive Emergency cases in human health (Johnson et al., 2011; Doarn et al., 2019).

4.1 Satellite-based positioning and navigation technology

The Global Navigation Satellite System (GNSS) corresponds to a network of Earth-orbiting sensors delivering satellite transmissions that relay positioning and monitoring method to its final-users, who measure surface locations through triangulation (Yu et al., 2020). The usage of video satellites has also had a major outcome on treatment of infections, from measuring the patient's accurate position to controlling the transport of products, as well as providing a massive disinfection of unmanned aircraft navigation systems based on the Global Positioning System (GPS). Digital Navigation Technology such as the Global Positioning System depend on Earth-orbiting satellites to deliver Global Positioning System location and time information to localization devices everywhere on Earth (Mbunge, 2020).

These navigation techniques are essential throughout a pandemic like COVID-19. They assist the vital transport and distribution of medical equipment and basic health necessities around the globe. Hence, urgent Respondents can indeed hit places quicker and more easily. In specific situations, digital control systems can help ensure real-time contact with drivers delivering vital goods (Suryaatmadja and Maulani, 2020). They will handle constantly changing traffic patterns while locking down and traffic limits are already in optimal position (Ting et al., 2020).

4.2 Unmanned aerial vehicle (UAV) for large-scale disinfection

By integrating a high-precision mapping and navigation facility, unmanned aerial vehicle sterilization will immediately track the bacterial region and execute at centimeter-level precision, demonstrating high versatility and elevated functional performance throughout aerial antiseptic spraying (Mohammed et al., 2020). Until being used by the disease disinfection teams, several drones were used as farm nozzles. By moving equipment from pesticide to antiseptic liquid, small aerial devices rapidly move their domain of action from agriculture to heavily populated city blocks, neighborhoods having elevated or reduced COVID-19 infections, and remote villagers with a fragile health infrastructure and complex sanitation conditions (Albert et al., 2021).

Drone-based airborne bioremediation is showing to be the best efficient means of providing public sanitation. Spray performance of unmanned aerial vehicles can indeed be nearly 50 times greater than conventional means, thus reducing the danger to staff who may be exposed to biosecurity risks; increased either disinfectant compounds or harmful viruses (Kumar et al., 2021).

4.3 Satellite remote sensing and earth observation

Satellite sensing is a science that recognizes the features of the earth's surface and tests its geo-biophysical properties using electromagnetic radiation as a satellite interaction with the primary objective of spatial mapping, earthquake, flood tracking, environmental monitoring and resource observation (Ennouri and Kallel, 2019). In addition, remote technology can be used in disease prevention operations. Remote application can encourage the creation of a geographic information system (GIS), which is a set of computer-based tools designed to capture, store, manipulate, manage and visualize all types of geographic data required for health mapping, and enable a collective analysis of geographical data obtained from satellites (Ennouri et al., 2021).

4.4 Artificial intelligence-driven satellite for a geographic information system-based health mapping

A geographic information system incorporates spatial visualization, mathematical simulation and advanced algorithms to promote the management and distribution of epidemiological data (Ennouri et al., 2020a) and to play an important role in interpreting an extensive picture of public health status. Mapping by Artificial Intelligence-based satellite support for infectious disease monitoring. Artificial Intelligence technology enables satellites to have more efficient information analysis capability and better spatial time resolution, allowing for programmed filtering and processing of recorded images to assess the information categories that should be sent back to Earth (Ottaviani et al., 2020). By integrating intelligent satellite data collection and dissemination technologies, semantic recognition methods and geocoding framework, the mapping of COVID-19 health displays specifically acquired satellite data with additional local functionality, and then users can see spatial distribution and direction of studied pathogens (Ennouri et al., 2020b).

4.5 Dissemination of public health information

When clusters of at-risk groups are recognized, successful distribution of information about public health becomes vital. By integrating numerous data streams, including geo-tagged household surveys and remote sensing data, Machine Learning models inform how societies absorb news. Data indicating Television watching, radio listening, newspaper readership, and mobile phone possession are now accessible within 1 km^2. This knowledge is used to pick platforms for personalized campaigns and additionally to improve the distribution of messages (Tornberg et al., 2021). In the aim to recognize the short, medium and long-term effects of a pandemic, it is important to make good usage of advanced Machine Learning data while preparing specific interventions.

By using local population statistics in an epidemiologic study, downstream impacts can be easily foreseen such as global shocks and food security issues. This form of sophisticated modeling helps businesses prioritize geography and make choices in the sense of dynamic situations. Machine Learning generates population data that is uniquely appropriate to address questions considered as essential to Coronavirus suppression (Pinter et al., 2020). The capacity of technology to take into account current learning, changing risk factors and recently discovered regions makes it a perfect informant during an progressing disaster (Gao et al., 2020).

5. Conclusion

Innovative technology promotes patient interaction, increases disinfection efficiency, tracks surgical devices, transmits sanitation and decreases COVID-19 infection. As the Coronavirus crisis increasingly becomes a global issue, no nation is doing enough to protect its population health. It is necessary to share expertise of disease management with different countries and to enhance international collaborations with the use of space technology in order to accelerate disease prevention. In addition to the digital analysis of satellite imagery, the usage of statistical and numerical tools as well as Machine Learning techniques has also broadened the potential for researching the spatial spread of diseases. These helpful devices allowed the possibility of generating risk maps based on multivariate and hierarchical models. Furthermore, the role of remote sensing technology is more and more confirmed in monitoring and detecting the localization of endemic diseases and then organizing therapeutic activities.

References

Ahani, A., Nilashi, M., 2020. Coronavirus outbreak and its impacts on global economy: the role of social network sites. J. Soft Comput. Decis. Support Syst. 7 (2), 19–22.
Albert, S., Amarilla, A.A., Trollope, B., Sng, J.D., Setoh, Y.X., Deering, N., Khromykh, A.-A., 2021. Assessing the potential of unmanned aerial vehicle spraying of aqueous ozone as an outdoor disinfectant for SARS-CoV-2. Environ. Res. 196, 110944.
Alphan, H., Çelik, N., 2016. Monitoring changes in landscape pattern: use of Ikonos and Quickbird images. Environ. Monit. Assess. 188 (2), 81.
Arora, S., Bhaukhandi, K.D., Mishra, P.K., 2020. Coronavirus lockdown helped the environment to bounce back. Sci. Total Environ. 742, 140573.
Bauer, A., Weber, E., 2020. COVID-19: how much unemployment was caused by the shutdown in Germany? Appl. Econ. Lett. 28, 1053–1058.
Chamola, V., Hassija, V., Gupta, V., Guizani, M., 2020. A comprehensive review of the COVID-19 pandemic and the role of IoT, drones, AI, blockchain, and 5G in managing its impact. IEEE Access 8, 90225–90265.
Claverie, M., Vermote, E.F., Franch, B., Masek, J.G., 2015. Evaluation of the Landsat-5 TM and Landsat-7 ETM+ surface reflectance products. Remote Sens. Environ. 169, 390–403.
Dang, L.M., Piran, M., Han, D., Min, K., Moon, H., 2019. A survey on internet of things and cloud computing for healthcare. Electronics 8 (7), 768.
De Alban, J.D.T., Connette, G.M., Oswald, P., Webb, E.L., 2018. Combined Landsat and L-band SAR data improves land cover classification and change detection in dynamic tropical landscapes. Remote Sens. 10 (2), 306.
Diffenbaugh, N.S., Field, C.B., Appel, E.A., Azevedo, I.L., Baldocchi, D.D., Burke, M., Wong-Parodi, G., 2020. The COVID-19 lockdowns: a window into the Earth System. Nat. Rev. Earth Environ. 1 (9), 470–481.
Ding, X., Clifton, D., Ji, N., Lovell, N.H., Bonato, P., Chen, W., Zhang, Y.T., 2020. Wearable sensing and telehealth technology with potential applications in the coronavirus pandemic. IEEE Rev. Biomed. Eng. 14, 48–70.
Djalante, R., Shaw, R., DeWit, A., 2020. Building resilience against biological hazards and pandemics: COVID-19 and its implications for the Sendai Framework. Prog. Disaster. Sci. 6, 100080.
Doarn, C.R., Travis, T.W., Currie-Gregg, N.K., Nicogossian, A.E., Weyland, M., Shepanek, M., Williams, R.S., 2019. Engineering, life sciences, and health/medicine synergy in aerospace human systems integration: the Rosetta stone project—an executive summary. New Space 7 (2), 110–113.
Dou, M., Chen, J., Chen, D., Chen, X., Deng, Z., Zhang, X., Wang, J., 2014. Modeling and simulation for natural disaster contingency planning driven by high-resolution remote sensing images. Futur. Gener. Comput. Syst. 37, 367–377.
Ennouri, K., Ottaviani, E., Smaoui, S., Triki, M.A., 2020a. Use of remote sensing technology and geographic information system for agriculture and environmental observation. In: Ennouri, K. (Ed.), Emerging Technologies in Agriculture and Food Science. Bentham Science Publishers, pp. 108–138 (31) https://doi.org/10.2174/9789811470004120010l.
Ennouri, K., Triki, M.A., Kallel, A., 2020b. Applications of remote sensing in pest monitoring and crop management. In: Keswani, C. (Ed.), Bioeconomy for Sustainable Development. Springer, Singapore, https://doi.org/10.1007/978-981-13-9431-7_5.
Ennouri, K., Kallel, A., 2019. Remote sensing: an advanced technique for crop condition assessment. Math. Probl. Eng. 2019, 9404565.
Ennouri, K., Smaoui, S., Triki, M.A., 2021. Detection of urban and environmental changes via remote sensing. Circ. Econ. Sustain. https://doi.org/10.1007/s43615-021-00035-y.

Fan, X., Kawamura, K., Guo, W., Xuan, T.D., Lim, J., Yuba, N., Wang, Z., 2018. A simple visible and near-infrared (V-NIR) camera system for monitoring the leaf area index and growth stage of Italian ryegrass. Comput. Electron. Agric. 144, 314–323.

Farmakis-Serebryakova, M., Hurni, L., 2020. Comparison of relief shading techniques applied to landforms. ISPRS Int. J. Geo Inf. 9 (4), 253.

Filippini, T., Rothman, K.J., Goffi, A., Ferrari, F., Maffeis, G., Orsini, N., Vinceti, M., 2020. Satellite-detected tropospheric nitrogen dioxide and spread of SARS-CoV-2 infection in Northern Italy. Sci. Total Environ. 739, 140278.

Fong, S.J., Dey, N., Chaki, J., 2020. Artificial Intelligence for Coronavirus Outbreak. Springer, Singapore.

Franch-Pardo, I., Napoletano, B.M., Rosete-Verges, F., Billa, L., 2020. Spatial analysis and GIS in the study of COVID-19. A review. Sci. Total Environ. 739, 140033.

Gandhi, G.M., Parthiban, S., Thummalu, N., Christy, A., 2015. Ndvi: vegetation change detection using remote sensing and gis—a case study of Vellore District. Procedia Comput. Sci. 57, 1199–1210.

Gao, Y., Cai, G.Y., Fang, W., Li, H.Y., Wang, S.Y., Chen, L., Gao, Q.L., 2020. Machine learning based early warning system enables accurate mortality risk prediction for COVID-19. Nat. Commun. 11 (1), 1–10.

Garni, R., Tran, A., Guis, H., Baldet, T., Benallal, K., Boubidi, S., Harrat, Z., 2014. Remote sensing, land cover changes, and vector-borne diseases: use of high spatial resolution satellite imagery to map the risk of occurrence of cutaneous leishmaniasis in Ghardaïa, Algeria. Infect. Genet. Evol. 28, 725–734.

Guo, X., Zhang, H., Wu, Z., Zhao, J., Zhang, Z., 2017. Comparison and evaluation of annual NDVI time series in China derived from the NOAA AVHRR LTDR and Terra MODIS MOD13C1 products. Sensors 17 (6), 1298.

Hiscott, J., Alexandridi, M., Muscolini, M., Tassone, E., Palermo, E., Soultsioti, M., Zevini, A., 2020. The global impact of the coronavirus pandemic. Cytokine Growth Factor Rev. 53, 1–9.

Houborg, R., Fisher, J.B., Skidmore, A.K., 2015. Advances in Remote Sensing of Vegetation Function and Traits.

Jamison, A., Tuttle, E., Jensen, R., Bierly, G., Gonser, R., 2015. Spatial ecology, landscapes, and the geography of vector-borne disease: a multi-disciplinary review. Appl. Geogr. 63, 418–426.

Jensen, J.R., 2016. Introductory Digital Image Processing: A Remote Sensing Perspective, fourth Ed. Pearson Series.

Johnson, S.B., Gormley, T., Kessler, S., Mott, C., Patterson-Hine, A., Reichard, K., Scandura Jr, P. (Eds.), 2011. System Health Management: With Aerospace Applications. John Wiley & Sons.

Kganyago, M., Odindi, J., Adjorlolo, C., Mhangara, P., 2018. Evaluating the capability of Landsat 8 OLI and SPOT 6 for discriminating invasive alien species in the African Savanna landscape. Int. J. Appl. Earth Obs. Geoinf. 67, 10–19.

Kumar, A., Sharma, K., Singh, H., Naugriya, S.G., Gill, S.S., Buyya, R., 2021. A drone-based networked system and methods for combating coronavirus disease (COVID-19) pandemic. Futur. Gener. Comput. Syst. 115, 1–19.

Lan, T., Shao, G., Tang, L., Xu, Z., Zhu, W., Liu, L., 2021. Quantifying spatiotemporal changes in human activities induced by COVID-19 pandemic using daily nighttime light data. IEEE J. Sel. Top. Appl. Earth Obs. Remote Sens. 14, 2740–2753.

Lausch, A., Borg, E., Bumberger, J., Dietrich, P., Heurich, M., Huth, A., Schaepman, M.E., 2018. Understanding forest health with remote sensing, part III: requirements for a scalable multi-source forest health monitoring network based on data science approaches. Remote Sens. 10 (7), 1120.

Mahabir, R., Croitoru, A., Crooks, A.T., Agouris, P., Stefanidis, A., 2018. A critical review of high and very high-resolution remote sensing approaches for detecting and mapping slums: trends, challenges and emerging opportunities. Urban Sci. 2 (1), 8.

Masrur, A., Yu, M., Luo, W., Dewan, A., 2020. Space-time patterns, change, and propagation of COVID-19 risk relative to the intervention scenarios in Bangladesh. Int. J. Environ. Res. Public Health 17 (16), 5911.

Maxwell, A.E., Warner, T.A., Fang, F., 2018. Implementation of machine-learning classification in remote sensing: an applied review. Int. J. Remote Sens. 39 (9), 2784–2817.

Mbunge, E., 2020. Integrating emerging technologies into COVID-19 contact tracing: opportunities, challenges and pitfalls. Diabetes Metab. Syndr. Clin. Res. Rev. 14 (6), 1631–1636.

Mohammed, M.N., Syamsudin, H., Hazairin, N.A., Haki, M., Al-Zubaidi, S., Sairah, A.K., Eddy, Y., 2020. Toward a novel design for spray disinfection system to combat coronavirus (Covid-19) using IoT based drone technology. Rev. Argent. Clín. Psicol. 29 (5), 240.

Obata, K., Tsuchida, S., Yamamoto, H., Thome, K., 2017. Cross-calibration between ASTER and MODIS visible to near-infrared bands for improvement of ASTER radiometric calibration. Sensors 17 (8), 1793.

Ogen, Y., 2020. Assessing nitrogen dioxide (NO2) levels as a contributing factor to coronavirus (COVID-19) fatality. Sci. Total Environ. 726, 138605.

Omrani, H., Omrani, B., Parmentier, B., Helbich, M., 2020. Spatio-temporal data on the air pollutant nitrogen dioxide derived from Sentinel satellite for France. Data Brief 28, 105089.

Ottaviani, E., Barelli, E., Ennouri, K., 2020. Machine learning for precision agriculture: methods and applications. In: Ennouri, K. (Ed.), Emerging Technologies in Agriculture and Food Science. Bentham Science Publishers, pp. 91–107, https://doi.org/10.2174/9789811470004120010l.

Pal, S., Das, P., Mandal, I., Sarda, R., Mahato, S., Nguyen, K.A., Saha, T.K., 2021. Effects of lockdown due to COVID-19 outbreak on air quality and anthropogenic heat in an Industrial belt of India. J. Clean. Prod. 297, 126674.

Pan, S.L., Zhang, S., 2020. From fighting COVID-19 pandemic to tackling sustainable development goals: an opportunity for responsible information systems research. Int. J. Inf. Manag. 55, 102196.

Pang, G., Wang, X., Yang, M., 2017. Using the NDVI to identify variations in, and responses of, vegetation to climate change on the Tibetan Plateau from 1982 to 2012. Quat. Int. 444, 87–96.

Pettorelli, N., Schulte to Bühne, H., Tulloch, A., Dubois, G., Macinnis-Ng, C., Queirós, A.M., Nicholson, E., 2018. Satellite remote sensing of ecosystem functions: opportunities, challenges and way forward. Remote Sens. Ecol. Conserv. 4 (2), 71–93.

Pimentel, R., Aguilar, C., Herrero, J., Pérez-Palazón, M.J., Polo, M.J., 2016. Comparison between Snow Albedo Obtained from Landsat TM, ETM+ Imagery and the SPOT VEGETATION Albedo Product in a Mediterranean Mountainous Site. Hydrology 3 (1), 10.

Pinter, G., Felde, I., Mosavi, A., Ghamisi, P., Gloaguen, R., 2020. COVID-19 pandemic prediction for Hungary; a hybrid machine learning approach. Mathematics 8 (6), 890.

Qadir, A., Mondal, P., 2020. Synergistic use of radar and optical satellite data for improved monsoon cropland mapping in India. Remote Sens. 12 (3), 522.

Rashid, M.T., Wang, D., 2021. CovidSens: a vision on reliable social sensing for COVID-19. Artif. Intell. Rev. 54 (1), 1–25.

Reddy, G.O., 2018. Satellite remote sensing sensors: principles and applications. In: Geospatial Technologies in Land Resources Mapping, Monitoring and Management. Springer, Cham, pp. 21–43.

Riva, M.J., Daliakopoulos, I.N., Eckert, S., Hodel, E., Liniger, H., 2017. Assessment of land degradation in Mediterranean forests and grazing lands using a landscape unit approach and the normalized difference vegetation index. Appl. Geogr. 86, 8–21.

Sepuru, T.K., Dube, T., 2018. An appraisal on the progress of remote sensing applications in soil erosion mapping and monitoring. Remote Sens. Appl. 9, 1–9.

Serrano, F., Kazda, A., 2020. The future of airport post COVID-19. J. Air Transp. Manag. 89, 101900.

Solberg, S., Walker, S.E., Schneider, P., Guerreiro, C., 2021. Quantifying the impact of the Covid-19 lockdown measures on nitrogen dioxide levels throughout Europe. Atmosphere 12 (2), 131.

Sun, H., Shimada, M., Xu, F., 2017. Recent advances in synthetic aperture radar remote sensing—systems, data processing, and applications. IEEE Geosci. Remote Sens. Lett. 14 (11), 2013–2016.

Sun, W., Zhu, L., De Smedt, I., Bai, B., Pu, D., Chen, Y., Yang, X., 2021. Global significant changes in formaldehyde (HCHO) columns observed from space at the early stage of the COVID-19 pandemic. Geophys. Res. Lett. 48, e2020GL091265.

Sun, X., Wandelt, S., Zhang, A., 2020. How did COVID-19 impact air transportation? A first peek through the lens of complex networks. J. Air Transp. Manag. 89, 101928.

Suryaatmadja, S., Maulani, N., 2020. Contributions of space technology to global health in the context of Covid-19. J. Admin. Kesehat. Indones. 8 (2), 60–73.

Tassa, A., 2020. The socio-economic value of satellite earth observations: huge, yet to be measured. J. Econ. Policy Reform 23 (1), 34–48.

Ting, D.S.W., Carin, L., Dzau, V., Wong, T.Y., 2020. Digital technology and COVID-19. Nat. Med. 26 (4), 459–461.

Tornberg, H.N., Moezinia, C., Wei, C., Bernstein, S.A., Wei, C., Al-Beyati, R., Diemert, D.J., 2021. Assessing the dissemination of COVID-19 articles across social media with Altmetric and PlumX metrics: correlational study. J. Med. Internet Res. 23 (1), e21408.

Tu, W., Hu, Z., Li, L., Cao, J., Jiang, J., Li, Q., Li, Q., 2018. Portraying urban functional zones by coupling remote sensing imagery and human sensing data. Remote Sens. 10 (1), 141.

Tuli, S., Tuli, S., Tuli, R., Gill, S.S., 2020. Predicting the growth and trend of COVID-19 pandemic using machine learning and cloud computing. Internet Things 11, 100222.

Vîrghileanu, M., Săvulescu, I., Mihai, B.A., Nistor, C., Dobre, R., 2020. Nitrogen dioxide (NO2) pollution monitoring with sentinel-5P satellite imagery over Europe during the coronavirus pandemic outbreak. Remote Sens. 12 (21), 3575.

Wang, Q., Su, M., 2020. A preliminary assessment of the impact of COVID-19 on environment—a case study of China. Sci. Total Environ. 728, 138915.

Wu, M., Li, H., Huang, W., Niu, Z., Wang, C., 2015. Generating daily high spatial land surface temperatures by combining ASTER and MODIS land surface temperature products for environmental process monitoring. Environ Sci Process Impacts 17 (8), 1396–1404.

Xiao, X., Zhang, T., Zhong, X., Shao, W., Li, X., 2018. Support vector regression snow-depth retrieval algorithm using passive microwave remote sensing data. Remote Sens. Environ. 210, 48–64.

Yagoub, H., Belbachir, A.H., Benabadji, N., 2014. Detection and mapping vegetation cover based on the Spectral Angle Mapper algorithm using NOAA AVHRR data. Adv. Space Res. 53 (12), 1686–1693.

Yu, J., Meng, X., Yan, B., Xu, B., Fan, Q., Xie, Y., 2020. Global navigation satellite system-based positioning technology for structural health monitoring: a review. Struct. Control. Health Monit. 27 (1), e2467.

Zhang, C., Marzougui, A., Sankaran, S., 2020b. High-resolution satellite imagery applications in crop phenotyping: an overview. Comput. Electron. Agric. 175, 105584.

Zhang, Y.F., Gong, J.W., Yin, M., 2020a. Influences and response measures of COVID-19 epidemic on shipping and port industry in China. J. Traffic Transp. Eng. 20 (3), 159–167.

Zhong, Y., Zhu, Q., Zhang, L., 2015. Scene classification based on the multifeature fusion probabilistic topic model for high spatial resolution remote sensing imagery. IEEE Trans. Geosci. Remote Sens. 53 (11), 6207–6222.

Zhou, C., Su, F., Pei, T., Zhang, A., Du, Y., Luo, B., Xiao, H., 2020. COVID-19: challenges to GIS with big data. Geogr. Sustain. 1 (1), 77–87.

Zhu, Z., Zhou, Y., Seto, K.C., Stokes, E.C., Deng, C., Pickett, S.T., Taubenböck, H., 2019. Understanding an urbanizing planet: strategic directions for remote sensing. Remote Sens. Environ. 228, 164–182.

CHAPTER 7

Potential constraint of rainfall availability on the establishment and expansion of agroforestry in the Joe Gqabi, Alfred Nzo and OR Tambo Districts, Eastern Cape in South Africa

Phokele Maponya[a], Casper Madakadze[b], Nokwazi Mbili[c], Zakheleni Dube[d], Thabo Nkuna[a], Meshack Makhwedzhana[a], Takalani Tahulela[e], and Kgosi Mongwaketsi[e]
[a]Agricultural Research Council-Vegetable, Industrial and Medicinal Plants, Pretoria, Gauteng, South Africa
[b]University of Pretoria, Faculty of Agriculture and Natural Science, Pretoria, South Africa
[c]University of KwaZulu-Natal, College of Agriculture, Engineering and Science, School of Agricultural, Earth and Environmental Science, Durban, South Africa
[d]University of Mpumalanga, Faculty of Agriculture and Natural Science, Mbombela, South Africa
[e]South Africa Forestry Company Limited (SAFCOL) Research, SAFCOL (Pty) LTD, Sabie, South Africa

1. Introduction

According to CCAFS (2019), forests are the biggest users of water worldwide and extensive forested areas have been lost or are undergoing conversion to agriculture, creating concerns about loss of hydrological functions and increasing the competition for scarce water between agriculture, urban centers, industries and wildlife. The challenge is to improve the sustainability and productivity of land and water use, especially for the growing populations of many developing countries hence agroforestry is recommended. Agroforestry is a land use system that includes the use of woody perennial and agricultural crops and animals in combination to achieve beneficial ecological and economical interactions for food, fiber and livestock production. According to Nair (1985), several criteria can be used to classify and group agroforestry systems (and practices). The following are the most commonly used criteria: the system's structure (composition and arrangement of components), its function, its socio-economic scale and level of management, and its ecological spread. According to Nair (1985) structurally, the system can

Agricultural Bioeconomy
https://doi.org/10.1016/B978-0-323-90569-5.00009-3

be grouped as agrosilviculture (crops—including tree/shrub crops—and trees), silvopastoral (pasture/animals + trees), and agrosilvopastoral (crops + pasture/animals + trees).

According to Maponya et al. (2020) agroforestry systems play a major role throughout human history in supporting livelihoods, assisting various communities generate income, create job opportunities, as well as meeting food security and nutritional needs in Limpopo Province. The researchers further indicated an evidence of the importance of agroforestry systems (and Timber Based Mixed Farming systems) especially silvipasture and agrosilvipasture for supporting food production and income generation in Limpopo Province. Some farmers in Limpopo Province highlighted that they are generating income through renting of farms for grazing, selling trees to the communities to build shelter, kraals, medicinal purposes, fuelwood, etc.

Furthermore, South Africa is considered a semi-arid country vulnerable to water stress, particularly drought (Hassan, 2013). In the previous years, South Africa has experienced one of the worst drought in history where some provinces were declared disaster areas. The threshold for rainfall agriculture is averaged at 250 mm annually and in terms of forestry, rainfall needs to be higher than 750 mm per annum to sustain commercial forestry. According to ARC (2020), OR Tambo district receives a median annual rainfall that ranges mostly between 800 and more than 1000 mm. Climatically the area is thus well suited to rainfed arable agriculture where slopes and soils permit. A moderate summer peak in rainfall is evident. Summer rains start in September and the wettest months are November and December, and the driest, May to August. Similar to OR Tambo, the Alfred Nzo district receives a median annual rainfall that ranges mostly between 800 and more than 1000 mm. Climatically the area is thus well suited to rainfed arable agriculture where slopes and soils permit. A moderately strong summer peak in rainfall is evident. Summer rains start in September. The wettest months are November and December, and the driest, May to August. Furthermore, Joe Gqabi district is vastly agricultural and it provides an ideal investment opportunity for agro-forestry, and its proximity to the ports of East London and Port Elizabeth, with links to major routes like the N6 and N10, is an advantage. Hence, the objective of the study was to determine the potential constraint of rainwater on the establishment and expansion of agroforestry in the Joe Gqabi, Alfred Nzo and OR Tambo Districts, Eastern Cape.

In addition, Alfred Nzo district is predominantly rural and most agriculture is at subsistence level (ADM, 2014). It is further noted that agricultural input costs are high and land claims are affecting development. According to

ADM (2014) there is nevertheless potential to develop the local economy through agriculture (cattle, goats, sheep; crops such as sorghum, maize, lucerne, vegetables, tropical fruit and nuts, pineapples and chicory), tourism (arts and crafts, scenery, wildlife, wetlands and cultural heritage), fishing (in the dams and rivers) and forestry. Oliver Tambo district has a relatively small formal economy compared to the rest of the province, providing 11% of value added (ECDC, 2014) Agriculture is the major private sector activity, contributing 13% of value added and 8% of formal employment. Forestry is the main formal agricultural enterprise, with large forests to the north and west of Mthatha, mainly leased by the private sector but with community plantations also present. According to ECDC (2014), there are significant numbers of small commercial farmers in the area, concentrating on mixed farming of livestock and crops (mostly maize). Some commercial farmers have invested in small-scale irrigation, with cabbage and potatoes the most popular crops. Subsistence agriculture makes a major contribution to household food security. Most production is for home consumption or limited local sale (ECDC, 2014).

2. Methodology

2.1 Study area

As indicated in Fig. 1, six forestry enterprises participated in the study, namely (1) Sinawo Forestry Enterprise (2) Mkambathi Forestry Enterprise (3) Izinini Forestry Enterprise (4) Gqukunqa Forestry Enterprise (5) Sixhotyeni Forestry Enterprise (6) Lusikisiki Forestry Enterprise.

2.1.1 Sinawo Forestry Enterprise

Sinawo is located in the Mfolozi, Hlolweni and Greenville villages on the R61 between Port Edward and Mbizana in the north-eastern part of the Eastern Cape about 15 km from Port Edward. It is located in the Mbizana Local Municipality in the Alfred Nzo District Municipality. With a 1100 ha plantation established in the 1970s/1980s, Sinawo offers fertile ground to revive the existing plantation. This meant the community could immediately draw benefits from the clear felling of the existing timber and agroforestry. The enterprise has 800 beneficiaries (ECRDA, 2020).

2.1.2 Mkambathi Forestry Enterprise

The Mkambathi forestry project is situated some 60 km from Flagstaff and 290 km from the SAPPI pulp mill at Mkomaas in KwaZulu-Natal. It is

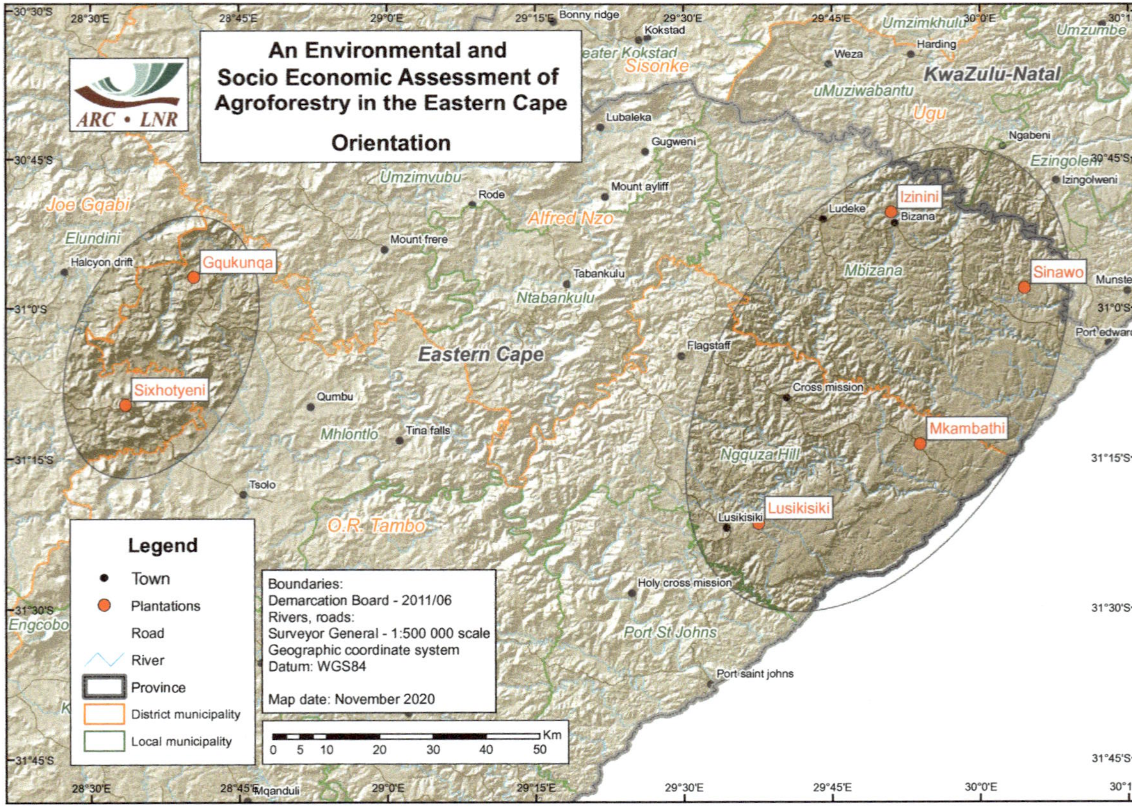

Fig. 1 Location of the Agroforestry Sites in the Eastern Cape, South Africa (ARC-SCW, 2020).

located at Ingquza Hill Local Municipality in the OR Tambo District Municipality. The Mkambathi community already had a gum plantation which was established by the then Transkei Agriculture Corporation (TRACOR) in 1980 making it easy to rehabilitate the existing plantation because the community would draw immediate benefit from the clear-felling of the existing timber and agroforestry. The project employs 127 people from the 8 villages (ECRDA, 2020). A total of 668.8 ha was planted with *Eucalyptus dunnii*, which is suitable for the pulp market (ECRDA, 2020).

2.1.3 Izinini Forestry Enterprise

It is located on the outskirts of Mbizana in the north-eastern region of the Eastern Cape; Izinini is situated in the Mbizana Local Municipality in the Alfred Nzo District Municipality. It is located some 200 km from the SAPPI cellulose plant at Mkomaas in KwaZulu-Natal and some 60 km from Flagstaff. Izinini has an existing plantation with wattle, which was established by the community. This made it easy to rehabilitate the plantation and introduce agroforestry. This also made for immediate economic benefits for the community from the clear felling of the existing timber and agroforestry. The project has 1237 beneficiaries. About 293 ha was planted with *Eucalyptus dunnii*, which is suitable for the pulp market (ECRDA, 2020).

2.1.4 Gqukunqa Forestry Enterprise

Gqukunqa is situated in Qumbu in the north of the Eastern Cape about 40 km from Maclear on the R56 Maclear/Mount Fletcher road. The project lies 60 km from the PG Bison board mill at Ugie. Its total area is 1578 ha in extent. It is located in the Mhlontlo Local Municipality in the OR Tambo District Municipality. The area planted is approximately 605 ha resulting in the creation of 211 jobs from 18 villages. The area was planted with *Eucalyptus nitens*, which is suitable for the pulp market and pole market (ECRDA, 2020).

2.1.5 Sixhotyeni Forestry Enterprise

Sixhotyeni is situated in the north Eastern Cape approximately 25 km from Maclear on the R396 road between Maclear and Mount Fletcher. The project is 45 km from the PG Bison board mill at Ugie and 226 km from the Harding Treated Timbers pole plant at Harding. It is located in Elundini Local Municipality in the Joe Gqabi District Municipality. A combined area of 980 ha was initially identified for forestry development by the community and a soil survey confirmed the suitability of the soils for commercial plantation establishment. The project has employed about 127 people from the villages bordering the Ntwenka plantation, was planted with 214 ha of *Eucalyptus nitens*, which is suitable for the pulp market and pole market (ECRDA, 2020).

2.1.6 Lusikisiki Forestry Enterprise

Lusikisiki lies in the Ingquza Hill Local Municipality in the Eastern Cape Province. Lusikisiki is 45 km inland from and north of Port St Johns. It is positioned along the R61 (future N2 Wild Coast Toll Route) leading to Mthatha to the west and Durban to the north. There is a series of rolling hills, lush tropical forests, wide open spaces, untouched beaches and an untamed mystery—supports the local people's lifestyle that has changed little over the years.

2.2 Research approach

According to Backeberg and Sanewe (2010), participatory action research is a most appropriate research method since people, especially farmers, benefit while the research is ongoing. Participatory action approach was also recommended by various researchers who emphasized that this approach is a good alternative to the traditional "transfer of technology" or "top-down approach" to agricultural research and extension. Coordinates were received from the Department of Environment, Forestry and Fisheries (DEFF) to plot the climate maps.

2.3 Sampling method

A purposive sampling technique was used to select five agroforestry sites namely (1) Sinawo Forestry Enterprise (2) Mkambathi Forestry Enterprise (3) Izinini Forestry Enterprise (4) Gqukunqa Forestry Enterprise (5) Sixhotyeni Forestry Enterprise (6) Lusikisiki Forestry Enterprise.

2.4 Data analysis

The following approach was used to determine average monthly rainfall (Malherbe and Tackrah, 2003): Decadal (10-day period) 1 km × 1 km surfaces were created from rainfall data (1920–99) downloaded from the AgroMet databank at the Agricultural Research Council-Soil, Climate and Water (ARC-SCW) (South African Weather Service and SCW weather stations) from stations with a recording period of 10 years or more. Regression analysis and spatial modeling were utilized taking into account topographic indices such as altitude, aspect, slope and distance to the sea during the development of the surface.

2.4.1 The following approach was used to determine average monthly rainfall (Malherbe and Tackrah, 2003)

Decadal (10-day period) 1 km × 1 km surfaces were created from rainfall data (1920–99) downloaded from the AgroMet databank at the Agricultural Research Council-Soil, Climate and Water (ARC-SCW) (South African Weather Service and SCW weather stations) from stations with a recording period of 10 years or more. Regression analysis and spatial modeling were utilized taking into account topographic indices such as altitude, aspect, slope and distance to the sea during the development of the surface.

2.4.2 The following approach was used to determine soil types (ARC-SCW, 2017)

Digital land type information—soil depth and the spatial component were used. Soil depth is recorded as a range for each soil entry. A weighted average was calculated for each land type unit. Land Type Survey Staff. 1972–2006. Land Types of South Africa: Digital map (1:250000 scale).

2.4.3 The following model was used to determine average monthly temperature (Malherbe and Tackrah, 2003)

Decadal (10-day period) 1 km × 1 km surfaces were created from temperature data (1920–99) downloaded from the AgroMet databank at the ARC-SCW (South African Weather Service and SCW weather stations) from stations with a recording period of 5 years or more. Regression analysis and spatial modeling were utilized taking into account topographic indices such as altitude, aspect, slope and distance to the sea during the development of the surface.

2.4.4 The following model was used to determine soil drainage (ARC-SCW, 2017)

See Table 1.

3. Results and discussions

3.1 Soil conditions

According to ARC (2017) in most parts of the Eastern Cape, Deep, loamy well-drained alluvial soils are found in along most drainage courses and on

Table 1 Digital land type information—soil type and the spatial component.

Class	Soil drainage	Qualifying soil forms	Percentage qualifying soil in land type
1	Poor	Ch, Rg, Wo, Ka, Kd, Es	>40
2	Impeded	Ch, Rg, Wo, Ka, Kd, Es, Lo, Wa, Cf, La	>40
3	Somewhat impeded	Ch, Rg, Wo, Ka, Kd, Es, Lo, Wa, Cf, La, Bo, Fw, Vf, We, Av, Gc, Pn, Bv, Ss, Va, Sw	>40
4	Other		

river terraces. Along the major rivers, these are almost all under irrigation. Moderately deep, mostly well-drained, loamy soils are fairly abundant in most valley situations and on plains, including the coastal plains. Shallow soils, however, dominate in mountainous areas and most karoo plains. Although soils suitable for irrigation are not abundant everywhere, soil availability and suitability is much less of an impediment to lateral expansion of irrigated agriculture than water availability (ARC, 2017).

As is the case in almost all of the seaboard areas, the landscape is dissected by numerous small and larger streams. The resultant topography and soil pattern may not be everywhere conducive to large-scale modern mechanized agriculture, but can, in places, be conducive to the small scale, production of vegetables under supplementary (e.g., by treadle pump) irrigation. The soils, however, are mostly not very deep (601–900 mm) except in the Lusikisiki agroforestry site (901–1200 mm) (Fig. 3). Protection of fields against farm and other animals would be crucial. The soils, except those on dolerite, are erodible and organic matter needs to be conserved.

Most agroforestry site has well drained soil except Mkambathi agroforestry site (Fig. 2). There is an abundance of soils with moderately good to good physical properties. This is particularly true for the high-lying, undissected parts of plains (where, unfortunately, it is difficult to supply irrigation water). As is the case everywhere else, on-site confirmation of soil depth and drainage as well as sampling for fertility analysis and lime requirement should precede any project implementation.

According to Maponya et al. (2021) agroforestry systems respect structural and ecological diversity of species in ecosystem and a well-managed trees have a great capacity of maintaining soil fertility by building up organic matter and mediating nutrient cycling. In fact, soils in agroforestry systems

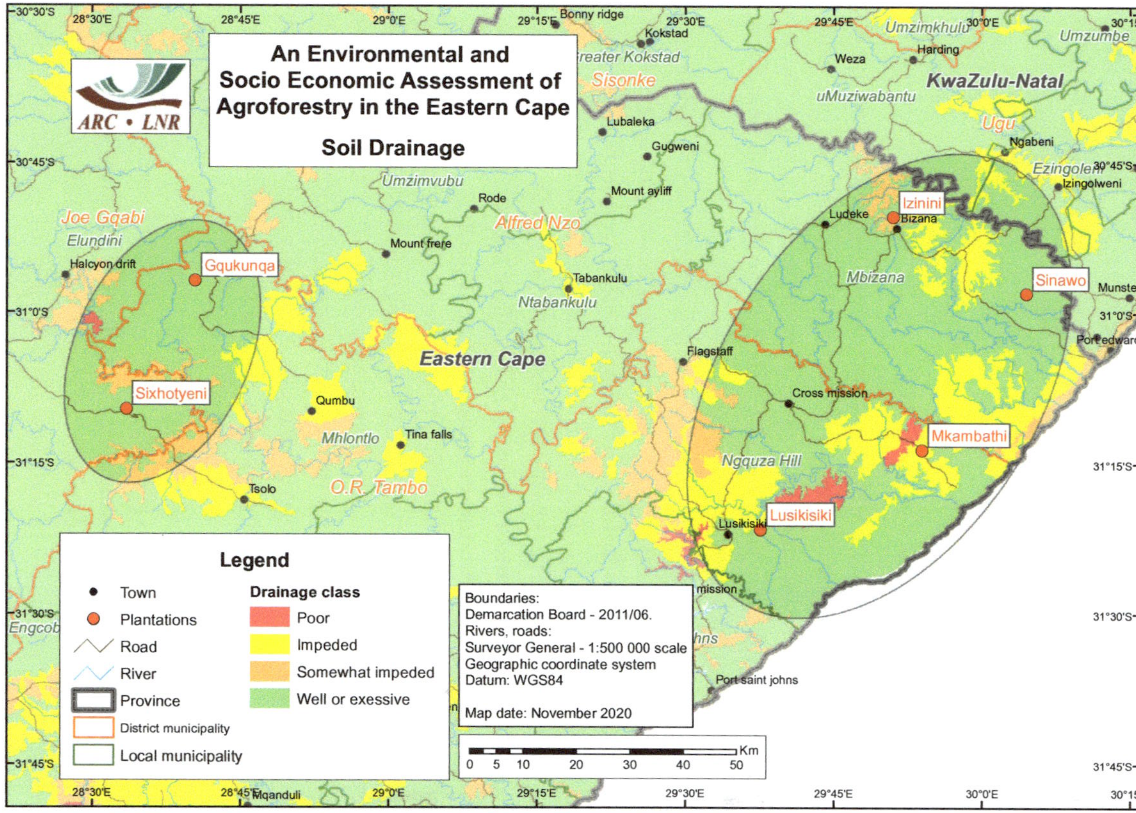

Fig. 2 Eastern Cape Province soil drainage (ARC-SCW, 2020).

are richer in nitrogen, phosphorus, potassium, calcium and organic carbon (ARC, 2019; Hillbrand et al., 2017). As a study of fertile lands in Brazil found out, presence of trees on farms managed by indigenous farmers led up to 20 times higher content of phosphorus and 2.5 higher level of potassium in soils (Ospina, 2016). This is because trees are more efficient at extracting nutrients from deeper soil layers than annual crops with shallow roots. In general, deeper and stronger root systems of trees reach nutrients that would be unavailable to other plants (Fig. 3).

3.2 Rainfall availability

Timber-based mixed farming and agroforestry systems that have commercial timber species such as Eucalyptus as the woody component, are only suited to areas that receive sufficient rainfall to support production of the trees (Maponya et al., 2021). In such areas, it is likely that there will be sufficient rainfall to support the other components of the system.

As indicated in Fig. 4, the average rainfall for September is fairly good (75–100 mm) for most parts of the Eastern Cape Province agroforestry sites as compared to the agricultural open fields areas. As seen in Fig. 5, the situation improves during October (101–150 mm) with an increase in rainfall in the agroforestry sites. This situation is not surprising, as the Department of Environment, Fisheries and Forestry (DEFF) have identified those sites as good for the establishment and expansion of agroforestry.

As indicated in Figs. 6 and 7, long-term increases in rainfall from November to December are experienced in the agroforestry sites. The agroforestry sites were experiencing increased rainfall at 125–175 mm and 125–175 mm respectively. This trend offers a good platform for the establishment and expansion of agroforestry as rainwater is not a constraint. The results are also in line with Maponya et al. (2021) who found that rainfall is not a constraint in the establishment of agroforestry in the Limpopo and Mpumalanga Provinces identified agroforestry sites.

The last three rainfall maps (Figs. 8–10) show the 33rd percentile, median (50th percentile) and 67th percentile. To explain what these maps depict, one can consider the 33rd percentile. *If there were 100 years of recorded data arranged in sequence from dry to wet, then the 33rd percentile would be the value of the 33rd year. In other words, the chances are good to exceed this rainfall, or the chances are small that you will have less rain.* The model estimated annual rainfalls for the broad study area at 801–1000+mm for the 33rd percentile; 901–1000+ at the 50th percentile and +1000 mm at the 67th percentile.

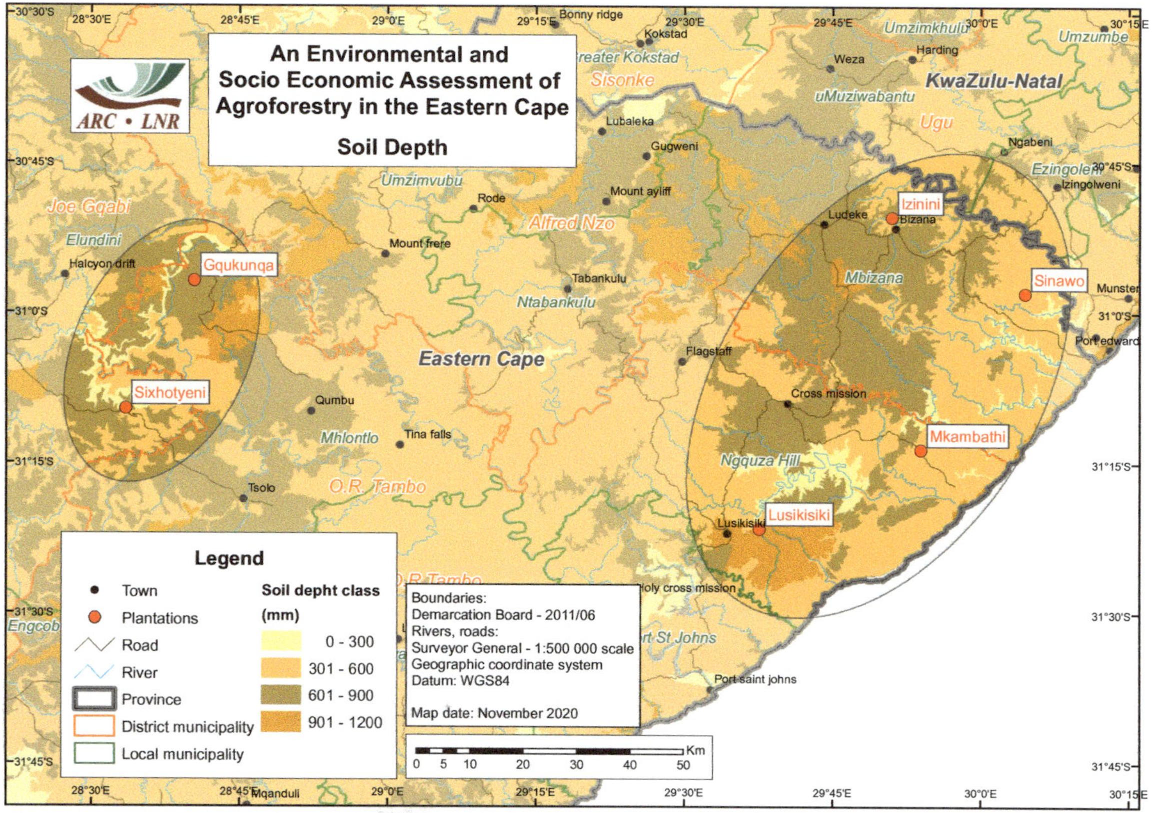

Fig. 3 Eastern Cape Province soil depth (ARC-SCW, 2020).

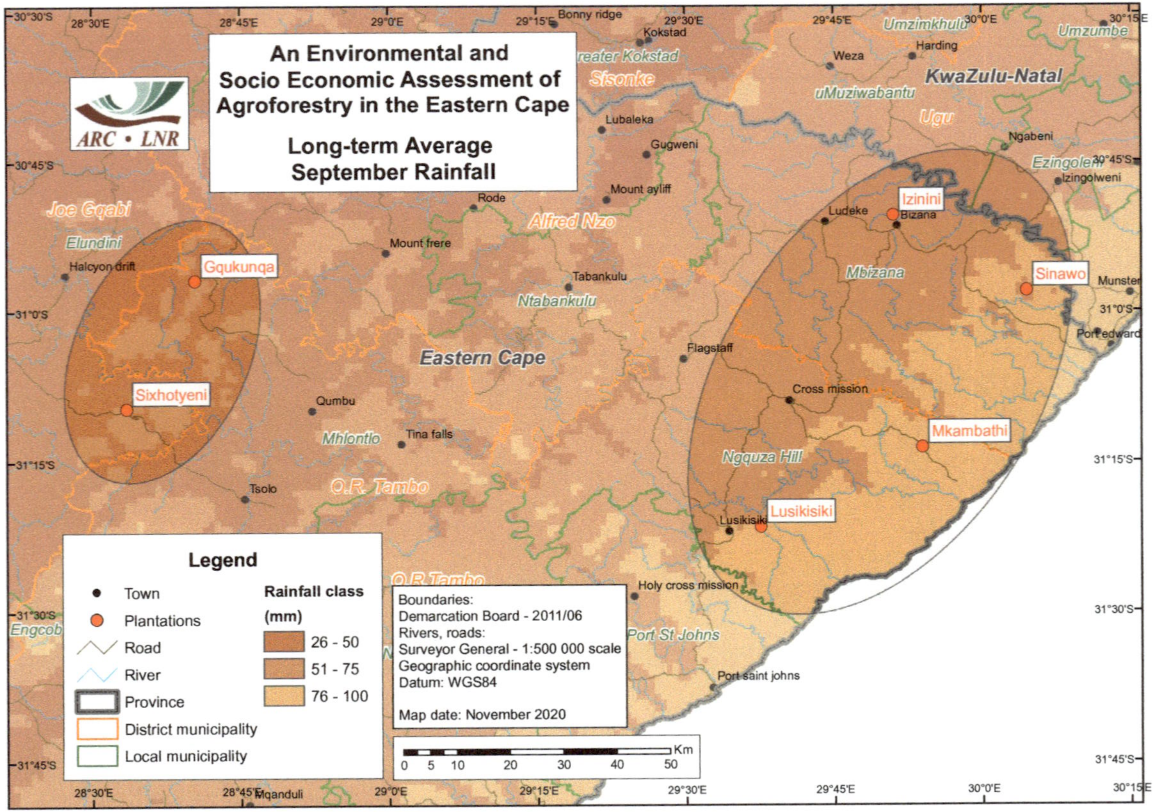

Fig. 4 Eastern Cape Province long-term average September rainfall (ARC-SCW, 2020).

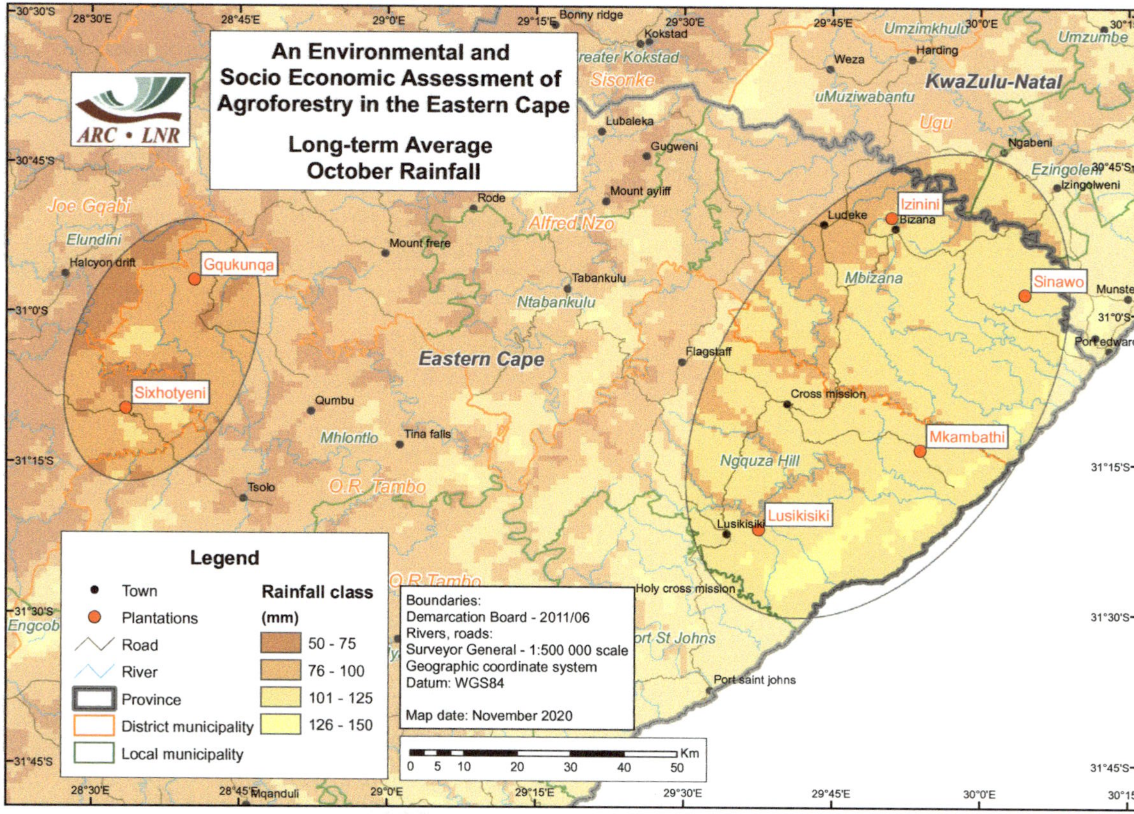

Fig. 5 Eastern Cape Province long-term average October rainfall (ARC-SCW, 2020).

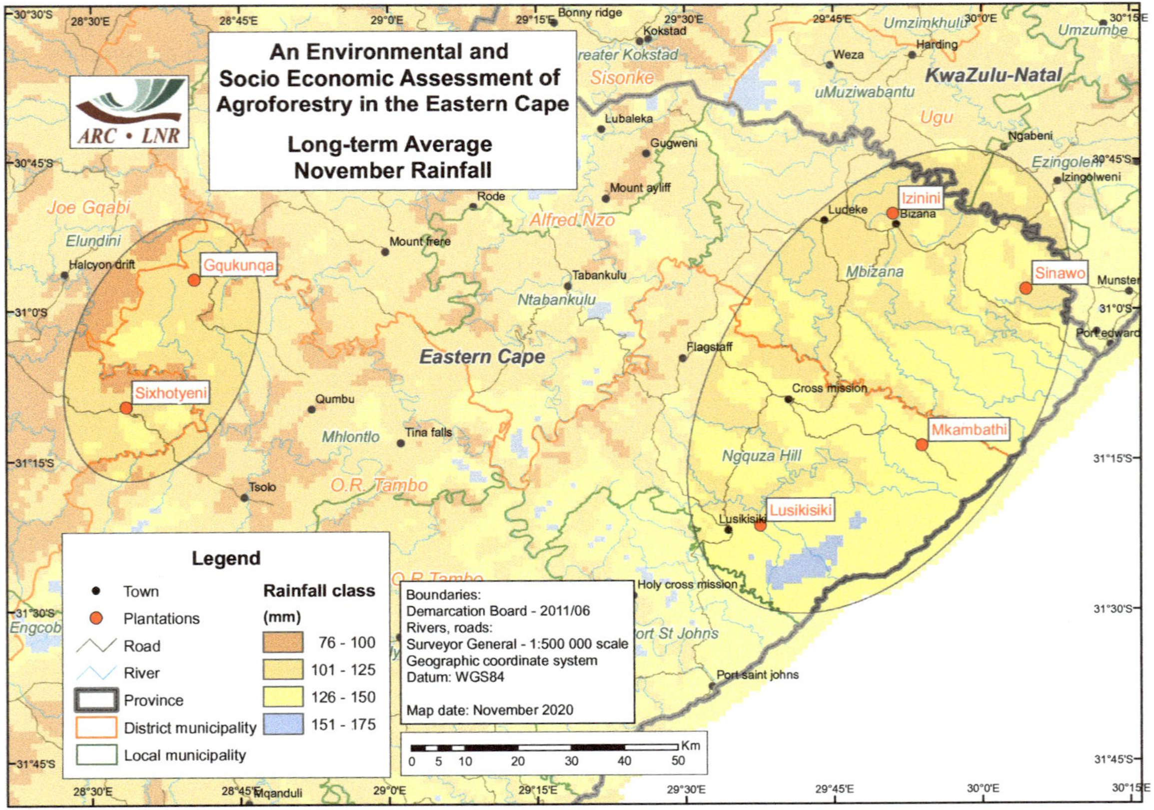

Fig. 6 Eastern Cape Province long-term average November rainfall (ARC-SCW, 2020).

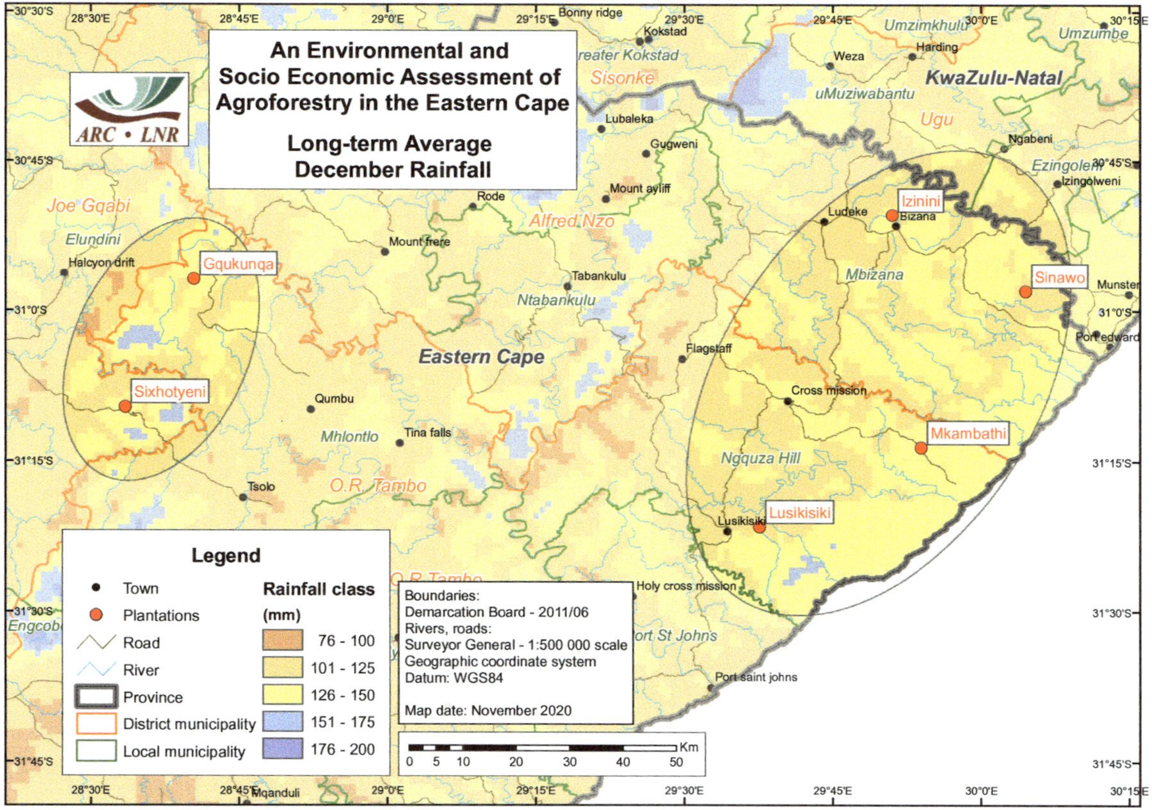

Fig. 7 Eastern Cape Province long-term average December rainfall (ARC-SCW, 2020).

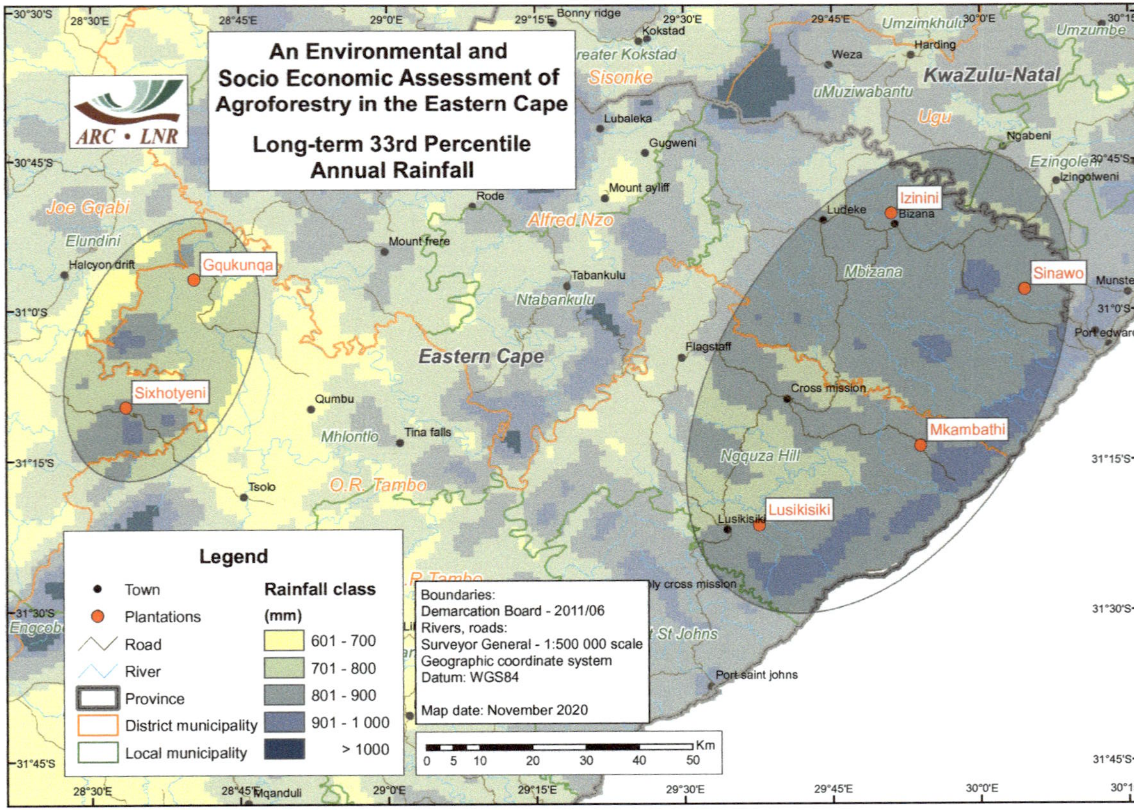

Fig. 8 Eastern Cape Province long-term 33rd percentile annual rainfall (ARC-SCW, 2020).

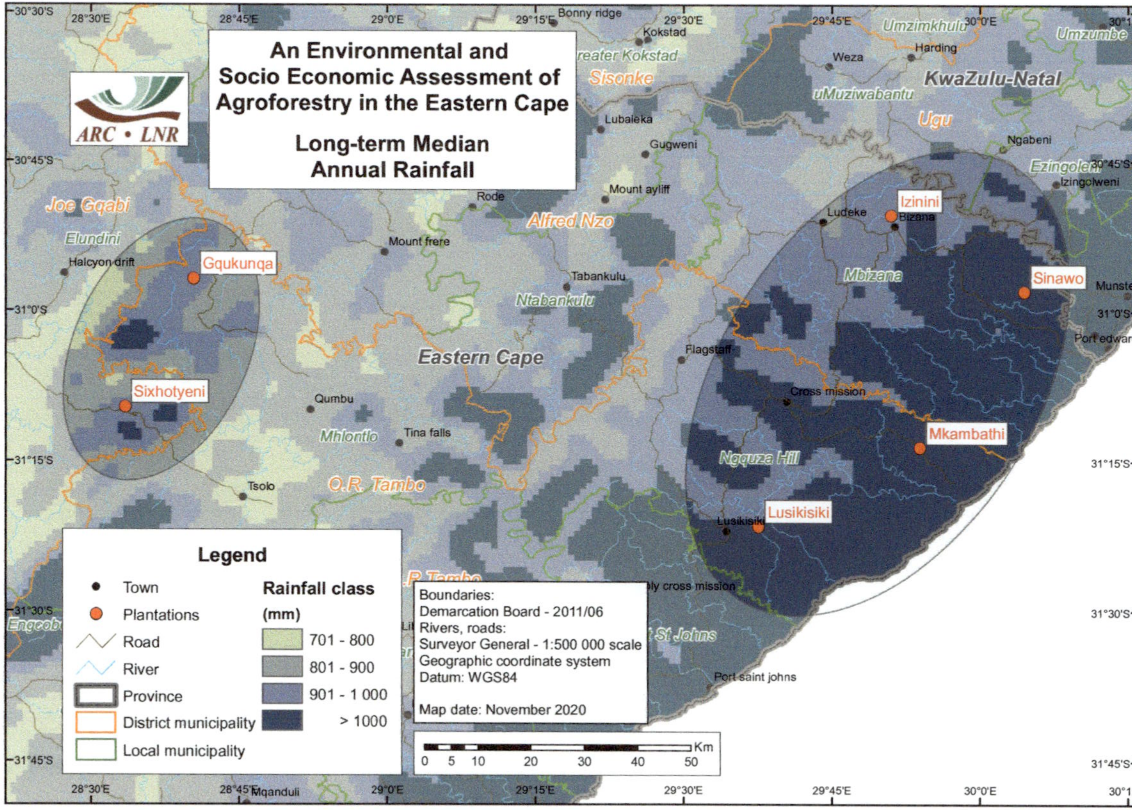

Fig. 9 Eastern Cape Province long-term median percentile annual rainfall (ARC-SCW, 2020).

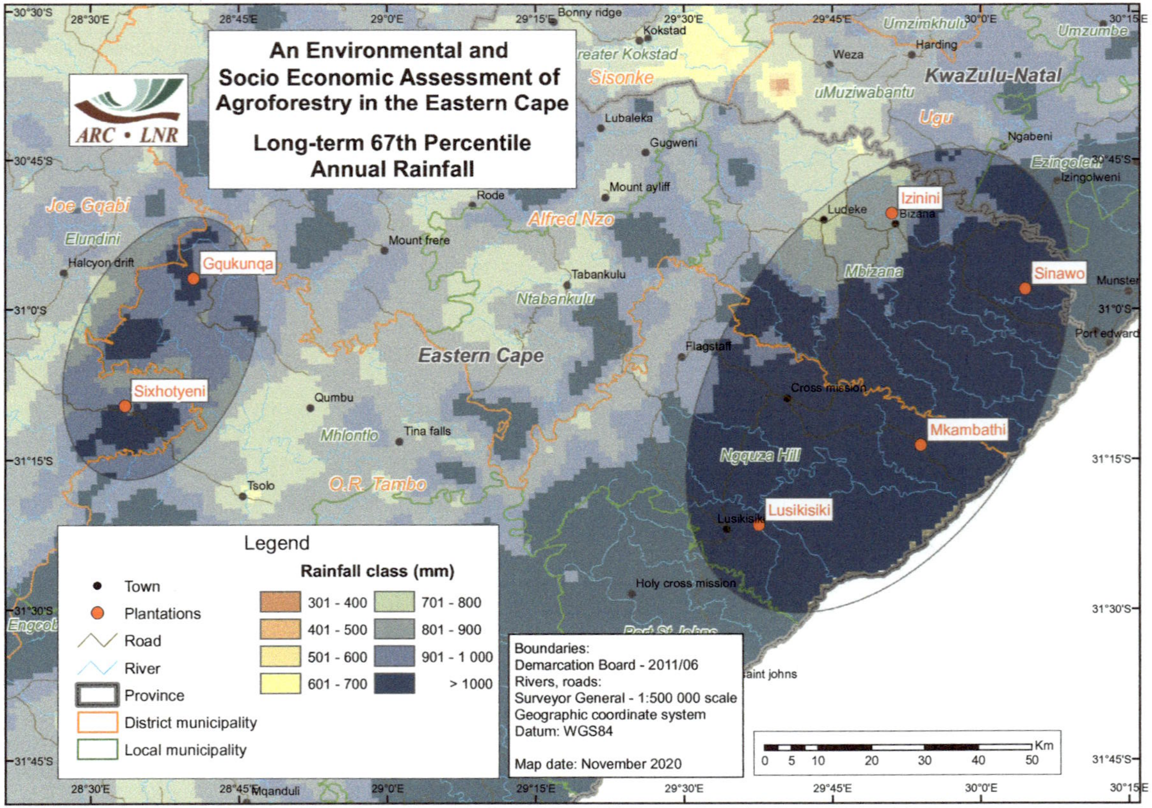

Fig. 10 Eastern Cape Province long-term 67th percentile annual rainfall (ARC-SCW, 2020).

These agroforestry sites will allow for timber production and will thus support agroforestry/timber based mixed farming systems.

4. Conclusion and recommendations

In conclusion, it must be emphasized that trees do so much for water. Their role in water management is irreplaceable as incorporating trees into the food production system is in many cases one of the best strategies one can choose to make sure that soils have optimum moisture levels to sustain crops. In addition, trees are especially helpful for farmers without access to irrigation. Soils then remain moist for longer periods of time, allowing crops to have better access to water from upper soil layers where their roots are. The monthly average rainfall results confirms that rainwater is not a constraint for agroforestry establishment and expansion in the agroforestry sites. The rainfall belt is also more evident in the agroforestry sites as seen in Figs. 8–10. These interesting results are in line with the similar studies conducted in the agroforestry sites in Limpopo and Mpumalanga Provinces (Maponya et al., 2018, 2019, 2020, 2021). Currently the profitable commodities were recommended to the Department of Environment, Fisheries and Forestry (DEFF) for agroforestry integration in the Eastern Cape Province. The study recommends that the establishment and expansion of agroforestry be carried out in the identified suitable areas and in line with the Department of Agriculture, Forestry and Fisheries Agroforestry 2017 Implementation strategy.

Declaration statement

Funding (information that explains whether and by whom the research was supported)

The research was supported by Agricultural Research Council, South African Forestry Company Limited (SAFCOL), Mountain to Ocean (MTO) Department of Environment, Forestry & Fisheries (DEFF), Universities of Kwazulu Natal, Pretoria & Mpumalanga.

Conflicts of interest/competing interests (include appropriate disclosures)

No conflict of interests that are directly or indirectly related to the work submitted for publication.

Availability of data and material (data transparency)

Not applicable.

Code availability (software application or custom code)

Not applicable.

Authors' contributions (optional: Please review the submission guidelines from the journal whether statements are mandatory)

The author contributed to the study as a project leader and main author.

Additional declarations for articles in life science journals that report the results of studies involving humans and/or animals

Not applicable.

Ethics approval (include appropriate approvals or waivers)

Ethical approval was received from the Department of Environment, Fisheries and Forestry (DEFF) to use coordinates for the Eastern Cape Province agroforestry sites.

Consent to participate (include appropriate statements)

Coordinates were received from the Department of Environment, Fisheries and Forestry (DEFF).

Consent for publication (include appropriate statements)

The author gives consent for publication.

Acknowledgments

The authors wish to acknowledge the following: Agricultural Research Council, South African Forestry Company Limited (SAFCOL), Mountain to Ocean (MTO), Department of Environment, Forestry & Fisheries (DEFF), Universities of Kwazulu Natal, Pretoria & Mpumalanga. A special thanks to Ms. Marjan Van Der Walt (ARC ISCW) for her excellent & professional work in drawing the climate maps.

References

Alfred Nzo District Municipality (ADM), 2014. The Agriculture Status in the Alfred Nzo District, Eastern Cape Province, 2014/15 Integrated Development Plan.

ARC (Agricultural Research Council-SCW), (ARC-SCW), 2017. Pretoria, South Africa.

ARC (Agricultural Research Council-SCW), (ARC-SCW), 2019. Pretoria, South Africa.

ARC (Agricultural Research Council-SCW), (ARC-SCW), 2020. Pretoria, South Africa.

Backeberg, G.R., Sanewe, A.J., 2010. Towards productive water use and household food security in South Africa. In: Paper Presented at the 6th Asian Regional Conference of ICID, 10–16 October 2010, Indonesia.

CCAFS (Climate Change, Agriculture and Food Security), 2019. The 5th Global Science Conference on Climate—Smart Agriculture—Transforming Food Systems Under Changing Climate, 8–10 October, 2019, Bali, Indonesia.

ECDC (Eastern Cape Development Corporation), 2014. Promoting Sustainable and Shared Economic Growth and Development, Eastern Cape Province.

ECRDA (Eastern Cape Rural Development Agency), 2020. Forest Development, Eastern Cape Province.

Hassan, R., 2013. Drought management strategies in South Africa and the potential for economic policy instruments. In: Drought in Arid and Semi-Arid Regions., https://doi.org/10.1007/978-94-007-6636-5-21.

Hillbrand, H., Borelli, S., Conigliaro, M., Olivier, A., 2017. Agroforestry for Landscape Restoration—Exploring the Potential of Agroforestry to Enhance the Sustainability and Resilience of Degraded Landscapes.

Malherbe, J., Tackrah, A., 2003. Long-Term Average Ten Daily 1km X 1km Temperature, Rainfall and Evaporation Grid Surfaces Modelled From Weather Station Data With a Long-Term Recording Period. ARC-Institute for Soil, Climate and Water, Pretoria.

Maponya, P., Venter, S.L., Du Plooy, C.P., Backeberg, G.R., Mpandeli, S.N., Nesamvuni, E., 2018. Oral presentation & full paper publication, perceptions on the constraints to agroforestry competitiveness: a case of smallholder farmers in Limpopo Province. In: 9th International Scientific Agriculture Symposium, 04th–07th October 2018, Bosnia—Herzegovonia.

Maponya, P., Venter, S.L., Du Plooy, C.P., Backeberg, G.R., Mpandeli, S., Nesamvuni, E., 2019. Evaluation of the timber based mixed farming/agroforestry systems: a case of farmers in Limpopo Province, South Africa. In: 4th World Agroforestry Congress, 20–22 May 2019 Le Corum, Montpellier, France.

Maponya, P., Venter, S.L., Du Plooy, C.P., Backeberg, G.R., Mpandeli, S.N., Nesamvuni, E., 2020. Timber based mixed farming/agroforestry benefits: a case study of smallholder farmers in Limpopo Province, South Africa. In: Venkatramanan, V., Shah, S., Prasad, R. (Eds.), Global Climate Change and Environmental Policy: Agriculture Perspectives. Global Climate Change and Environmental Policy, Springer, pp. 275–302.

Maponya, P., Madakadze, I.C., Mbili, N., Dube, Z., Isaacs, L., Makhwedzhana, M., Nkuna, T., Tahulela, T., 2021. Potential constraint of rainwater availability on the establishment and expansion of agroforestry in the Ehlanzeni District, Mpumalanga Province in South Africa. In: Abstract Accepted & Full Paper Submitted, 71st IEC Meeting and 24th ICID International Congress on Irrigation and Drainage 2021, 06–12 July 2021. International Convention Centre, Sydney, Australia.

Nair, P.K.R., 1985. Classification of agroforestry systems. Agrofor. Syst. 3, 97–128.

Ospina, C., 2016. Climate and Economic Benefits of Agroforestry Systems. Climate Institute, Washington, DC.

CHAPTER 8

Agricultural biotechnology in Bangladesh: The way forward

Abdullah Mohammad Shohael[a] **and Kathleen L. Hefferon**[b]
[a]Department of Biotechnology and Genetic Engineering, Jahangirnagar University, Savar, Bangladesh
[b]Department of Microbiology, Cornell University, Ithaca, NY, United States

1. Introduction

The term "agricultural biotechnology" refers to a set of experimental strategies for improving plants, animals, and microbes. Knowledge of genetics has led to innovations that boost crop yields. Farmers' potential for developing plants and animals is bolstered by biotechnology, beginning with the ability to detect genes that may bestow benefits on particular crops and continuing with the ability to work with such traits very accurately. The conventional breeding methods for species improvement have their limits, but biotechnology has opened up enormous possibilities that were before unattainable. Agricultural biotechnology has the potential to boost food and nutrition security and increase agricultural output. There is mounting evidence that the origins of agriculture are inextricably linked to climate events. Climate constraints may have slowed the long transition from gathering crops in the wild to the beginnings of agriculture (Darlington, 1969). A small number of species provide a disproportionately large share of the world's food and feed. It also has the massive issue of feeding 10 billion people by 2050 as environmental conditions continue to deteriorate. Although cereals are indeed an excellent supply of carbohydrates in the form of starch, they are relatively deficient in several essential nutrients, including proteins, minerals, and vitamins. This results in the condition known as "hidden hunger," which occurs when a person's daily caloric demands are satisfied however the mineral content, notably iron, zinc, and vitamins, but also phytonutrients with antioxidant potential are insufficient to ensure optimal physical and mental development (Zsögön et al., 2022).

Since the mid-1990s, commercial cultivation of genetically modified (GM) food and feed crops has been expanding, and farmers currently grow over 190 million hectares of biotech crops (Brookes and Barfoot, 2012). Despite this, GM food crops have not been commercially cultivated in most

Agricultural Bioeconomy
https://doi.org/10.1016/B978-0-323-90569-5.00012-3

EU countries (except for Spain, Portugal, the Czech Republic, Slovakia, and Romania). While a negative attitude toward GM crops is perpetuated, it is often associated with insufficient knowledge of modern breeding technologies, a lack of trust in regulators, inadequate communication regarding the risks and/or benefits, and personal values. The cumulation of these factors can seriously impact food security issues, particularly in the global South. Biotechnology can improve plant nutritional value, increase crop yields, confer resistance to pests and increase tolerance to environmental changes created warmings such as extreme temperatures, flooding, and drought.

Bangladesh is an agricultural country and its economy is based on agricultural production contributing 19.6% to the national GDP and providing employment for 63% of the population. A good number of people completely depended on their livelihoods on agricultural production and products are the key source of their, food, income, and employment. Rice is the main staple food in Bangladesh. After the major crops (rice, wheat, and maize), fruits, vegetables, fisheries, and livestock are produced to provide the major food supply for the huge population. Agriculture plays a vital role in providing food and nutritional supports as well and elevating national GDP (Sarkar et al., 2020). Though the contribution of agriculture to the growth of GDP is very crucial, several challenges are ahead, including population growth, urbanization, shrinking of arable land, climate change and threats of abiotic and biotic factors that hinder agricultural production. Bangladesh Delta Plan (BDP) 2100 incorporating agriculture, water resources, support for food security, and mitigating climate change risks will ensure effective and sustainable development of these sectors.

The development and adoption of modern biotechnological tools and techniques for the improvement of genetic resources are necessary to enhance the efficiency of plant genetic resources to the alteration of new crop varieties. Implementation of functional genomic tools, to enable the relationship between gene and phenotype, has allowed the development in breeding for crop improvement. For the development of potential crop varieties, new breeding techniques (NBTs) are targeted to isolate specific genes of interest and their precise editing and transfer into the targeted crop hosts for sustainable agriculture (He et al., 2014; Lusser et al., 2012; Varshney et al., 2014; Gentile and La Malfa, 2019).

Bangladesh is one of the major victims of climate change. Climate change followed by global warming is creating huge threats to agricultural production through the pool of abiotic stressors. Continuous increases in drought, heat, cold, flood, salinity and tropospheric ozone reduce crop yield yearly. On the other hand, the changing climates are altering the growth and characteristics of insects and pathogens toward yield losses ((Rivero et al., 2022). To ensure food security and eliminate zero hunger it is important to study the physiological and molecular mechanisms of different crops under different levels of multifactorial stresses and the need to develop resilient climate-smart crops that are suitable for upcoming threats using modern biotechnological tools and techniques (Fig. 1). This chapter analyzes the existing landscape of agricultural biotechnology in Bangladesh and forecasts its future effects. Additionally, it exemplifies the potential of genome editing and genetic modification as methods for advancing Bangladesh's agricultural sector in the future. This chapter offers a condensed analysis of the key contribution and present status of GM crops to national food security and bioeconomy, as well as the significant challenges and concerns for GM acceptability associated with the introduction of new technology.

2. Agriculture in Bangladesh

Bangladesh is a small but promising country in South Asia covering 148,460 km^2 of land and positioned as the largest Delta in the world. It is one of the most densely populated countries in the world with a population of over 165 million and an estimated population, will be over 200 million by 2050. Bangladesh is an agriculture-based country and the contribution of agriculture to the national GDP of Bangladesh is tremendous. In 2019, the share of agriculture in Bangladesh's gross domestic product was

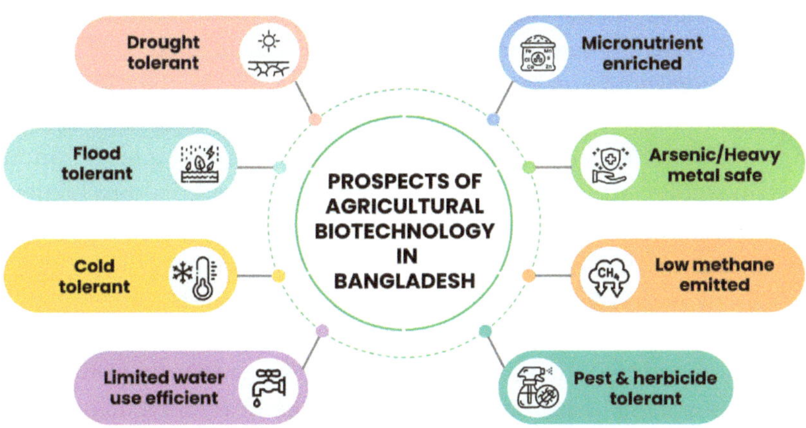

Fig. 1 Opportunities for Agricultural biotechnology in Bangladesh to battle zero hunger, climate change, and hidden hunger.

12.68%. The agricultural development of this country is alleviated by the fertile alluvial soil of the Ganges Meghna-Brahmaputra delta coupled with high rainfall and easy cultivation (Choudhury and Islam, 2002). Along with staple food rice, it grows potato, sugarcane, jute, wheat, and corn as major crops. Other vegetables, fruits, spices, oilseeds are also grown as minor crops. According to the World Bank, the total arable land in Bangladesh is 61.2% of the total land area (down from 68.3% in 1980). The impact of the growing population, decrease of arable land, and climate change has adversely affected agricultural production and is a challenge to food security. After gaining its independence in 1971, Bangladesh's agricultural industry quickly emerged as the country's primary engine of economic growth. Its contribution to GDP was perhaps in the vicinity of sixty percent. We are all aware that agriculture in Bangladesh plays an essential role in terms of people's ability to make a living, employment opportunities, and overall contribution to GDP. For the past ten years, its contribution has decreased, falling from 17% in 2010 to 12.6% in 2020 (BBS). The sector is at the heart of our economy, helping to alleviate poverty while also maintaining the safety of our food supply. Despite the expanding population (from 147.6 million in 2010 to 164.7 million in 2020) and the negative effects of the global pandemic and climate change, the industry has shown remarkable resilience in terms of profitability and production (World Bank).

Nearly fifty percent of Bangladesh's population is largely occupied in agriculture, while more than seventy percent of the country's land is devoted to crop cultivation. Rice, jute, wheat, tea, lentils, oil seeds, vegetables, and fruits are the most important crops. Despite agriculture's significant contribution to Bangladesh's overall economy, it has remained mostly subsistence-based, with variable crop yields due to climate change, lack of scientific understanding, rudimentary agricultural methods, less genetically modified product promotion, and poor infrastructure preventing farmers from completely commercializing their crops (Fig. 2). According to the findings of certain studies, the amount of inputs used in agricultural production in Bangladesh is significantly higher than the average amount. Recognizing this context, an In-Depth Country Assessment conducted as part of the Global Strategy to Improve Agricultural and Rural Statistics concluded that the Bangladesh Bureau of Statistics needs to improve internal and external coordination and that an agriculture-specific plan is necessary to complement and supplement national strategies for the development of statistics (FAO). Bangladesh's staple food crop, rice, has been the primary focus of agricultural studies during the Green revolution. A significant rise in rice

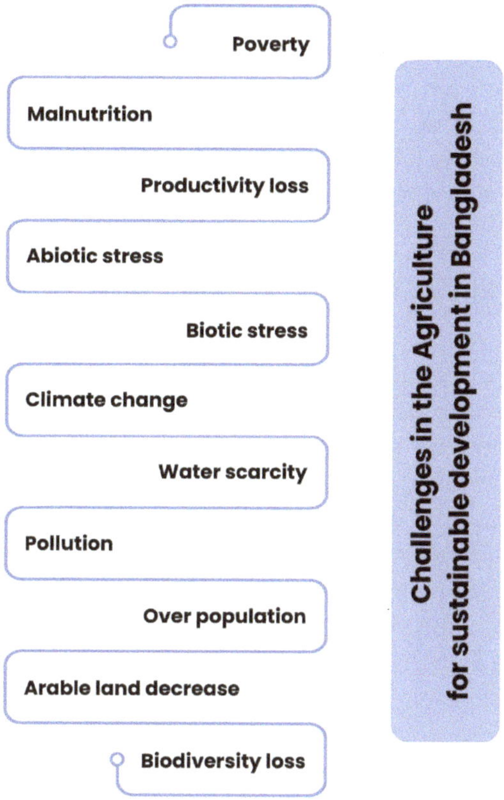

Fig. 2. Major challenges stand in the way of Bangladesh establishing sustainable agricultural output.

output can be attributed to the efforts of the research community. It has also proven successful to diversify rice crops, with new varieties emerging that can withstand both salty and dry soil and underwater. In addition to rice, the government plans to develop 21 new types of agricultural technology and 22 new crop varieties. Research on developing environmentally benign and disaster-resistant cultivars of jute as a non-cereal rice alternative is already underway. Many other crops, however, have great potential in the country. Among the vegetables; potatoes and brinjal, cotton, onions, and other non-cereal crops have been found to provide economic and private returns comparable to or higher than current rice farming. In addition to rice, these crops should be the focus of cutting-edge biotechnology research and development. Genetically modified organisms (GMOs) are the result of modern biotechnology's ability to alter or modify a living organism's original genetic structure in a laboratory to introduce novel and useful genes (GMOs). By boosting food production and availability while simultaneously improving food quality, genetically modified organisms (GMOs) can play a critical role in achieving food security. Assuring the availability of micronutrients and other nutraceuticals in GM crops could improve farmers' economic and social status (Qaim and Kouser, 2013).

3. Adopting agricultural biotechnology for circular bioeconomy

The Food and Agriculture Organization (FAO) estimates that more than 800 million people around the world are affected by food insecurity, and another two billion are suffering from a lack of essential micronutrients. The increasing need for food around the world is a result of factors such as a growing population and increased competition for crops as renewable energy sources. Environmental stressors, such as climate change, a lack of available water, and intense competition for land, pose a threat to the production of food and the security of food supplies worldwide (Antoine et al., 2021). A high premium is placed on food and energy security in the United Nations (UN) sustainable development goals (SDGs) 2030 agenda, and agricultural biotechnology will play a momentous role in achieving these goals and facilitating the transition to a circular bioeconomy (Aguilar et al., 2019). Building bridges between biotechnology and the economy, as well as between research, business, and society, is what the field of bioeconomy is all about. It examines economic systems that are founded on the responsible utilization of renewable biological resources. The development of new biomaterials and bioenergy sustainably and cost-effectively for an ever-increasing global population are only two of the many issues that face bioeconomics today. Other challenges include the efficient production of products at a large scale using minimal resources. In comparison to traditional breeding, molecular plant breeding is regarded as a more effective and trustworthy method for studying the relationship between genotype and phenotype. This is done to improve crop varieties so that they are more resistant to the effects of environmental, nutritional, or other stresses. The use of contemporary plant biotechnologies, such as genetic modification (GM) and genome editing (GE), has the potential to significantly speed up the creation of better crops for farmers. The increased precision that is now possible in plant breeding as a result of the use of GE techniques represents a significant departure from GM, which has relied on unpredictable insertions of isolated genes into the plant genome, as well as from conventional breeding approaches, which, for the most part, rely on random, uncontrolled chemical- or radiation-induced mutagenesis. GE techniques have made it possible for plant breeding to achieve this level of precision (Bujnicki et al., 2017). The uses of biotechnology in industries such as medicine and agriculture, as well as the consumption of biotechnology products by humans and animals, are all a part of our everyday life (Moshelion and Altman, 2015).

4. Recent breakthroughs in the cultivation of GM crops in Bangladesh

Plants that have had their DNA sequences altered through genetic engineering to produce desirable features are known as genetically modified (GM) crops. These crops are a relatively recent addition to the vast array of crop types that have been generated by more conventional processes such as breeding, selection, and mutation. Traditional breeding methods have been used for years to alter the genomes of plants and animals. The diversity of organisms that have emerged from artificial selection for certain, desirable features is just the consequence of natural variation. Technology advancements in genetic engineering have made it possible to precisely regulate the introduction of genetic modifications into an organism, allowing for the transfer of genes from one species to another that is completely unrelated. This has the potential to increase agricultural yields while also accelerating the production of profitable pharmaceutical compounds. Crop plants, farm animals, and soil microorganisms are examples of some of the most well-known organisms that have been modified via the use of genetic engineering (Phillips, 2008). Scientists are genetically modifying crops to make them more nutritious to assist communities that do not have access to particular micronutrients that are essential for living a healthy life. GM crops are superior to conventional varieties in several respects, including their capacity to endure adverse environmental conditions, their superior defense mechanisms against insects and other unwanted organisms, their capacity to increase crop yields, and the essential vitamins and minerals that they provide. The use of GM crops in Bangladesh with the goals of increasing agricultural output and mitigating the impacts of global warming may be a model for other emerging regions dealing with similar challenges (Table 1).

Table 1 Recent developments and prospects for genetically modified crops in Bangladesh.

	Name of the targeted gene	Developed crops	Current status	Year of introduction	Local developer/ collaborator
1.	cry1AC	Brinjal (Bt Begun)	Approved and Cultivated nationwide	2013	Bangladesh Agricultural Research Institute (BARI)
2.	Crt1/PSY	Rice (Golden Rice)	Under review for Biosafety approval	–	Bangladesh Rice Research Institute (BRRI)
3.	3R LBR (Rpiblb2/ Rpi-vnt1/ Rpimcq1)	Potato (LBR Potato)	Under trial	–	Bangladesh Agricultural Research Institute (BARI)
4.	cry1AC	Cotton (Bt Cotton)	Progressing for Approval	2022	The Cotton Development Board (CBD)
5.	OsNAS2/ SoyferH1	Rice (High Iron Zinc Rice)	Under trial	–	Bangladesh Rice Research Institute (BRRI)

4.1 Vegetables

Vegetables contain a high amount of vitamins and minerals and common source of human diets around the world. Among the arable land in Bangladesh, around 1.25 million hectares were used for vegetable production, where around 26.7 million tons of vegetables were produced in the 2018–19 fiscal year (Department of Agricultural Extension). That has shown a significant rise—37.63%—between 2013–14 and 2018–19. Farmers are interested in vegetable farming due to the increase in demands with a good return than other crops. Eggplant, potato, tomato, cabbage, cauliflower, pumpkin are the most popular vegetables in the country. Eggplant (known as brinjal or begun in Bangladesh) is a versatile used cheap and popular vegetable cash crop in Bangladesh grown widely after potato. In total vegetable, production eggplants were attributed 4.7% in winter and 9.6% in summer (Bangladesh Bureau of Statistics (BBS), 2018).

4.1.1 Eggplant

Fruit and shoot borer (FSB) is a devastating pest that causes major production loss in brinjal. To protect the eggplant from FSB, farmers are frequently spraying various pesticides even daily to the field (Shelton et al., 2018). This pesticide dependency strategy is alarming for both personal health and hazard to the environment. To protect the eggplants from deadly FSB and to save humans and the environment from the residues of pesticides, plant biotechnology employed a solution by developing genetically modified Bt plants. *Bacillus thuringiensis* containing a gene (Cry1Ac) that produces crystal protein can destroy the killer FSB larvae into the plant cell. Transgenic plants harboring Cry1Ac have long been safely used to control the plants from FSB (ILSI CERA, 2010). Bt Brinjal was first developed by the India-based Maharashtra Hybrid Seed Company (Mahyco) (Shelton et al., 2017). Bangladesh Agricultural Research Institute (BARI) was officially involved in the field trial and release through a public-private partnership between Mahyco, Cornell University, Sathguru Management Consultants, BARI, and the United States Agency for International Development (USAID) under the Agricultural Biotechnology Support Project II cooperative agreement (ABSPII; http://absp2.cornell.edu). Four Bt Brinjal varieties were officially released for seed production and commercialization on October 30, 2013 and till 2018–19, Bt brinjal was grown by 20,695 farmers on 1213.3 ha (Shelton et al., 2020).

4.1.2 Potato

Potato is a widely consumed important tuber crop ranked third in the world after rice and wheat (International Potato Center). The total world potato production is estimated at 388,191,000 ton in 2017 (Source: (Faostat, 2019)). Except for diversified industrial products, more than a billion people eat potatoes. It was mostly grown and consumed in Europe, North America, and the former Soviet Union until the early 1990s. After that demand in Asia, Africa, and Latin America, increased 165 million tonnes in 2007 (FAO). Potato is a very popular cash crop in Bangladesh widely consumed all over the country, though it has grown in winter but consumed over the year. Bangladesh secured the top seventh position around the world for potato production and now has the surplus after its yearly consumption. The northern areas of Bangladesh are the potato-growing hot spot due to the favorable environment and land capacity. The potato was cultivated in a total of 0.172 million hectares of land of eight districts in two agriculture regions of the Rangpur and Dinajpur divisions in 2021 (The Department of

Agricultural Extension) where the targeted production is assuming 42.47 lakh tons. Potato Late blight (PLB) disease caused by *Phytophthora infestans*, was responsible for the 19th-century Irish potato famine that had led to 1 million deaths from starvation, still affects more than 3 million hectares of potato crops globally, and causes economic losses estimated at $2.75 billion a year. In 1922 it was first reported PLB in Bangladesh and still, now it is one of the major diseases that cause about 20% yield loss yearly (Hossain et al., 2009). To overcome this disease genetic engineers were trying to find out a solution by transferring the disease-resistant genes into the potato. The transfer of three resistance (R) genes from wild relatives (RB, Rpi-blb2 from *Solanum bulbocastanum* and Rpi-vnt1.1 from *S. venturii*) into the potato. Transgenic potato lines, stacking of the three R genes produced a high frequency of resistance to late blight in the field over several seasons (Ghislain et al., 2019). To get a sustainable solution of producing LB resistant potato, the Bangladesh government relies on modern agricultural biotechnology after the success of FSB resistant genetically engineered Bt Brinjal and finally allowed the Bangladesh Agricultural research Institute (BARI) to import 3R blight resistant (RB) potato lines, developed at the Michigan State University (MSU) harboring three blight-resistant genes derived from three wild potato varieties—Mexican origin *Solanum bulbocastanum*, Argentine origin *Solanum venturii* and a Peruvian origin *Solanum mochiquense*, and go for a field trial. According to Dr. David Douches (Feed the Future Biotechnology Potato Partnership, a US government food security initiative), BARI is getting two blight-resistant yellow skin high yielding cultivar Diamant (DIA-MSU-UB #1 and DIA-MSU-UB #2) lines in early 2021 for research and development out of 10 best blight-resistant potato lines developed and evaluated at confined field trials at MSU (www.potapro.com).

4.2 Rice

In Asia and Africa, thousands of people are suffering micronutrient deficiencies. Limited income, food habits, rice-based staple food consumption, and socio-economic patterns are the limiting factors to uptake adequate nutrition in Bangladesh. Supplementation and biofortification programs are implementing in different areas to address the micronutrient deficiencies but the success is limited (Sarkar et al., 2020). Rice is consumed three times a day by most of the peoples in Bangladesh is a good platform for providing nutrition through it. Elimination of hidden hunger is one of the most important agendas to achieve the SDGs by 2030. To achieve the goal and fulfill the demands of required micronutrients like Vitamin A, Zinc, Iodine to people, multiple approaches should be taken. Childhood blindness, anemia, and weakened host resistance to infections are increasing disease risks and death in a low-income country due to Vitamin A deficiency (VAD) where food-based nutrition is not available or affordable. Ingo Potrykus, A German scientist, and Peter Beyer have developed an endosperm-specific biosynthetic pathway of provitamin A enriched Golden rice that has promising advancement of supplying Vitamin A through Rice in late 1990 to save millions of children from preventable xeropthalmia. Efficiently vitamin A producing (37 μg/g in the uncooked rice) Golden rice event was produced through the transformation of psy (from *Zea mays*) and crtI (from *Erwinia uredovora*) in a tropical japonica background Keybonnet (Paine et al., 2005). Finally, transgenic golden rice GR2E donor event of the popular Indica rice variety was selected after precise selection by the International Rice Research Institute (IRRI) for the South and Southeast Asian countries to introgression of β-carotene biosynthesis locus into the widely grown rice varieties (Humanitarian Board, 2009; Beyer, 2010; Dubock, 2014, 2019). At this moment an average of 10.6 ± 1.16 μg/g total carotenoid content was yielded in the GR2-E BRRI dhan29 Golden Rice in the multi-location trials in Bangladesh by Bangladesh Rice Research Institute (BRRI) is sufficient to deliver approximately 66% and 80% of the recommended daily intake of vitamin A for children and women through the alternation of white rice to golden rice (Biswas et al., 2021).

4.3 Cotton

Cotton (*Gossypium hirsutum*) is an important fiber crop being using for a long period to meet the demands of clothing to cover and protect human beings. Due to its high oil (18–24%) and protein content (20–40%) cotton seeds also have industrial potentiality and using as vegetable cooking oil and feeds. About 350 million people are engaged in the cotton industry that has a tremendous role in economic and social welfare (Siddiqui et al., 2019). Bangladesh is now one of the top garments exporters in the world where cotton is the major raw material. The annual requirement of raw cotton for this consumption is estimated at 8.3 million bales and currently, the country is spending about $4.5 billion a year for the cotton import. Historic "Muslin" was made from the local finest cotton balls and its fame was circulated around the world. This glory revealed that the weather and soil conditions were favorable for exclusive fine cotton production in Bangladesh.

Gradually cotton production is reducing due to the abiotic and biotic stressors. Pink bollworm (*Pectinophora gossypiella*), armyworm (*Spodoptera littoralis*), cotton bollworm (*Helicoverpa armigera*) and spotted bollworm (*Earias insulana*) are the major pests that limit cotton production around the globe (Naqvi et al., 2017). Neurotoxic pesticides are widely used agrochemicals to protects the insects in cotton are harmful to humans, biodiversity, and the environment. Biotechnologists were trying to solve this problem and to minimize the pesticides uses, genetically modified crops expressing Cry1Ac and Cry2Ab genes from *Bacillus thuringiensis*, were developed and widely adopted by farmers in the world (Ferry et al., 2006). Bt cotton expressing Cry1Ac and Cry2Ab genes possess inbuilt protection against the insects played an important role for sustainable cotton production (Devare et al., 2004). To increase cotton production Bangladesh government has decided to start the cultivation of GM cotton in 2017. For this venture Hyderabad-based, JK Agri Genetics Ltd. is collaborating with the Bangladesh government to hold field trials of GM cotton. Cotton Development Board (CDB) began trials of the genetically-modified Bt cotton with two varieties, JKCH 1947 Bt and JKCH 1950 Bt, which can resist bollworm and Fall armyworm in the plant. CDB has successfully finished the Contained trial in greenhouse condition on March 2020 at BARI greenhouse and confined the trial in the open space by the approval of Bangladesh's National Committee on Biosafety (NCB). In the coming season, there will be a regional trial in four different places in Bangladesh. After successful completion of the regional trial with the approval of NCB, it could be released at the farmer's level for cultivation. For this initiative total of 60,000 farmers have been trained for and 3050 demonstration plots have been set for cotton production and are hopeful to produce a million bales of cotton on 200,000 ha of land and save on some 15% of import costs (https://textilefocus.com).

5. Genome editing as an emerging technology for the agricultural improvement in Bangladesh

Genome editing (GEd) allows precise modifications to DNA, just a few nucleotides among billions in live cell genomes. Researchers are using this method to investigate how the genome functions and to create crops that are resistant to pests, have improved nutritional value, and can flourish in dry environments (Voytas and Gao, 2014). CRISPR (Clustered Regularly Interspaced Short Palindromic Repeats) with the CRISPR-associated protein Cas9 (CRISPR-Cas9) has changed the area of genome engineering and emerged as a revolutionary technique for genome editing in recent years. When compared to other genome-editing techniques like zinc finger nucleases and transcriptional activator-like effector nucleases (TALENs), this approach is more adaptable and straightforward (Cong et al., 2013). This ground-breaking technique has emerged to provide effective breeding tools without the need for exogenous DNAs that cause genetic differences, such as natural, spontaneous mutations. The method has been demonstrated to provide several elite features that traditional breeding either cannot produce or can only produce via a time-consuming and tedious procedure. The production of enough food to feed the globe by 2050 is not achievable with the state of plant breeding and agricultural technology today. Climate change and the decline in available farmland as a consequence of increasing urbanization provide further challenges to food production. To create novel traits/cultivars that may increase yield or maintain food supply under adverse circumstances, it is necessary to apply a revolutionary plant breeding approach such as GEd (Van Vu et al., 2022).There are two elements involved in the use of the CRISPR-Cas system. These are gRNAs that are unique to the target sequence, as well as Cas enzymes that induce double-strand breaks (Jinek et al., 2012). To connect to the DNA, it is looking for, it needs a sequence known as a protospacer adjacent motif (PAM). This particular PAM sequence is unique to every Cas protein. The choice of a target sequence is considerably restricted when using the most popular form of SpCas9 since NGG is required. Despite the fact that each genome has a sufficient number of NGG sites, there may be instances in which one is required to make a cut at a particular location owing to the presence of a functional amino acid or protein domain. For gene knock-out or knock-in, DNA double-strand breaks are followed by nonhomologous end-joining (NHEJ) or homology-directed repair (HDR). The CRISPR-Cas9 system has a number of benefits that set it apart from other common platforms for editing the genome, such as zinc finger nucleases (ZFNs) and transcription activator-like nucleases (TALEN). To begin, in the CRISPR-Cas9 system, sgRNAs that are approximately 100 base pairs in length can be easily customized for targeted genes. This makes the system significantly easier to use when compared to tools based on ZFN or TALEN, which require complex design and the synthesis of a cumbersome guiding protein (zinc finger DNA-binding domain and transcription activator-like effector). Second, when compared to the other two technologies, the CRISPR-Cas9 technology provides a better level of accuracy in the editing of genomes. Third, the CRISPR-Cas9 system enables the possibility of editing at many genomic loci at the same time, which is known as multiplexing. This is something that other currently available technologies are unable to do (Fang et al., 2022). The widespread use of breakthrough technologies is proof. There has not been any other technology in the life sciences that have been introduced in recent years that has been utilized so extensively and effectively in such a short amount of time. The only exception to this is the advent of next-generation sequencing technology. The reason for this is not only that it can be used in any organism that has been examined to this point, but more importantly, it is extremely easy to produce the essential reaction reagents. The transformation of a basis vector that carries expression cassettes expressing a CRISPR-Cas protein and a guide RNA (gRNA) is often adequate for the most typical use of this technology. In addition, the use of procedures that are based on type-II restriction enzymes makes the production of complicated structures a simple and easy process (Engler et al., 2008). In addition, in recent years, commercial suppliers have been able to include CRISPR items in their product portfolios at rates that are reasonable, which has further fueled the adoption of the CRISPR-Cas technology.

Genome editing does not require the introduction of new gene sequences; rather, it may direct only one or a few nucleotide changes within a plant genome (Mao et al., 2016; Rani et al., 2016). Genome editing technologies such as CRISPR-Cas9 make use of various defense strategies derived by bacteria to target specific sequences of DNA and cleave them at these sites with Cas nucleases. The technology is then able to take advantage of DNA repair mechanisms found already in the cells of all organisms. By repairing the sites of cleavage, specialized changes that will be carried through the genome of the "edited" organism to subsequent generations can be produced. Genome editing is an extremely promising new tool for plant breeding, as many complex traits such as crop yield, photosynthesis, and drought resistance involve multiple genes, and thus would be difficult to achieve through conventional breeding alone.

The ease of genome editing makes it a viable option for countries such as Bangladesh. First, a sgRNA (guide RNA) that targets a specific nucleotide sequence is designed, using algorithms that avoid off-target effects. The sgRNA and Cas protein gene sequences are constructed into a vector along with a suitable promoter. The vector is inserted into plant cells using standard techniques such as *Agrobacterium tumefaciens*, PEG transformation, or biolistic delivery. Plants are regenerated from plant cells using tissue culture technologies and edited plants are confirmed using standard nucleotide sequence analysis, polymerase chain reaction, etc.

Plant genome editing was explored as a new breeding technique as early as 2012, with the generation of blight-resistant rice, powdery mildew-resistant wheat, and a variety of herbicide-tolerant crops. Shortly after, other valuable traits such as high yield waxy corn, soybean with reduced trans fats, non-browning mushrooms, and potatoes with altered starch content became available. Next-generation genome-edited crops will contain traits such as improved photosynthetic ability, tolerance to heat and cold stresses, improved nitrogen use efficiency, tolerance to salinity, and drought resistance. Improving photosynthesis via genome editing will ensure that plants are more productive under heat stress conditions anticipated by climate change, especially since improvements using traditional breeding methods have reached their limits. Genome-edited crops with different fruit and seed traits such as size, shape, number, yield, time of ripening, and regulation of senescence are also under development. In addition to this, crops with improved protein yield, amino acid content, and enhanced micronutrient content are currently being generated using genome editing technologies. The next section describes the progress of a few of these crops.

Tolerance to abiotic stresses such as high temperatures will be needed to improve yield in the face of climate change. CRISPR-Cas9 has been used to knock out the MAP kinase gene SlMAPK in tomatoes, and the resulting products were able to resist heat stress (Yu et al., 2019). Similarly, Shi et al. (2017) showed that CRISPR created variants of an ARGOS8 gene in corn that could increase crop yields in drought conditions. CRISPR also is opening the door to crops with a better nutritional profile, providing nitrogen more efficiently and reducing food waste. For example, Li et al. (2018) have demonstrated that a single nucleotide polymorphism in the NRT1.1B gene that is responsible for the improved nitrogen use efficiency in indica rice can be moved into japonica rice, using CRISPR/Cas9 gene-editing technology. The NRT1.1B gene encodes for a nitrate transporter protein. This took place within a single generation and at a frequency of 6.72%, indicating that it is completely feasible to replace alleles for agriculturally important traits quickly, inexpensively and effectively. Without the use of genome editing technology, the transfer of this trait would have taken a plant breeder years through crossing/backcrossing to introduce the trait into intended cultivars.

Another way yield can be increased is by reducing crop waste due to browning. For example, the polyphenol oxidase (PPO) gene is responsible for oxidative browning of several fruits and vegetables including potatoes and apples. Genome editing has been used to block oxidation by polyphenols and reduce browning in both of these crops (Waltz, 2015; Mellidou et al., 2014). A trait that has importance to Bangladesh is the prevention of browning of brinjal (Maioli et al., 2020). The researchers made use of the polyphenol oxidases (PPOs), which are responsible for catalyzing the oxidization of polyphenols, causing eggplant flesh to brown after cutting. There are 10 different PPO genes in eggplant, 3 were knocked out using CRISPR-Cas technologies. These three genes displayed elevated transcript levels after cutting, suggesting they played an active role in the browning process. The induced mutations were stably inherited in several generations of progeny and conferred with a lack of browning flesh cutting.

Salt and drought tolerance are also important traits for tomorrow's crops. The wild barley *Hordeum brevisubulatum*, which prefers salty environments in Asia, produces a transcription activation gene (HbMBG1a) that can enhance salt tolerance in the common mustard plant, *Arabidopsis thaliana*. The same DREB gene used to enhance drought tolerance also boosted salt tolerance in soybean plants. Genome-edited drought resistance has also been explored through osmoregulation (Ozturk et al., 2020; Zia et al., 2021). Osmoregulation involves the synthesis and accumulation of osmoprotectants including soluble proteins, sugars, and amino acids such as proline. Other stress-responsive genes conferring drought tolerance in plants have also been identified. This provides several avenues for genome editing but also highlights the complexity of drought resistance as a trait in general. For example, genome editing was used to introduce novel alleles into the gene OPEN STOMATA 2 (OST2), a prominent plasma membrane H^+ ATPase responsible for stomatal response in Arabidopsis (Osakabe et al., 2016). Similarly, ethylene resistance was provided to maize plants by editing ARGOS8 gene expression, thus improving grain yield under drought conditions (Barnawal et al., 2017). As another example, the transcription factor PtoMYB170 was altered using CRISPR/Cas9, to confer drought tolerance in Arabidopsis (Li et al., 2017a,b). Other genome technologies, including zinc-finger nucleases (ZFN) and transcription activator-like effector nucleases (TALEN) have been used for targeted gene editing to confer drought tolerance in plants (Xu et al., 2008; Xue et al., 2011).

Cadmium-tolerance is desirable in several crops. The Arabidopsis defensin gene AtPDF2.5 can be induced when the plant is subjected to Cd stress (Luo et al., 2019). AtPDF2.5 contains eight cysteine residues, which play a role in intolerance through the chelation of Cd. When plants overexpressing AtPDF2.5 were generated, tolerance to Cd was displayed and Cd efflux from the cytoplasm was observed.

Improvements in yield have also been explored using CRISPR-Cas9 technologies by selecting a variety of target genes (Ricroch, 2019). Li et al. (2017a,b) edited rice (Zhonghua 11) genes responsible for grain number, panicle architecture, grain size and plant architecture (Gn1a, DEP1, GS3 and IPA1). The first exon of Gn1a and GS3 and the third exon of DEP1 and IPA1 were chosen as targets for the sgRNAs because mutations within these regions have been proven to result in higher yield. The DEP1 mutants increased the number of flowers per panicle (about 50% more than the control) IPA1 mutations result in improved yield phenotypes as well. Xu et al. (2017) edited three genes related to grain weight through genome editing, which resulted in significant increases. By deleting 625 bp within the DEP1 gene using CRISPR/Cas9, erect panicles with a higher grain number and lower plant height were produced (Huang et al., 2009).

Genome editing has been used to generate glyphosate-resistant plants. Other means to confer resistance include altering target-effector interactions, knocking out genes that make the host plant susceptible, and uncoupling defense hormones (Bisht et al., 2019). Zhang et al. (2019) used CRISPR-Cas9 to create a variety of stackable herbicide tolerance traits against several popular herbicides. This nontransgenic approach provides a powerful tool for future weed management. This task would be virtually impossible to carry out by traditional breeding due to the polyploid nature of wheat, as individual mutations would confer a relatively low level of resistance. Genome editing may become the method of choice in the future due to its ease of use across crop species for important agronomic traits such as this. The CRISPR–Cas system is the most flexible genome-editing technology in the history of molecular biology due to its precise ability to modify genomes. GEds have the potential to be used in Bangladesh to improve plant growth and biomass production, control heavy metal transport/metabolization/ compartmentalization, root system development, and also increase their phytoremediation capacity. The versatility of CRISPR allows for the simultaneous targeting of numerous genes or phenotypes, resulting in more efficient genome editing, reduced costs, and provide less environmental consequences.

6. The regulatory framework for GMOs adoption in Bangladesh

According to the Department of Environment (DOE), under the Ministry of Environment and Forest (MOEF) of the People's Republic of

Bangladesh, the national biosafety framework (NBF) is a combination of policy, legal, administrative and technical instruments that are developed to ensure an adequate level of protection in the field of the safe transfer, handling and use of GMOs resulting from modern agricultural biotechnology that may have adverse effects on the conservation and sustainable use of biological diversity, taking also into account risks to human health. At present, there are some existing policies in relevant sectors such as (i) Environment Policy, 1992 (ii) National Biodiversity Strategy and Action Plan for Bangladesh, 2004 (iii) National Biotechnology Policy, 2006 (iv) Biosafety Guidelines of Bangladesh, 2006 (v) National Guidelines for Fish and Animal Biotechnology, 2006 (vi) National Guidelines on Medical Biotechnology, 2006. To monitor and regulate the GMO related biosafety issues, National Committee on Biosafety (NCB), Biosafety Core Committee (BCC), National Technical Committee on Biosafety (NTCB), Institutional Biosafety Committee (IBC), Field Level Biosafety Committee (FBC), Bangladesh Biosafety Clearing House (BBCH) are working (NBF, 2007).

7. Consumer acceptance and future regulation of new breeding technologies in Bangladesh

Globally, consumer acceptance of plant biotechnology has been varied. One of the major problems regarding consumer acceptance has been the possibility that transgenes may have adverse effects on human health or the environment. Despite the scientific evidence supporting the safety of transgenic crops, the debate continues to this day. Genome editing technologies, such as the CRISPR/Cas9 system, can efficiently achieve transgene-free gene modifications in plants. It is unclear whether public attitudes will initially be able to accept these plants over transgenic plants. It will be necessary for governments the world over to reassess their current regulatory frameworks and establish new guidelines to include genome-edited crop technologies. At the same time, improvements in communication between the public and scientists regarding the risks and benefits of these crops will be necessary to integrate these crops into society. In the past, more emphasis has been placed on the farmer's not the consumer's benefit, for example, relief from the labor of weeding or pesticide application in the case of herbicide-tolerant or insect-resistant crops, respectively. More recently, however, GM crops with improved consumer benefits have become available, such as the Arctic Apple, a non-browning apple cultivar that is genetically engineered with a transgene that produces RNAi which significantly reduces the expression level of polyphenol oxidase (PPOs). Another example is the Innate Potato, which has two RNAi systems that lower specific transcript levels and thus limit the formation of acrylamide. Other consumer benefits of genome-edited crops include potatoes (produced by TALENS) with a lower accumulation of reducing sugars during cold storage, rice seed with enhanced longevity or increased fragrance, soybean with increased oleic acid, and tomatoes with a longer shelf life (Clasen et al., 2015; Ma et al., 2015; Shan et al., 2015; Haun et al., 2014).

Regulatory models for genome-edited crops have been proposed, and currently, the EU and New Zealand are the only legislations where genome-edited plants are regulated as GMOs. Many other countries move to exempt genome-edited crops, and this will have significant implications for countries found within the global South (Schmidt et al., 2020). There could be consequences for farmers who are prevented from taking advantage of technological advances, and the impacts could be felt on food security in countries such as Bangladesh. By focusing on traits representing consumers' benefits, the social integration of crops produced via genome editing will be accelerated.

8. Conclusions

To attain comprehensive food and health security, political promises must be implemented, and quantitative objectives for boosting food production must be established (Koul, 2022). In the fields of plant science and agriculture, the use of plant biotechnology represents a significant step forward. To solve the challenges of sustainable development, the government of Bangladesh has shown a readiness to embrace and execute a contemporary agricultural set of policies and standards that encourage the secure and ethical use of science and technology, including biotechnology. Improved agricultural biotechnology might help Bangladesh satisfy its food requirements, decrease poverty, eliminate malnutrition deficiencies, and protect the environment. Increased revenue, reduced pesticide use, and strong market demand for pest-free brinjals are just a few of the positive social effects that have resulted from Bt brinjal cultivation. As a signatory to the Cartagena Protocol on Biosafety and the creator of a suite of legislative and enabling frameworks for managing genetically modified crops, Bangladesh is a prospective South Asian country for biotechnology crop Commercialization shortly. Capacity building, seed production, stewardship, and outreach communications were all prioritized before the biotech crop adoption. Golden Rice, High Iron and Zinc Rice (HIZR), LBR potatoes, and Bt cotton are just a few of the other genetically modified (GM) crops in the pipeline for commercialization that has the potential to boost health, production, and economic outcomes in Bangladesh.

References

Aguilar, A., Twardowski, T., Wohlgemuth, R., 2019. Bioeconomy for Sustainable Development. Biotechnol. J. 14, 1800638. https://doi.org/10.1002/biot.201800638.

Bangladesh Bureau of Statistics (BBS), 2018. Bangladesh Agricultural Statistics Yearbook. Government of Bangladesh. Available from: https://databd.co/resources/agricultural-statistics-yearbook.

Antoine L. Harfouche, Vasiliki Petousi, Richard Meilan, Jeremy Sweet, Tomasz Twardowski, Arie Altman (2021). Promoting Ethically Responsible Use of Agricultural Biotechnology, Trends in Plant Science, 26:(6) 546-559.

Barnawal, D., Bharti, N., Pandey, S.S., Pandey, A., Chanotiya, C.S., Kalra, A., 2017. Plant growth-promoting rhizobacteria enhance wheat salt and drought stress tolerance by altering endogenous phytohormone levels and TaCTR1/TaDREB2 expression. Physiol. Plant. 161 (4), 502–514.

Beyer, P., 2010. Golden rice and 'Golden' crops for human nutrition. Nat. Biotechnol. 27, 478–481. https://doi.org/10.1016/j.nbt.2010.05.010.

Bisht, D.S., Bhatia, V., Bhattacharya, R., 2019. Improving plant-resistance to insect pests and pathogens: the new opportunities through targeted genome editing. Semin. Cell Dev. Biol. 96, 65–76. https://doi.org/10.1016/j.semcdb.2019.04.008. Epub 2019 May 8.

Biswas, P.S., Swamy, B.P.M., Kader, M.A., Hossain, M.A., Boncodin, R., Samia, M., Hassan, M.L., Wazuddin, M., MacKenzie, D., Reinke, R., 2021. Development and field evaluation of near-isogenic lines of GR2-EBRRI dhan29 golden rice. Front. Plant Sci. 12, 619739. https://doi.org/10.3389/fpls.2021.619739.

Brookes, G., Barfoot, P., 2012. The income and production effects of biotech crops globally 1996-2010. GM Crops Food 3, 265–272. https://doi.org/10.4161/gmcr.20097.

Bujnicki, J, Dykstra, P, Wegener, H, 2017. New techniques in agricultural biotechnology. Publications Office of the European Union, Luxembourg.

Choudhury, N., Islam, M.S., 2002. Biotechnology in Bangladesh. Research and Information System for Developing Countries, New Delhi, India, pp. 39–59. http://www.ris.org.in/abdr_nov4.pdf. (Accessed 15 October 2010).

Clasen, B.M., Stoddard, T.J., Luo, S., Demorest, Z.L., Li, J., Cedrone, F., Tibebu, R., Davison, S., Ray, E.E., Daulhac, A., Coffman, A., Yabandith, A., Retterath, A., Haun, W., Baltes, N.J., Mathis, L., Voytas, D.F., Zhang, F., 2015. Improving cold storage and processing traits in potato through targeted gene knockout. Plant Biotechnol. J. 14, 169–176.

Cong, L., Ran, F.A., Cox, D., Lin, S., Barretto, R., Habib, N., et al., 2013. Multiplex genome engineering using CRISPR/Cas systems. Science 339, 819–823. https://doi.org/10.1126/science.1231143.

Darlington, C.D., 1969. The silent millennia in the origin of agriculture. In: Ucko, P.J., Dimbleby, G.W. (Eds.), The Domestication and Exploitation of Plants and Animals. Routledge, London, pp. 67–72.

Devare, M., Jones, C., Thies, J., 2004. Effect of Cry3Bb transgenic corn and tefluthrin on the soil microbial community: biomass, activity, and diversity. J. Environ. Qual. 33, 837–843.

Dubock, A., 2014. The present status of Golden Rice. J. Huazhong Agric. Univ. 33, 69–84.

Dubock, A., 2019. Golden rice: to combat vitamin A deficiency for public health. In: Zepka, L.Q., Jacob-Lopes, E., De Rosso, V.V. (Eds.), Vitamin A. Intechopen, London, https://doi.org/10.5772/intechopen.84445.

Engler, C, Kandzia, R, Marillonnet, S, 2008. A one pot, one step, precision cloning method with high throughput capability. PLoS ONE 3 (11), e3647. https://doi.org/10.1371/journal.pone.0003647.

Fang, T., Cao, X., Ibnat, M., et al., 2022. Stimuli-responsive nanoformulations for CRISPR-Cas9 genome editing. J Nanobiotechnol 20, 354. https://doi.org/10.1186/s12951-022-01570-y.

Faostat (2019) Food and Agriculture Organization of the United Nations, 2019. Production: Crops. http://faostat.fao.org.

Ferry, N., et al., 2006. Transgenic plants for insect pest control: a forward-looking scientific perspective. Transgenic Res. 15, 13–19.

Gentile, A., La Malfa, S., 2019. New breeding techniques for sustainable agriculture. In: Farooq, M., Pisante, M. (Eds.), Innovations in Sustainable Agriculture. Springer, Cham, https://doi.org/10.1007/978-3-030-23169-9_13.

Ghislain, M., Byarugaba, A.A., Magembe, E., Njoroge, A., Rivera, C., Román, M.L., Tovar, J.C., Gamboa, S., Forbes, G.A., Kreuze, J.F., et al., 2019. Stacking three late blight resistance genes from wild species directly into African highland potato varieties confers complete field resistance to local blight races. Plant Biotechnol. J. 17, 1119–1129.

Haun, W., Coffman, A., Clasen, B.M., Demorest, Z.L., Lowy, A., Ray, E., Retterath, A., Stoddard, T., Juillerat, A., Cedrone, F., Mathis, L., Voytas, D.F., Zhang, F., 2014. Improved soybean oil quality by targeted mutagenesis of the fatty acid desaturase 2 gene family. Plant Biotechnol. J. 12, 934–940.

He, J., Zhao, X., Laroche, A., Lu, Z.X., Liu, H., Li, Z., 2014. Genotyping-by-sequencing (GBS), an ultimate marker-assisted selection (MAS) tool to accelerate plant breeding. Front. Plant Sci. 5, 484.

Hossain, M., Dey, T.K., Iqbal Hossain, M., Begum, S.N., Kadian, M.S., 2009. Research experience on potato late blight disease management in Bangladesh. Acta Hortic. 834, 175–186. https://doi.org/10.17660/ActaHortic.2009.834.19.

Huang, X., Qian, Q., Liu, Z., Sun, H., He, S., Luo, D., Xia, G., Chu, C., Li, J., Fu, X., 2009. Natural variation at the DEP1 locus enhances grain yield in rice. Nat. Genet. 41 (4), 494–497. https://doi.org/10.1038/ng.352. Epub 2009 Mar 22.

Humanitarian Board, 2009. Humanitarian Board for golden rice. In: Golden Rice Network Meeting, 9–10 March, 2009. International Rice Research Institute, Los Baños.

ILSI CERA, 2010. A Review of the Environmental Safety of the Cry1Ac Protein. ILSI CERA.

Jinek, M, Chylinski, K, Fonfara, I, Hauer, M, Doudna, JA, Charpentier, E, 2012. A programmable dual-RNA-guided DNA endonuclease in adaptive bacterial immunity. Science 337 (6096), 816–821. https://doi.org/10.1126/science.1225829.

Koul, B., 2022. Future Prospects of GM Plants. In: Cisgenics and Transgenics. Springer, Singapore. https://doi.org/10.1007/978-981-19-2119-3_6.

Li, J., Cang, Z., Jiao, F., Bai, X., Zhang, D., Zhai, R., 2017a. Influence of drought stress on photosynthetic characteristics and protective enzymes of potato at seedling stage. J. Saudi Soc. Agric. Sci. 16 (1), 82–88.

Li, H., Hu, B., Chu, C., 2017b. Nitrogen use efficiency in crops: lessons from Arabidopsis and rice. J. Exp. Bot. 68, 2477–2488.

Li, J., Zhang, X., Sun, Y., Zhang, J., Du, W., Guo, X., Li, S., Zhao, Y., Xia, L., 2018. Efficient allelic replacement in rice by gene editing: a case study of the NRT1.1B gene. J. Integr. Plant Biol. 60 (7), 536–540. https://doi.org/10.1111/jipb.12650. 29575650.

Luo, J.S., Yang, Y., Gu, T., Wu, Z., Zhang, Z., 2019. The Arabidopsis defensin gene AtPDF2.5 mediates cadmium tolerance and accumulation. Plant Cell Environ. 42, 2681–2695. https://doi.org/10.1111/pce.13592.

Lusser, M., Parisi, C., Plan, D., Rodriguez-Cerzo, E., 2012. Deployment of new biotechnologies in plant breeding. Nat. Biotechnol. 30, 231–239.

Ma, L., Zhu, F., Li, Z., Zhang, J., Li, X., Dong, J., Wang, T., 2015. TALEN-based mutagenesis of lipoxygenase LOX3 enhances the storage tolerance of rice (*Oryza sativa*) seeds. PLoS ONE 10, e0143877.

Maioli, A., Gianoglio, S., Moglia, A., Acquadro, A., Valentino, D., Milani, A.M., Prohens, J., Orzaez, D., Granell, A., Lanteri, S., Comino, C., 2020. Simultaneous CRISPR/Cas9 editing of three PPO genes reduces fruit flesh browning in *Solanum melongena* L. Front. Plant Sci. 11, 607161. https://doi.org/10.3389/fpls.2020.607161.

Mao, Y., Zhang, Z., Feng, Z., Wei, P., Zhang, H., Botella, J.R., et al., 2016. Development of germ-line-specific CRISPR-Cas9 systems to improve the production of heritable gene modifications in Arabidopsis. Plant Biotechnol. J. 14, 519–532. https://doi.org/10.1111/pbi.12468.

Mellidou, I., Buts, K., Hatoum, D., Ho, Q.T., Johnston, J.W., Watkins, C.B., Schaffer, R.J., Gapper, N.E., Giovannoni, J.J., Rudell, D.R., Hertog, M.L., Nicolai, B.M., 2014. Transcriptomic events associated with internal browning of apple during postharvest storage. BMC Plant Biol. 14, 328. https://doi.org/10.1186/s12870-014-0328-x.

Moshelion, Menachem, Altman, Arie, 2015. Current challenges and future perspectives of plant and agricultural biotechnology. Trends Biotechnol 1879-3096. 33 (6), 337–342. https://doi.org/10.1016/j.tibtech.2015.03.001.

Naqvi, R.Z., et al., 2017. Development of a triple gene Cry1Ac-Cry2Ab-EPSPS construct and its expression in *Nicotiana benthamiana* for insect resistance and herbicide tolerance in plants. Front. Plant Sci. 8, 55.

National Biosafety Framework, June 2007. "National Biosafety Framework" of the Government of the People's Republic of Bangladesh. Department of Environment Ministry of Environment and Forest Government of the People's Republic of Bangladesh. http://old.doe.gov.bd/publication_images/9_nationalbiosafetyframeworkbd.pdf.

Osakabe, Y., Watanabe, T., Sugano, S.S., Ueta, R., Ishihara, R., Shinozaki, K., Osakabe, K., 2016. Optimization of CRISPR/Cas9 genome editing to modify abiotic stress responses in plants. Sci. Rep. 6, 26685.

Ozturk, M., Turkyilmaz Unal, B., García-Caparrós, P., Khursheed, A., Gul, A., Hasanuzzaman, M., 2020. Osmoregulation and its actions during the drought stress in plants. Physiol. Plant. https://doi.org/10.1111/ppl.13297. Epub ahead of print. 33280137.

Paine, J.A., Shipton, C.A., Chaggar, S., Howells, R.M., Kennedy, M.J., Vernon, G., et al., 2005. Improving the nutritional value of Golden Rice through increased pro-vitamin A content. Nat. Biotechnol. 23, 482–487. https://doi.org/10.1038/nbt1082.

Phillips, T., 2008. Genetically modified organisms (GMOs): Transgenic crops and recombinant DNA technology. Nature Education 1 (1), 213.

Qaim, M., Kouser, S., 2013. Genetically modified crops and food security. PLoS ONE 8 (6), e64879. https://doi.org/10.1371/journal.pone.0064879.

Rani, R., Yadav, P., Barbadikar, K.M., Baliyan, N., Malhotra, E.V., Singh, B.K., Kumar, A., Singh, D., 2016. CRISPR/Cas9: a promising way to exploit genetic variation in plants. Biotechnol. Lett. 38 (12), 1991–2006. https://doi.org/10.1007/s10529-016-2195-z.

Ricroch, A., 2019. Global developments of genome editing in agriculture. Transgenic Res. 28 (Suppl 2), 45–52. https://doi.org/10.1007/s11248-019-00133-6.

Rivero, Rosa M, Mittler, Ron, Blumwald, Eduardo, Zandalinas, Sara I, 2022. Developing climate-resilient crops: improving plant tolerance to stress combination. Plant J 1365-313X. 109 (2), 373–389. https://doi.org/10.1111/tpj.15483.

Sarkar, B., Akter, F., Johora, F.T., Md. Ullah, A., Shohael, A.M., 2020. Micronutrient biofortification in rice through new breeding techniques (NBTs): Bangladesh perspective. Curr. Nutr. 1, 1. https://doi.org/10.2174/2665978601999201202114714.

Schmidt, S.M., Belisle, M., Frommer, W.B., 2020. The evolving landscape around genome editing in agriculture: many countries have exempted or move to exempt forms of genome editing from GMO regulation of crop plants. EMBO Rep. 21 (6), e50680. https://doi.org/10.15252/embr.202050680. Epub 2020 May 19 32431018.

Shan, Q., Zhang, Y., Chen, K., Zhang, K., Gao, C., 2015. Creation of fragrant rice by targeted knockout of the OsBADH2 gene using TALEN technology. Plant Biotechnol. J. 13, 791–800.

Shelton, A.M.K.E., Hokanson, D.M., Hautea, M.J., Hossain, M.A., Hossain, V., et al., 2017. Bt Eggplant: A Genetically Engineered 'Minor' Crop Comes of Age in Bangladesh and the Philippines. ISB Report.

Shelton, A.M., Hossain, M.J., Paranjape, V., Azad, A.K., Rahman, M.L., Khan, A.S.M.M. R., et al., 2018. Bt Eggplant project in Bangladesh: history, present status, and future direction. Front. Bioeng. Biotechnol. 6, 106. https://doi.org/10.3389/fbioe.2018.00106.

Shelton, A.M., Sarwer, S.H., Hossain, M.J., Brookes, G., Paranjape, V., 2020. Impact of Bt brinjal cultivation in the market value chain in five districts of Bangladesh. Front. Bioeng. Biotechnol. 8, 498. https://doi.org/10.3389/fbioe.2020.00498.

Shi, J., Gao, H., Wang, H., Lafitte, H.R., Archibald, R.L., Yang, M., Hakimi, S.M., Mo, H., Habben, J.E., 2017. ARGOS8 variants generated by CRISPR-Cas9 improve maize grain yield under field drought stress conditions. Plant Biotechnol. J. 15 (2), 207–216. https://doi.org/10.1111/pbi.12603. Epub 2016 Aug 17 27442592.

Siddiqui, H.A., Asif, M., Asad, S., et al., 2019. Development and evaluation of double gene transgenic cotton lines expressing Cry toxins for protection against chewing insect pests. Sci. Rep. 9, 11774. https://doi.org/10.1038/s41598-019-48188-z.

Van Vu, T., Das, S., Hensel, G., et al., 2022. Genome editing and beyond: what does it mean for the future of plant breeding? Planta 255, 130. https://doi.org/10.1007/s00425-022-03906-2.

Varshney, R.K., Terauchi, R., McCouch, S.R., 2014. Harvesting the promising fruits of genomics: applying genome sequencing technologies to crop breeding. PLoS Biol. 12, e1001883.

Voytas, DF, Gao, C, 2014. Precision genome engineering and agriculture: opportunities and regulatory challenges. PLOS Biol 12 (6), 1–6. https://doi.org/10.1371/journal.pbio.1001877.

Waltz, E., 2015. USDA Approves Next-Generation GM Potato. Nature Publishing Group.

Xu, D.-Q., Huang, J., Guo, S.-Q., Yang, X., Bao, Y.-M., Tang, H.-J., et al., 2008. Overexpression of a TFIIIA-type zinc finger protein gene ZFP252 enhances drought and salt tolerance in rice (*Oryza sativa* L.). FEBS Lett. 582 (7), 1037–1043.

Xu, R., Wei, P., Yang, J., 2017. Use of CRISPR/Cas genome editing technology for targeted mutagenesis in rice. Methods Mol. Biol. 1498, 33–40. https://doi.org/10.1007/978-1-4939-6472-7_3.

Xue, G.-P., Way, H.M., Richardson, T., Drenth, J., Joyce, P.A., McIntyre, C.L., 2011. Overexpression of TaNAC69 leads to enhanced transcript levels of stress up-regulated genes and dehydration tolerance in bread wheat Mol. Plant 4 (4), 697–712.

Yu, W., Wang, L., Zhao, R., et al., 2019. Knockout of SlMAPK3 enhances tolerance to heat stress involving ROS homeostasis in tomato plants. BMC Plant Biol. 19, 354. https://doi.org/10.1186/s12870-019-1939-z.

Zhang, R., Liu, J., Chai, Z., et al., 2019. Generation of herbicide tolerance traits and a new selectable marker in wheat using base editing. Nat. Plants 5, 480–485. https://doi.org/10.1038/s41477-019-0405-0.

Zia, R., Nawaz, M.S., Siddique, M.J., Hakim, S., Imran, A., 2021. Plant survival under drought stress: implications, adaptive responses, and integrated rhizosphere management strategy for stress mitigation. Microbiol. Res. 242, 126626. https://doi.org/10.1016/j.micres.2020.126626. 33189069. Epub 2020 Oct 18.

Zsögön, A., Peres, L.E.P., Xiao, Y., Yan, J., Fernie, A.R., 2022. Enhancing crop diversity for food security in the face of climate uncertainty. Plant J. 109 (2), 402–414. https://doi.org/10.1111/tpj.15626.

CHAPTER 9

Strategic tropical natural resources management and household socio-economic resilience due to the Covid-19 pandemic

S. Andy Cahyono[a], Cahyono Agus[b,c], S. Agung Sri Raharjo[d], Agung Wahyu Nugroho[a], and Yonky Indrajaya[a,]*
[a]Research Center for Ecology and Ethnobiology, National Research and Innovation Agency (BRIN), Cibinong, Indonesia
[b]Faculty of Forestry, Universitas Gadjah Mada, Yogyakarta, Indonesia
[c]PKBTS Tamansiswa Alumni Board, Yogyakarta, Indonesia
[d]Institute for Implementation Standard of Environment and Forestry Instrument, Surakarta, Indonesia
*_Corresponding author: yonky.indrajaya@brin.go.id_

1. Introduction

Indonesia is traversed by the equator, with abundant natural resources and a high risk of disaster. Its strategic location includes (1) the equator, (2) the ring of fire, and (3) the Eurasia, Pacific, and Indo-Australian earth plates. Indonesia is the largest archipelagic country globally, with 16,056 islands located between the Indian and Pacific Oceans in Southeast Asia. The country plays a crucial role in geography, politics, economics, sociology, defense, and international security. Indonesia has a tropical ecosystem with a wealth of natural resources, flora, fauna, ethnic diversity, mega-diversity, and abundant natural beauty. Development and economic growth are multiplying but are still based on resource-based development even though Indonesia's economic growth relies on a resource base averages more than 5% a year.

The novel Coronavirus disease (Covid-19) epidemic is a global health emergency triggered by the Coronavirus-2 that causes Severe Acute Respiratory Syndrome (SARS-CoV-2) (Parvin et al., 2020). Covid-19 was declared a public health emergency on January 30, 2020, and a pandemic on March 11, 2020, by the World Health Organization (WHO)

Agricultural Bioeconomy
https://doi.org/10.1016/B978-0-323-90569-5.00007-X

(Ullah et al., 2020). In the last two decades, novel viruses such as H1N1SAR, H5N1, and MERS, have caused worldwide health crises (Djalante et al., 2020). Since sufferers infected with Covid-19 in Indonesia were first confirmed on March 2, 2020, the public can access direct data about the impact of Covid-19 in Indonesia online.

As in many other nations, social distancing policies have been shown to help minimize the number of cases of Covid-19. Starting April 2020, the Indonesian government applied the Large-scale Social Restriction (LSSR), and the newest is Enforcement of Restrictions on Community Activities (ERCA). The policy restricts non-essential activities, including the closure of schools and workplaces, sharply curtailing religious activities, halting non-essential public activities, promoting public safety campaigns, encouraging social distancing, or implementing any stay-at-home directive (Bloem and Salemi, 2021). The government has also issued many policies and programs to reduce the impact of Covid-19, including the "Family of Hope Program" for 10 million "Beneficiary Families," essential food assistance, pre-employment cards, electricity subsidies for 450 VA (23.8 million customers), and 900 VA (7.2 million customers), housing incentives for low-income people, etc.

The Covid-19 pandemic has caused great havoc in many life spheres. Although this havoc has been experienced globally, the effects of the Covid-19 outbreak have been more severe in low and low-middle-income countries. In Indonesia's case, Covid-19 caused Indonesia's economic growth to reach 2.97% in the first quarter of 2020. However, in the second quarter, Indonesia's economic growth became minus 5.32%. Developed countries, such as the United States, Italy, France, Germany, and South Korea, even had their economic growth reaching minus 17–20%. The Ministry of Finance has released news that economic growth in the third quarter of 2020 was around minus 2.9% to minus 1.1%. This phenomenon indicates that Indonesia will enter an economic recession.

The Covid-19 pandemic caused a world economic recession, as economic growth declined sharply; unemployment increases, exports decline, investment decreases, purchasing power decreases, poverty, and economic uncertainty. Developing and emerging countries with weak access to technology and health have been particularly affected. All sectors of the Indonesian economy experienced a decline, except for the agricultural industry, which grew slightly positive. This paper discusses the role of an integrated bio-cycle system in household socio-economic resilience and management of tropical natural resources in the Covid-19 and post-Covid-19 era.

2. The effects of Covid 19 pandemic: From global to households

2.1 The impact on agricultural production and income generation

During the Covid-19 pandemic, the lockdown and movement restrictions led to a labor shortage in agricultural production (Ilesanmi et al., 2021). Owing to the lack of mechanized farming equipment in Indonesia, agrarian labor is performed manually daily. Annually, peak agricultural activity necessitates a high demand for labor, but farmers and laborers are prohibited from accessing farmland due to the lockdown and movement restriction. As a result, the number of people available to cultivate farmland has decreased, and agricultural production has fallen across the region. The seed supply chain, agrochemicals, and fertilizer importation were disrupted, resulting in fewer viable seeds and lower agricultural yields. Due to the travel restrictions imposed, transportation of this perishable produce to market was hampered, resulting in increased food wastage.

It can be implicitly interpreted that there are activities in the agricultural sector that have been negatively affected, but other activities can reap benefits during the Covid-19 pandemic. For example, many vegetable farmers and breeders have decreased their income after terminating contracts with hotels, restaurants, catering, and quiet markets. However, many vegetable farmers have increased their income, such as immune-boosting foods, ginger, herbs, and healthy foods.

2.2 The impact on household food security

Covid-19 not only posed a threat to human health, but it also had the potential to cause a global food crisis (Balkan, 2020). One-third of the world's population has been confined due to the Covid-19 pandemic (Galanakis, 2020). The effect and duration of the economic crisis on individual households resulting from the pandemic are challenging to forecast since many unknowns surround the crisis's duration, such as the length of "stay-at-home" orders, affected sectors post-crisis consumption, and recovery (Martin et al., 2020). Prices of agricultural commodities have fallen by 20% due to a global decline in demand from hotels and restaurants (Nicola et al., 2020). Consumers' fears of a potential food shortage and reduced agricultural production due to farm labor restrictions (Glauber et al., 2020).

2.3 The impact socio-economic activities

World economic conditions and households depend on the interaction and activities of the community. High community interaction and movement will improve the world economy. Covid-19 forces many countries to carry out lockdowns and restrictions on interactions and activities to prevent transmission of the virus. As a result, economic activity has stalled, and world economic growth has fallen sharply. Lockdown and physical distancing make people keep their distance from each other, be alert to each other, and change the social order of people's lives.

Activities and social interactions, eco, and culture, as well as households, change. This change in social interaction results in changes in the behavior of agricultural entrepreneurs. Changes in behavior include optimizing the use of household labor. If agricultural business actors can use agricultural labor, limiting social interaction has forced agricultural business actors to optimize the workforce in their family environment. Changes in social interaction resulting from the LSSR policy have forced agricultural business actors to take advantage of their existing agricultural production facilities, such as fertilizers and plant seeds. Besides, many agricultural business actors reduce fertilizers and pesticides due to the limited availability of agricultural production facilities on the market (Sunarsih et al., 2020).

2.4 The impact prices of agricultural products and households

Three factors caused the decline in agricultural commodity prices during the Covid-19 pandemic. First, the restrictions on transportation and economic activities, such as the PSBB policy, impacted the slow distribution process of agricultural commodities. As a result, a decrease in the quality of product freshness and a decline in prices. Second, massive layoffs of employment in various business sectors have reduced purchasing power and market demand, resulting in further pressure on agricultural commodity prices. Third, cessation of social activities such as celebrations, parties, and hospitality cause foodstuffs' demand to decline. However, agricultural commodities have also increased rapidly during Covid-19, including immune-boosting foods (fruits, ginger, rhizome, honey, etc.) and healthy foods (Sunarsih et al., 2020).

According to the Smeru (2021) survey, three-quarters of households (74.3%) experienced a decrease in income from January 2020 to January 2021. The proportion of households with a more significant income reduction occurred in households with children (75.3%) and those living in urban areas (78.3%). On average, urban households also experienced a more significant decrease in income than rural households. Besides, nearly a quarter (24.4%) of households reported an increase in expenditure, primarily from an increase in the cost of spending on food and other necessities. The proportion of households with children who spent more on the internet and mobile phone costs was significantly more significant (65%) than households without children (28.9%).

On the other hand, only a few of the primary breadwinners (14%) have changed jobs due to Covid-19. However, almost half (47.3%) of those who changed jobs changed from jobs in the formal sector to jobs in the informal sector where labor protection was low. Half of all households (51.5%) have no precautionary savings. Nearly a third (27.3%) pawned possession of goods to survive. Another finding showed that a quarter of them (25.3%) borrowed money informally from family or friends. Nearly a third (30%) of respondents are worried that they will not feed their families. The proportion of households facing moderate and severe food insecurity increased to 11.7% in 2020. Decreased income and disruption of the food supply system are the main factors causing food insecurity.

2.5 The impact on forest and environmental

The Covid-19 pandemic has been cited as a significant factor in the ongoing global economic downturn. It has been beneficial to the environment (Roy et al., 2021; Golar et al., 2020). However, Covid-19 has also hampered forest and environmental afforestation, rehabilitation, reforestation, and recovery programs due to difficulties in human resources, funding, perceptions, health protocols, etc., so that many are neglected (Mohan et al., 2021). The LSSR makes it difficult for people to get acceptance, so they use the land to survive by cultivating land in the forest and taking forest products to get a fast income by cutting wood (Mockrin et al., 2019; Ambarwati et al., 2018). Forest area managers must optimize the use of environmental services and non-timber forest products such as resin, sap, honey, rattan, fruit, and medicinal plants. Area managers must be wise in providing access support to use these resources so that the decline in household income around the forest does not trigger forest destruction in the Covid-19 era.

3. Covid-19: Disruption on socio-economic and nature

Covid-19 has changed our thinking from resource extraction to resource efficiency, technological innovation, e-commerce, zero waste, and product safety. Food commodities, frozen food, biopharmaceuticals, non-timber forest products, and immune-enhancing foods have increased in demand during the Covid-19 pandemic, followed by changes in consumption, production, and distribution patterns. The pandemic has resulted in a relaxation of pressure on the earth's environment because almost all business, industrial, school, government, social, and transportation activities are locked down. Tropical ecosystems can be recovering and rapidly healing because they are supported by high rainfall, temperature, humidity, and organic cycles throughout the year.

Covid-19 has made a change in mindset in most households, life activities, education, and works. Shifts in perspective from resource extraction to resource efficiency, from comparative advantage to competitive advantage, from linear economic to a circular economy, from high economic growth to steady stage economy, from national food security to household food security, from open cycle to closed cycle, changes in production patterns, consumption patterns, and distribution patterns. While agricultural businesses saw a drop in exports, some agricultural products, particularly grain and oil, remained stable or even increased, indicating the pandemic's strong demand for staple foods. Exports of the medicinal plant, not surprisingly, significantly increased during the pandemic.

The pandemic has caused the export demand for plantation products to fall a lot. However, horticultural products have received a blessing from the pandemic. Gunawan and Nida (2020) report that the increase in fruit exports in January–May 2020 for seasonal fruit exports increased by 308.73% with a value of 91.04%, and for annual fruit exports increased by 31.84% with a value of 73.39% compared to the same period in 2019. Data on domestic fruit consumption during and after the pandemic is estimated to increase along with the public's increasing awareness about the importance of consuming fruits and vegetables to increase endurance. Horticultural products are also rich in health benefits, so they are known as superfoods, such as berries (blueberries, cranberries, strawberries), pomegranates, and papayas (CBI, 2019). Another trend developed during Covid-19 is that consumers tend to look for low-carb, low-fat products and be low in sugar. Also, non-wood forest products that increase endurance such as honey, ginger, and eucalyptus oil. Consumers are also increasingly concerned about the origin of the food they consume. During a pandemic, the application of post-harvest technology to horticultural commodities plays an important role. In addition to increasing shelf life and reducing potential losses for farmers, distributors, and consumers, post-harvest technology can also maintain the functional components of fruits and vegetables, namely vitamins, minerals, fiber, and other active compounds that can increase immunity.

Covid-19 disrupts a healthier, more empathetic, and caring lifestyle. The movement to help people affected by Covid-19 spread to other communities with the viral social media aspect. The concept of *Jogo Tonggo* (take care of the neighbor) in Central Java and the form of concern among the community has emerged in various forms, including social and security safety nets and economic networks. The *Jogo Tonggo* Task Force is stationed in each RW (i.e., a group of about 150 households). The task force was created to guard and monitor each other's neighbors. Members of this movement's task are to ensure that assistance and support from outside the region that enters their area is right on target and effective.

Household consumption patterns changed after people became aware of Covid-19 and the existence of social/physical distancing. There were panic buying and scarcity of masks, hand sanitizers, flue-preventing drugs, multivitamins, disinfectants, spices, herbs, frozen food, processed products biopharmaceuticals during the Covid-19 pandemic (Mujiyono and Soenjoto, 2020). Frozen and processed food products have increased purchases by households since March 2020 (Morrison, 2020). The Covid-19 pandemic has also changed housewives' behavior in preparing menus for their families. Homemakers are more concerned about the quality of food for their families; this is related to increasing family immunity (Bahagia et al., 2020).

Yuswohady (2020) states that a business order in a "new normal" condition[a] must adapt to changes in people's behavior, especially consumers. There are five significant changes in people's behavior related to Covid-19, namely (1) Online shopping has begun to shift from products of needs to products. This change will be beneficial because most agricultural products are necessities. (2) Consumers begin to eat less at restaurants and switch to delivery services. (3) The work from home (WFH) policy encourages homemakers to cook their meals at home regularly. This situation has the potential to increase the demand for fresh agricultural products such as vegetables and meat. (4) Millennial families with a practical mindset will cook more of their food, but with ingredients ready to cook and can be cooked at any time (frozen food). (5) Online shopping patterns are expected to develop more rapidly during the new normal and in the future.

[a] New normal is a change in behavior or habits to continue carrying out activities as usual but by continuously implementing health protocols during the Covid-19 pandemic.

LSSR, which forces people to reduce travel, including shopping for daily necessities, is one factor in developing an online marketing system. Some startups are shopping for food and household products that connect traditional markets in cities with households, e.g., okesayur, online market, go-shop, pasarmu, tumbas.in, LaLettuce, pono.co.id, etc.

4. Innovation: A new paradigm of knowledge-based development in Covid-19 era

Covid-19 has accelerated the Industrial Revolution 4.0 in Indonesia. The Covid-19 pandemic has succeeded in forcing the world to pause and look for new patterns for its life. A new paradigm of knowledge-based development through the empowerment of natural resources must be developed more innovative, more focused, more global, and futuristic for sustainable development after the Covid-19 era. Covid-19 forces us to be more creative and take technological leaps that prioritize efficiency, safety, low waste, renewable, and meet human life needs.

The integrated bio-cycle system is an innovation that plays a vital role in the socio-economic resilience of households whose activities are increasingly limited. Household economic resilience with an integrated bio-cycle system can withstand Covid-19. Management of land and living resources based on natural nature is intended to harmonize added value to the environment, economic, socio-cultural, and health aspects. This integrated farming system has a multi-functional and multi-product that is conservative productive to change consumer behavior and technology disruption. Information technology, the Internet of Things, artificial intelligence, big data, blockchain, and networks can support a biological cycle system that is more interconnected, accurate, upscale, and based on an integrated biological cycle system for sustainable and dignified management of tropical natural resources.

After the post-Covid-19 period, Internet and Communication Technologies, blockchain in the food supply chain, and other Industry 4.0 applications, as well as methods that redefine the way we consume food (e.g., lab-grown meat, plant-based meat alternatives, and valorization of a wide variety of bioresources), are the most promising developments (Galanakis et al., 2021). There is also a pressing need to use social marketing to understand better the behaviors, expectations, and obstacles that affect customer and agri-food industry behavior change. As a result, this shift would help adapt to new norms forged by the Covid-19 pandemic, where there is a significant information gap for decision-making.

With technical advances, Covid-19 has rendered food protection and socio-economic resilience more conscious of households. Intelligent and active packaging, modern biosecurity structures (e.g., fostering a food safety culture in food processing facilities and farmlands), bio pesticides in agriculture, and Industry 4.0 (e.g., block chain technology) are all expected to rise rapidly in the coming years. The ultimate aim is to protect consumers by ensuring the protection of food and the food supply chain and minimizing food waste and the food sector's environmental effects. These developments can pave the way for new business models that disrupt the food supply chain and market in a techno-socioeconomic way (Galanakis et al., 2021).

The primary linear economy method, which is primarily focused on growing demand to meet rising consumption, has proven ineffective for long-term resource management (Stegmann et al., 2020). The shift from the existing development model to a circular bio-economy model may boost resilience (Mak et al., 2020; Farcas et al., 2021). As a result, a stronger push for innovation in this area would help offset the food supply chain's environmental effect by allowing for the development of less energy-intensive, environmentally sustainable processes, goods, and services (O'Neill et al., 2019). Food waste can be valorized using bio refineries that generate bioenergy, electric power, bio surfactants, and bio fertilizers, among many other things, in the circular bio-economy (Galanakis et al., 2021). Furthermore, aquaculture is a new high-protein, low-carbon-emission technology gaining popularity worldwide to ensure intensive food sustainability (Tahar et al., 2018a,b).

5. Integrated bio-cycles farming system in post Covid-19 era: The next level

A solely economic red economy has resulted in the loss of life and environmental degradation. While it was risky and expensive, a green economy is concerned with health and environmental value. Gunter Pauli came up with the idea of the blue economy in 2009 (Agus, 2018) by accelerating natural cycle processes and enabling land (land, water, and mineral) and biological (plants, animals, and humans) resources with the added value of the climate, economy, technology, socio-cultural, and sustainable management. The blue economy encourages productive investment, job growth, increased financing, increased creativity, social capital production, and entrepreneurship. Implemented by using waste and abandoned goods to be energy, food, and employment thus turn poverty to be sustainable development and insufficiency to be available (Bosman and Rotmans, 2016).

Tropical natural resources are a significant potential source, but their economic value is poor because they are still mismanaged under natural norms (Agus, 2018). Deforestation and open-pit mining overexploitation have been the main factors to land loss and significant local-global environmental harm (Agus, 2018). We must change our growth model from natural resource-based to knowledge-based development. A change from resource extraction to resource empowerment will pose a new challenge in moving from a red-green to a blue economic concept for sustainable development that is more intelligent, concentrated, global, and futuristic (Agus, 2018).

While tropical bio-geo-resources have high biomass productivity, their economic values are still low. The Integrated Bio-cycle Farming System (IBFS) is an ecological landscape management system that harmoniously combines agricultural and non-agricultural aspects. The 9R (reuse, reduce, recycle, refill, restore, repair, replant, rebuild, reward) system was used to manage the cycle of electricity, organic matter and carbon, water, nutrient, harvest, crop, and money to get the best results for the farmer, population, agriculture, and global climate (Agus, 2018). The system is multipurpose and can handle various products (food, feed, fiber, fuel, pharmacy, fertilizer, edutainment, ecotourism). IBFS was intended to provide additional benefits for rural, medium, and large-scale farmers by converting waste products into renewable resources to produce high-value items like organic fertilizer (liquid and solid), animal feed, and biogas energy sources (Agus, 2018).

The IBFS is an alternative farming system that brings together agricultural, fisheries, plantation, livestock farming, forestry sectors, and non-agricultural aspects like tourism, agroindustry, towns, and industry into one integrated area. Agus (2018) argue that IBFS was created by combining ICM (Integrated Crop Management), INM (Integrated Nutrient Management), IMM (Integrated Soil Moisture Management), and IPM (Integrated Pest Management).

The forest-based sector is vital in a bio-economy because it provides material (both wood and non-wood), bioenergy, and various other ecosystem-controlling and cultural services (Wolfslehner et al., 2016). The emergence of the bio-economy has heightened interest in forest resources from a variety of perspectives. First, the bio-economy raises demand for forest products and services, thus expanding the sector's economic opportunities. Bioenergy, wood construction, packaging materials, chemicals, textiles, and other industries are among the many possibilities.

Second, there is an increase in forest land demand for other environmental resources, such as biodiversity, carbon sequestration, tourism, and human health impact. The next level of IBFS makes more use of innovations and technologies that make IBFS more scalable, effective, efficient, integrated, and mitigate process errors.

On a household scale, the implementation can even be done in households with narrow land in implementing IBFS, producing their vegetables without buying them (Masduki, 2018). The hydroganic system is a hydroponic farming system, but it is done organically, often called aquaponics, which combines the hydroponic system with the fish farming pond underneath. Budikdamber or fish farming in buckets is a simple hydroganic model. This system uses buckets for fish cultivation and is planted with vegetables using fish manure as fertilizer or plant nutrition (Nursandi, 2018). This system helps provide the need for animal and vegetable protein and makes it easier for people to get fish and vegetables. At the time of Covid, the hydroganic and circular economy system went viral and met economic needs, and improved the mentalities of the people who were not allowed anywhere.

6. Household social economic resilience in Covid-19 era

Covid-19 shows an enormous uncertainty in living systems where anticipation and resilience are major problems in development (Leach et al., 2021). Folke (2006) defines resilience as the need for versatility or flexibility and adaptability and absorbing market and environmental shocks. Throughout the current Covid-19 era, resilience is a multidimensional concept that can be expressed across critical factors to minimize distress and promote adaptation (Verger et al., 2021). Financial assets determine resilience capacities in a household (e.g., savings, productive assets), social capital (e.g., connections), and human capital (e.g., education, knowledge).

Household socio-economic resilience is carried out using a farmer livelihood strategy that is economically in the form of savings and increased income, while the social livelihood strategy is the use of social networks in the community. Farmers reduce unneeded expenses except for basic needs and fulfillment of social needs (donations and others), ensuring their social network. This social network needs to be maintained even though it is burdensome for household expenses. This activity is done so that farmers are not considered unconventional, anti-social and guarantee social support when the economy becomes more difficult. The livelihood strategy in the form of savings is carried out according to economic stages. First, reducing expensive protein sources minimizes the frequency of eating and looking for conventional foods that are rarely eaten. In the next step, all family members (children, parents, and women) work whatever they can to get cash. In the next stage, if it is not sufficient, the assets owned will be sold. If this is not possible, they will try to take advantage of social networks with debt and take advantage of the culture of helping help in the community (Cahyono and Lastiantoro, 2014; Constas et al., 2014).

In Indonesia, household socio-economic resilience is carried out by optimizing planting in the yard for food availability (Pambudi, 2020), a family medicinal garden to increase endurance with *empon empon* plants for herbal medicine (Derajat et al., 2020; Nasrulloh et al., 2020; Pertiwi et al., 2020). People believe in treating shortness of breath and increasing body immunity by consuming ginger (Aryanta, 2019) in various traditional drinks such as ginger, ginger milk, ginger tea, *bandrek*, *wedang uwuh* (Mangifera et al., 2018; Sagala et al., 2020).

Social/physical distancing policies and Large-Scale Social Restrictions (LSSR) have begun to be implemented in several regions. It is just that its effectiveness in reducing the escalation of Covid-19 transmission has not been proven. Anthropologically, the Covid-19 outbreak concerns the biological aspects of the virus and biosocial elements related to efforts to change human behavior to comply with the rules. Non-compliance with the LSSR is also due to the socio-economic resilience of the household. To maintain the household's existence, household members work outside the house through the LSSR. Apart from the household's economic resilience, knowledge can regulate human behavior and depend on human internalization in the household's cultural context.

7. Conclusion

This paper examines the importance of an integrated bio-cycle system in household socio-economic resilience and tropical natural resource management during the Covid and post-Covid eras. Covid-19 has shifted our focus away from resource exploitation and resource quality, technical innovation, e-commerce, waste minimization, and product protection. During the Covid-19 pandemic, demand for food commodities, frozen food, biopharmaceuticals, non-timber forest products, and immune-boosting foods increased, accompanied by changes in consumption, processing, and distribution patterns. Since almost all industry, manufacturing, school, government, social, and transportation activities have been halted due to the pandemic, pressure on the earth's ecosystem has been relieved. Tropical habitats can quickly regenerate and heal because they are sustained all year by high rainfall, temperature, humidity, and organic cycles. For sustainable development after the Covid-19 period, a new paradigm of knowledge-based development by natural resource empowerment must be built that is more creative, global, cantered, and futuristic. Covid-19 can be withstood by household economic resilience with an integrated bio-cycle system. Land and living resource management based on natural nature is intended to balance added value to the climate, economy, socio-cultural aspects, and health. This integrated farming system has a multi-functional and multi-product that is conservatively productive to transform customer behavior and disrupt technology.

References

Agus, C., 2018. Development of blue revolution through integrated bio-cycles system on tropical natural resources management. In: Leal Filho, W., Pociovălişteanu, D.M., Borges de Brito, P.R., Borges de Lima, I. (Eds.), Towards a Sustainable Bioeconomy: Principles, Challenges and Perspectives. Springer International Publishing, Cham, pp. 155–172.

Ambarwati, M.E., Sasongko, G., & Therik, W.M. (2018). Dynamics of the tenurial conflict in state forest area (case in BKPH tanggung KPH semarang). Sodality: Jurnal Sosiologi Pedesaan, 6(2).Ambarwati, M.E., Sasongko, G., Therik, W.M., 2018. Dynamics of the tenurial conflict in state forest area (case in BKPH tanggung KPH semarang). Sodality 6 (2).

Aryanta, I.W.R., 2019. Manfaat jahe untuk kesehatan. J. Widya Kesehat. 1 (2), 39–43. https://doi.org/10.32795/widyakesehatan.v1i2.463.

Bahagia, Nurrahmawati, D., Nurhayati, I., 2020. Resilience of household mother in dealing with covid-19. Tunas Geografi 9 (2), 129–136.

Balkan, B.A., 2020. Effects of COVID-19 on food & agricultural systems. In: System Dynamics Society A&F SIG Meeting., https://doi.org/10.5281/zenodo.3780013.

Bloem, J.R., Salemi, C., 2021. COVID-19 and conflict. World Dev. 140, 105294. https://doi.org/10.1016/j.worlddev.2020.105294.

Bosman, R., Rotmans, J., 2016. Transition governance towards a bioeconomy: a comparison of finland and the Netherlands. Sustainability 8 (10), 1017. https://doi.org/10.3390/su8101017.

Cahyono, S.A., Lastiantoro, C.Y., 2014. Strategi nafkah petani di Kecamatan Bulu dan Weru Kabupaten Sukoharjo, Provinsi Jawa Tengah. In: Cahyadi, A., Argadyanto, N., Tivianton, T.A., Nugraha, H. (Eds.), Ekologi Lingkungan Kawasan Karst Indonesia: Menjaga asa Kelestarian Kawasan Karst Indonesia. Deepublish, Yogyakarta.

CBI, 2019. Exporting Fresh Pomegranates to Europe. https://www.cbi.eu/market-information/fresh-fruit-vegetables/pomegranates/europe. (Accessed 5 April 2021).

Constas, M., Frankenberger, T., Hoddinott, J., 2014. Resilience Measurement Principles: Toward an Agenda for Measurement Design. Resilience Measurement Technical Working Group. FSIN, Roma.

Derajat, A.R., Pamungkas, J., Setiawan, H.T., Hilmy, F., 2020. Pengembangan usaha jamu herbal untuk meningkatkan imunitas tubuh dalam menghadapi pandemi COVID-19. Civitas Ministerium 4 (01).

Djalante, R., Lassa, J., Setiamarga, D., Sudjatma, A., Indrawan, M., Haryanto, B., et al., 2020. Review and analysis of current responses to COVID-19 in Indonesia: period of January to March 2020. Prog. Disaster Sci. 6, 100091. https://doi.org/10.1016/j.pdisas.2020.100091.

Farcas, A.C., Galanakis, C.M., Socaciu, C., Pop, O.L., Tibulca, D., Paucean, A., et al., 2021. Food security during the pandemic and the importance of the bioeconomy in the new era. Sustainability 13 (1). https://doi.org/10.3390/su13010150.

Folke, C., 2006. Resilience: the emergence of a perspective for social–ecological systems analyses. Glob. Environ. Chang. 16 (3), 253–267. https://doi.org/10.1016/j.gloenvcha.2006.04.002.

Galanakis, C.M., 2020. The food systems in the era of the coronavirus (COVID-19) pandemic crisis. Foods 9 (4). https://doi.org/10.3390/foods9040523.

Galanakis, C.M., Rizou, M., Aldawoud, T.M.S., Ucak, I., Rowan, N.J., 2021. Innovations and technology disruptions in the food sector within the COVID-19 pandemic and post-lockdown era. Trends Food Sci. Technol. 110, 193–200. https://doi.org/10.1016/j.tifs.2021.02.002.

Glauber, J., Laborde Debucquet, D., Martin, W., Vos, R., 2020. COVID-19: trade restrictions are worst possible response to safeguard food security. In: IFPRI Book Chapters, pp. 66–68.

Golar, G., Malik, A., Muis, H., Herman, A., Nurudin, N., Lukman, L., 2020. The social-economic impact of COVID-19 pandemic: implications for potential forest degradation. Heliyon 6 (10), e05354. https://doi.org/10.1016/j.heliyon.2020.e05354.

Gunawan, E., Nida, F.S.H., 2020. Peluang dan strategi pengembangan ecommerce produk pertanian merespon dampak pandemic Covid19. In: Suryana, A., Rusastra, I.W., Sudaryanto, T., Pasaribu, S.M. (Eds.), Dampak Pandemic Covid-19: Perspektif Adaptasi dan Resiliensi Social Ekonomi Pertanian. IAARD Press, Bogor.

Ilesanmi, F.F., Ilesanmi, O.S., Afolabi, A.A., 2021. The effects of the COVID-19 pandemic on food losses in the agricultural value chains in Africa: the Nigerian case study. Public Health Pract. (Oxf.) 2, 100087. https://doi.org/10.1016/j.puhip.2021.100087.

Leach, M., MacGregor, H., Scoones, I., Wilkinson, A., 2021. Post-pandemic transformations: how and why COVID-19 requires us to rethink development. World Dev. 138, 105233. https://doi.org/10.1016/j.worlddev.2020.105233.

Mak, T.M.W., Xiong, X., Tsang, D.C.W., Yu, I.K.M., Poon, C.S., 2020. Sustainable food waste management towards circular bioeconomy: policy review, limitations and opportunities. Bioresour. Technol. 297, 122497. https://doi.org/10.1016/j.biortech.2019.122497.

Mangifera, L., Isa, M., Wajdi, M.F., 2018. Faktor-Faktor yang Mempengaruhi Konsumen Dalam Pemilihan Kuliner di Kawasan Wisata Alam Kemuning. Jurnal Manajemen Dayasaing 20 (1).

Martin, A., Markhvida, M., Hallegatte, S., Walsh, B., 2020. Socio-economic impacts of COVID-19 on household consumption and poverty. Econ. Disaster Clim. Chang. 4 (3), 453–479. https://doi.org/10.1007/s41885-020-00070-3.

Masduki, A., 2018. Hidroponik Sebagai Sarana Pemanfaatan Lahan Sempit di Dusun Randubelang, Bangunharjo, Sewon, Bantul. Jurnal Pemberdayaan: Publikasi Hasil Pengabdian Kepada Masyarakat 1 (2), 185–192.

Mockrin, M.H., Locke, D.H., Stewart, S.I., Hammer, R.B., Radeloff, V.C., 2019. Forests, houses, or both? Relationships between land cover, housing characteristics, and resident socioeconomic status across ecoregions. J. Environ. Manag. 234, 464–475. https://doi.org/10.1016/j.jenvman.2018.12.001.

Mohan, M., Rue, H.A., Bajaj, S., Galgamuwa, G.A.P., Adrah, E., Aghai, M.M., et al., 2021. Afforestation, reforestation and new challenges from COVID-19: thirty-three recommendations to support civil society organizations (CSOs). J. Environ. Manag. 287, 112277. https://doi.org/10.1016/j.jenvman.2021.112277.

Morrison, O., 2020. Coronavirus: Consumer Demand for Staple Foods "High but Stabilizing". https://www.foodnavigator.com/Article/2020/04/08/Coronavirus-Consumer-demand-for-staple-foods-high-butstabilising.

Mujiyono, A., & Soenjoto, W.P.P. (2020). Fenomena panic buying dan Scarcity di Masa pandemi Covid 19 Tahun 2020 (Kajian Secara Ekonomi Konvensional dan Syariah). Jurnal Istiqro, 6(2), 126–139, https://doi.org/10.30739/istiqro.v6i2.568.Mujiyono, A., Soenjoto, W.P.P., 2020. Fenomena panic buying dan Scarcity di Masa pandemi Covid 19 Tahun 2020 (Kajian Secara Ekonomi Konvensional dan Syariah). J. Istiqro. 6 (2), 126–139. https://doi.org/10.30739/istiqro.v6i2.568.

Nasrulloh, N., Pramono, A., Amar, M.I., Yulianti, R., 2020. Mengatasi dampak ekonomi pandemi covid-19 bagi masyarakat Kampung kb Gandul Cinere melalui pengolahan jamu herbal. LOGISTA-Jurnal Ilmiah Pengabdian kepada Masyarakat 4 (2), 443–449.

Nicola, M., Alsafi, Z., Sohrabi, C., Kerwan, A., Al-Jabir, A., Iosifidis, C., et al., 2020. The socio-economic implications of the coronavirus pandemic (COVID-19): a review. Int. J. Surg. 78, 185–193. https://doi.org/10.1016/j.ijsu.2020.04.018.

Nursandi, J., 2018. Budidaya ikan dalam ember (budikdamber) dengan aquaponik di lahan sempit. In: Prosiding Seminar Nasional Pengembangan Teknologi Pertanian, Lampung, October 8, 2018. Politeknik Negeri Lampung.

O'Neill, E.A., Rowan, N.J., Fogarty, A.M., 2019. Novel use of the alga Pseudokirchneriella subcapitata, as an early-warning indicator to identify climate change ambiguity in aquatic environments using freshwater finfish farming as a case study. Sci. Total Environ. 692, 209–218. https://doi.org/10.1016/j.scitotenv.2019.07.243.

Pambudi, P.A., 2020. Pandemi Covid-19: refleksi pentingnya optimasi lahan pekarangan sebagai penyokong kemandirian pangan dan kesehatan keluarga. EnviroScienteae 16 (3), 408–423.

Parvin, G.A., Ahsan, R., Rahman, M.H., & Abedin, M.A. (2020). Novel coronavirus (COVID-19) pandemic: the role of printing media in Asian countries. Front. Commun., https://doi.org/10.3389/fcomm.2020.557593.Parvin, G.A., Ahsan, R., Rahman, M.-H., Abedin, M.A., 2020. Novel coronavirus (COVID-19) pandemic: the role of printing media in Asian countries. Front. Commun. https://doi.org/10.3389/fcomm.2020.557593.

Pertiwi, R., Notriawan, D., & Wibowo, R.H. (2020). Pemanfaatan tanaman obat keluarga (TOGA) meningkatkan imunitas tubuh sebagai pencegahan COVID-19. Dharma Raflesia: Jurnal Ilmiah Pengembangan dan Penerapan IPTEKS, 18(2), 110–118.Pertiwi, R., Notriawan, D., Wibowo, R.H., 2020. Pemanfaatan tanaman obat keluarga (TOGA) meningkatkan imunitas tubuh sebagai pencegahan COVID-19. Dharma Raflesia 18 (2), 110–118.

Roy, S., Saha, M., Dhar, B., Pandit, S., Nasrin, R., 2021. Geospatial analysis of COVID-19 lockdown effects on air quality in the South and Southeast Asian region. Sci. Total Environ. 756, 144009. https://doi.org/10.1016/j.scitotenv.2020.144009.

Sagala, S.H., Maifita, Y., Armaita, A., 2020. Hubungan pengetahuan dan sikap masyarakat terhadap Covid-19: a literature review. Menara Medika 3 (1).

Smeru, 2021. Ringkasan Eksekutif: Dampak Sosial Ekonomi Covid-19 Terhadap Rumah Tangga dan Rekomendasi Kebijakan Strategis Untuk Indonesia. Smeru, Prospera, UNDP, Unicef, Jakarta.

Stegmann, P., Londo, M., Junginger, M., 2020. The circular bioeconomy: its elements and role in European bioeconomy clusters. Resour. Conserv. Recycl. Adv. 6, 100029. https://doi.org/10.1016/j.rcrx.2019.100029.

Sunarsih, Septanti, K.S., Suharyono, S., 2020. Perubahan perilaku di tingkat pelaku utama dan pelaku usaha pertanian pada era pandemi covid-19 (Dampak Pandemi Covid-19: Perspektif Adaptasi dan Resiliensi Sosial Ekonomi Pertanian). IAARD Press, Bogor.

Tahar, A., Kennedy, A., Fitzgerald, R.D., Clifford, E., Rowan, N., 2018a. Full water quality monitoring of a traditional flow-through rainbow trout farm. Fishes 3 (3). https://doi.org/10.3390/fishes3030028.

Tahar, A., Kennedy, A.M., Fitzgerald, R.D., Clifford, E., Rowan, N., 2018b. Longitudinal evaluation of the impact of traditional rainbow trout farming on receiving water quality in Ireland. PeerJ 6, e5281.

Ullah, M.A., Moin, A.T., Araf, Y., Bhuiyan, A.R., Griffiths, M.D., Gozal, D., 2020. Potential effects of the COVID-19 pandemic on future birth rate. Front. Public Health 8. https://doi.org/10.3389/fpubh.2020.578438.

Verger, N.B., Urbanowicz, A., Shankland, R., McAloney-Kocaman, K., 2021. Coping in isolation: predictors of individual and household risks and resilience against the COVID-19 pandemic. Soc. Sci. Humanit. Open 3 (1), 100123. https://doi.org/10.1016/j.ssaho.2021.100123.

Wolfslehner, B., Linser, S., Pülzl, H., Bastrup-Birk, A., Camia, A., Marchetti, M., 2016. Forest Bioeconomy—A New Scope for Sustainability Indicators. From Science to Policy, 4, https://doi.org/10.36333/fs04.

Yuswohady, 2020. 30 Prediksi Perilaku Konsumen di New Normal. Yuswohady.com.

CHAPTER 10

The healing and bioeconomic prospect of tropical ecosystems due to the Covid-19 pandemic

Cahyono Agus[a], Alue Dohong[b,c], Dian Yuanita Wulandari[d], Meilania Nugraheni[e,f], Catur Sugiyanto[f], Enggal Primananda[g], and Zahra Zafira Mutiara[h]

[a]Faculty of Forestry, Universitas Gadjah Mada, Yogyakarta, Indonesia
[b]The Environmental Science, Post Graduate Programme at Palangka Raya University, Palangkaraya, Indonesia
[c]Ministry of Environment and Forestry RI, Jakarta, Indonesia
[d]Faculty of Agriculture, Universitas Gadjah Mada, Yogyakarta, Indonesia
[e]Research and Development, Sinarmas Forestry, Riau, Indonesia
[f]Faculty of Economics and Business, Universitas Gadjah Mada, Yogyakarta, Indonesia
[g]Research Center for Plant Conservation and Botanic Gardens, Indonesian Institute of Sciences, Bogor, Indonesia
[h]Deutsche Gesellschaft für Internationale Zusammenarbeit (GIZ), Jakarta, Indonesia

1. Introduction

United Nation (2017) has forecasted that the world population is expected to reach 9.8 billion in 2050, increasing about 25% from the current figure. Average income levels worldwide will be multiples of what they are now, projected to drive the skyrocketing of food demand (Ayaz et al., 2019). Food production will see an increase by 70% to feed the entire world population to keep pace with explosive population growth and continual rise in income (FAO, 2009).

The tropical ecosystems have rich natural resources, mega-biodiversity, and the world's highest net primary productivity. By supporting high temperatures, rainfall, moisture, light intensity, and rapid organic throughout the year, they could facilitate a sustainable and dignified environment and life (Agus et al., 2021). Tropical natural resources have great potential resources but are still of less economic value because they are still under management that is not on nature-based development (Agus, 2019).

Thurston (1969) emphasized that tropical agriculture is a key to solving the world food crisis. Approximately 40% of the world's population is in the tropics, with growth rates much faster than in temperate regions (Mongabay, 2014). Those mentioned above, burgeoning human population growth

Agricultural Bioeconomy
https://doi.org/10.1016/B978-0-323-90569-5.00004-4

have been identified to lead to massive and tragic land conversion in the tropics (Achard, 2002), particularly for the expansion of agricultural purposes (Laurance et al., 2014). A study indicated that one billion ha of additional land would need to be converted to agriculture by 2050 to meet projected demands when global food production doubled. Studies have shown that land conversion often severely impacts the environment (Hao et al., 2017). Hence, it surges the need to produce more and better-quality food and tackle environmental issues (Loures et al., 2020). If agriculture can be achieved more efficiently, the amount of additional land required may be much smaller.

Bio-economy has been recognized as the critical action to establish future food production systems (Wesseler, 2015; Pigford et al., 2018; Kristinsson and Jörundsdóttir, 2019), and been pointed out that innovation and modernization processes are crucial to overcoming the challenges (Pigford et al., 2018). The critical importance of bio-economy development is the need to utilize and optimize any existing technologies to ensure improved food production and food security (Aguilar et al., 2019). Adopting agricultural innovation is indeed a drive to achieving food security in the face of a rise in population growth (Ajayi and Akinbobola, 2019).

The Covid-19 pandemic has resulted in widespread and tragic human and life disasters but has healed the earth's environment. Many countries have taken steps to calm down, soft down, and lockdown, to reduce slowly or drastically to lock down their activities. Observations on the air quality improvement after the Covid-19 lockdown in China show that the air quality index (AQI) experiences a 12.2% larger reduction in the cities with a lockdown policy. Besides, the reduction effect varies with different types of air pollutants (PM2.5, PM10, SO_2, NO_2, and CO decreased by 13.1%, 15.3%, 4%, 3.3%, and 3.3%, respectively (Wang et al., 2020). The pandemic continues to evolve despite the seasonal change, primarily La-Nina extreme climate phenomenon.

The Indonesian government is very concerned about the negative impact of La-Nina that is likely to exacerbate the pandemic situation. Various policies and strategies have been formulated to anticipate the worst. While experiences of coping with La-Nina events might be considerable, the current situation is unusual since response actions must include specific health protocols to prevent the spread of Covid-19 during evacuations.

The implementation of large-scale social restrictions (LSSR) in Indonesia and the policy to reduce industrial operations have been highlighted in many ways due to the contribution in improving environmental quality and reducing pollution. Nevertheless, they have implications for the economic sector, unemployment, and inequality issues due to workers' layoffs, reduction in hours, or pay cut, thereby reducing employees' income. The most affected sectors are the "green sector," consisting of forestry, agriculture, livestock, and plantation. This sector is always inversely proportional to the "ecological economics and environmental economics."

The forestry and agricultural sectors have the highest poverty rate of the week and vulnerable workers (20%), followed by the livestock sector (16%) and the plantation sector (14%) (Halimatussadiah et al., 2020). Essentially, the reduction in production and industrial activities in the green sector, in which most of them are always inversely proportional to the economic aspect, on the contrary, have a good impact on the flora and fauna activities and the quality of the environment in general. This chapter summarizes the strategic issues of the positive effects of the Covid-19 on improving the environmental quality, despite some points are still contradicting one another.

2. Covid-19 and La-Nina phenomenon in Indonesia

2020 was a remarkably tumultuous year for all countries globally, including Indonesia, that the Covid-19 outbreak has heavily impacted since the beginning of the year. The cases rapidly spread across the globe and changed fundamental life structures, including health, social, economy, politics, education, culture, tourism, trade, and transportation (Rupani et al., 2020; Susilawati et al., 2020).

The Agency for Meteorology, Climatology, and Geophysics (BMKG) claimed that La-Nina anomalies were observed to occur in the Pacific Ocean at moderate strength at the end of 2020. The impacts of the La-Nina event with extreme weather include sea-level rise (up to 15 cm), increasing wave height, higher rainfall intensity (up to 40%) than average, natural disasters, and potential damage to coastal agriculture and fisheries (Supriatin and Martono, 2016).

Since the beginning of 2020, many regions worldwide have been experiencing the double hazard of severe flooding and pandemic. Approximately 70 countries experienced flood-related events during the outbreak of Covid-19 (Simonovic et al., 2021). Meteorological and climatological hazards will concur and exacerbate the spread of Covid-19. Of all natural hazards, storms, floods, and droughts show the most severe pandemic impact (Phillips et al., 2020).

According to the data obtained from the Task Force for Covid-19 (Bidang data and Satgas Penanganan Covid-19, 2021), there is a correlation between the impact of the La-Nina event (floods) and the number of positive cases of Covid-19 (Fig. 1). The intensity of floods increased during December-February 2021, which is relatively the same as the increase in the number of positive cases of Covid-19 in Indonesia. The peak of La-Nina occurred in January 2021. Nevertheless, the correlation between these parameters is under investigation, whether floods are the only cause or another contributing factor. A study in Bangladesh revealed that among seven parameters, only two (minimum temperature and average temperature) had a significant relationship with new Covid-19 cases.

The impact of Covid-19 transmission that co-occurs with seasonal hazards, namely the La-Nina event, elevates public anxiety and hinders the government from carrying out mitigation and evacuation measures (Simonovic et al., 2021). These two multi-hazard aspects are crucial and critical since they pose threats to health and the environment in which they live. They change the disaster response strategies, including evacuation procedures, transportation mobility, types of assistance, and volunteers/medical personnel. Therefore, to anticipate and cope with seasonal hazards, it is necessary to design a complex strategy and easy-to-implement policies to balance

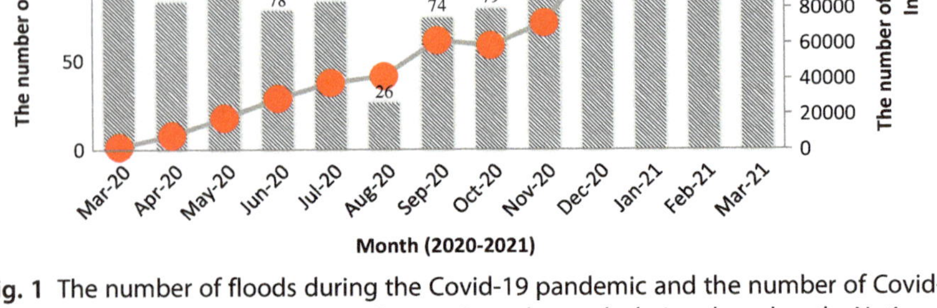

Fig. 1 The number of floods during the Covid-19 pandemic and the number of Covid-19 cases in Indonesia in 2020-2021. *Source: Open data, calculations based on the National Agency for Disaster Management/the BNPB Task Force for Covid-19 (Bidang data & Satgas Penanganan Covid-19, 2021. Analisis Data Covid-19 Indonesia. Retrieved from Indonesia: https://covid19.go.id/p/berita/analisis-data-covid-19-indonesia-update-28-maret-2021).*

Covid-19 transmission with the risk of natural hazard within the scheme of the evacuation-to-recovery process (Quigley et al., 2020).

3. The healing of tropical ecosystem in Indonesia

The Covid-19 pandemic has not only devastated the fundamental sectors such as health, economy, politics, and education but has specific, adverse impacts on the regeneration of tropical forest ecosystems. Adopting Zahawi et al. (2020), the measures to tackle the impacts of the Covid-19 pandemic on the regeneration and sustainability of tropical forests combined with the multi-hazard management priorities will be more complex.

The complexity reflects that many aspects are closely related and that there is a strong interaction between the natural environment, forest, and pandemic. Various diseases such as the Covid-19, ebola, SARS, and bird flu emerge as infectious diseases of animal origin or zoonotic diseases. Data shows that more than 60% of all types of infectious diseases that infect humans are zoonotic types, and ironically about 75% of all infectious diseases are newly emerging diseases (UNEP, 2016; Loh et al., 2015). The Covid-19 pandemic and other pandemics' risks are linked to deforestation and degradation of tropical forest ecosystems. Land-use changes (such as deforestation and forest conversion) are a significant driver of the emergence of new zoonotic diseases and are predicted to increase in the future (Loh et al., 2015).

Biodiversity conservation efforts are necessary to protect humans from the infectious disease since the high biodiversity of the intact forest ecosystem is able to control the spread of such diseases by inhibiting the spread of invasive pathogens. On the contrary, as forests are converted into monoculture plantations, pests, diseases, and pathogens are likely to develop rapidly (Keesing et al., 2010; Pongsiri et al., 2009; United Nations Environment Programme—UNEP, 2016).

Tropical forests are also very rich in diversity of plant species, including medicinal plants used as sources of medicines by communities living around the forest. A study claimed that secondary forests provided 2.7 times more than the number of identified medicinal plants found in primary forests, including herbs, weeds, cultivated, and exotic plants. The finding indicates that despite undergoing threats and habitat disturbances, medicinal plants' availability in secondary forests is still high. Meanwhile, the availability of medicinal plants in primary forests is increasingly scarce (Voeks, 1996).

Human population growth and technological developments accelerate the overexploitation of nature. The threats to tropical forests in Indonesia include forest conversion into oil palm plantation, monoculture for pulp and paper industry, agricultural expansion, open-pit mining, forest loss to residential development, and forest fires (Agus et al., 2020).

Types of pollution produced by industries are air pollution, water pollution, soil pollution, and noise pollution. Air pollution often occurs naturally (forest and land fires) and results from human activities such as the burning of fossil fuels for industries and motor vehicles (motorbikes, cars, planes, ships) (Noor, 2020). Air pollution's primary pollutants include CO_2, CO, SO_2, NOx, CFC, NH_3, volatile organic compounds (VOC), and particulate matter, which cause respiratory tract infections. Urban air pollution and other pollutions may trigger the spread of Covid-19 (Sharma et al., 2021). Indonesia has the world's highest mortality rate due to air pollution after China, India, and Pakistan in 2017 (Rupani et al., 2020).

According to Sharma et al. (2021), the positive effects of the Covid-19 lockdown on nature and the environment include:

a. They were stopping humans from performing all activities outside the house. The government's efforts to force lockdown and self-quarantine at home: Humans are reminded to pay attention to their health by taking a temporary hiatus.

b. Reducing industrial activity, particularly large-scale ones with many employees and small industries: Humans are reminded to stop for a moment from activities that cause pollution and industrial waste.

c. Reducing transportation mobility (planes, cars, motorbikes, etc.): Humans are reminded to stop for a moment from the mobility of motor vehicles which cause air pollution and traffic jams.

d. It was decreasing carbon emission rate: reduction of all fossil burning activities that lead to the temporary absence of pollutant sources, particularly compounds that can destroy the ozone layer.

e. Reducing air pollution and improving air quality level: in some areas, mountains that are commonly unseen due to vehicle fumes become visible, the sky is brighter than before, and satellite images of the atmosphere also show signs of reduced air pollution. In big cities, urban air emission indicators are getting better.

f. Reducing water pollution and improving water quality: by stopping industrial activities, mainly textile factories, the water quality level is better; the implication is that aquatic biota can be relieved without pollution in their habitat. The main rivers of India, such as the Ganga and Yamuna, are cleaner. Aquatic ecosystems usually regenerate, and beaches are cleaner than usual.

3.1 The effects of the Covid-19 pandemic on air quality index (AQI)

Regarding the observation of air pollution in Jakarta, Indonesia, the government implements the LSSR policy in big cities, including Jakarta, which has a very beneficial effect on the environment even in the short-term. This policy encourages individuals to minimize travel activities, decreasing significant air pollution levels. During the LSSR, the concentration of nitrogen dioxide in Jakarta fell by 40% from the level in 2019 (Halimatussadiah et al., 2020).

Air quality issues during the pandemic become the concern of many researchers, including Shareef and Hashmi (2020) on the air quality index (AQI) during the partial lockdown in Pakistan. Urban and industrial areas are the most significant pollutants since the highest pollutant concentrations, namely NO_x dan PM_{10}, are found in commercial and industrial areas. The study reveals that the air quality index during lockdown improved significantly. The PM10 concentration has decreased from to 50%, whereas the SO_2, NO_2, and CO compound decreased by 60–70% during the lockdown period (Shareef and Hashmi, 2020).

3.2 The effects of the Covid-19 pandemic on the terrestrial environment

In tropical countries, apart from reduced air pollution in urban areas as an implication of the LSSR policy, there is also an issue of air pollution caused by the destruction of tropical forests. It is crucial given the recent reports that increased air pollution can increase Covid-19 cases and deaths (Wu et al., 2020). Indonesia is a country with a high rate of forest fires due to anthropogenic activities, including agricultural conversion, timber harvesting, and peatland drainage (Harrison et al., 2020a). Forest and land fires are natural disasters that have disastrous effects, namely the apparent death of living organisms, both plants and animals, biodiversity loss, increased heart and smoke that causes air pollution even to other countries, and increases in greenhouse gas emissions (Page and Hooijer, 2016; Harrison et al., 2020a).

Syaharani and Tavares (2020) suggested the Indonesian commitment to reduce GHG emissions by 29% without international support and 41% with international support by 2030. Primary greenhouse gas reduction targets include forestry, energy, agriculture, industrial sectors, and waste. The most significant emissions are currently generated from the agricultural, forestry, land-use sectors, and energy sector. Therefore, in the National Action Plan

(NAP) on the Greenhouse Gases Emissions Reduction (GHGs) reduction target by 2020, the primary sector target is the Forestry and Peatlands sector, about 1039 $MtCO_2e$ or 87.38% (41% reduction target reducing scenarios) (Dohong et al., 2018a, b). Among the alternatives offered regarding efforts and techniques for effective tropical peatland restoration are small-scale trial activities, canal blocking, seedling transplantation, and seed dispersal. Nevertheless, it is highly recommended to simultaneously implement peatland restoration techniques through meaningful land-use policies and governance reform (Dohong et al., 2018a, b). It has been questioned whether the policy of Greenhouse Gases Emissions Reduction (GHGs) can be adequately achieved if the Covid-19 pandemic is identified as a potential threat to the conservation of tropical peatlands and local communities.

Weisse and Goldman (2021) reported that during the Covid-19 pandemic, Indonesia's tropical forests are relatively safe. Although the global deforestation rate is critical, primary forest loss has decreased for four consecutive years from 2017 to 2020. It is the first time Indonesia is not included in the top three countries based on primary forest loss. The government's efforts to protect forests are attained by temporarily stopping the issuance of new permits for oil palm plantations and imposing a permanent moratorium on converting primary forests and peatlands, social forestry, and agrarian reform.

Forest and land rehabilitation during the pandemic is exceptional since, in addition to protecting forests, it also contributes to improving the community welfare through labor-intensive mangrove planting programs. The rank of Indonesia 2020 will increase significantly to 24 or categorized as a moderate category. Efforts to reduce greenhouse gases are also inseparable from the success of controlling forest and land fires. The Greenhouse effect was successfully reduced by 82.01% in 2019 from 1,352,501 hectares of the burned area to 1.3 million hectares in 2020. Besides, in an attempt to recover the economy due to the pandemic, 39,970 farmers involved in this labor-intensive program.

3.3 The effects of the Covid-19 pandemic on the marine environment

Indonesia is a tropical maritime country with the longest coastline globally, rich in marine resources, fishery, marine, and underwater tourism. During the lockdown period, the Covid-19 pandemic was considered to impact coral reefs in Indonesia positively. Wisuda (2020) reported that the Covid-19 pandemic has heavily impacted the tourism industry, including marine tourism and diving tourism. Ebed de Rosary (2020) reaffirmed that reducing natural tourism activities in Indonesian waters during the Covid-19 pandemic becomes natural healing through a natural recovery process. It has improved coral reef health, coral fish abundance and population, and the presence of other protected reef organisms (Table 1).

Table 1 Significant reasons for pollutant reduction and environmental improvement indicators during the Covid-19 pandemic in Indonesia.

Significant reasons for pollutant reduction	Contradictory aspects	Problem-solving solution	Media sources
1) Tourist reduction 2) Diving activities closure 3) Diving spots closure 4) Reduced waste in tourist destinations 5) Reduced noise pollution (from ships) and tourist facilities 6) Affected bus and hotel industry 7) The presence of marine organisms 8) Improved water quality	1) Termination of workers, layoffs 2) Reduction of community income 3) Closure of fish market and nutrient supply deficiency 4) The exploitation of ornamental fish and excessive illegal fishing 5) Reduced export value of fisheries 6) The high amount of additional medical waste 7) The collapse of the bus and hotel industry	1) The policy on tax incentives for the community 2) Construction of cleanliness facilities at tourist attractions 3) Creativity in tourist destinations with the new usual concept 4) Improvement in the quality of marine processed products	https://www.mongabay.co.id/2020/05/12/hancurnya-industri-wisata-selam-indonesia-di-tengah-wabah-corona/ https://www.ugm.ac.id/id/berita/19576-kondisi-laut-membaik-selama-pandemi-covid-19 https://ekonomi.bisnis.com/read/20200425/12/1232674/-industri-pariwisata-siap-sambut-tren-positif-pasca-covid-19 http://sdgcenter.unpad.ac.id/seminars-and-conferences/tantangan-pencapaian-sdg14-dengan-adanya-pandemi-covid-19/

Coll (2020) demonstrated that the lockdown policy during the Covid-19 pandemic had a significant and positive effect on improving the marine ecosystem's quality. The pandemic allows coastal animals such as seabirds, shoreline birds, elasmobranches, and marine turtles to perform different activities. The emergence, movements, and behaviors of wildlife are the natural response to changes in their habitats due to reduced human activity that regularly disrupts their habitats but do not necessarily multiply their size of population and number (Bates et al., 2020).

4. Bio-economy of nature-based resources management

The sustainability of the whole nature-based resource utilization, encompassing land, water, air, minerals, forests, fisheries, wild flora, and fauna, defined nature-based resource management (Iyyanki et al., 2017). Depletion of nature-based resources might lead to the critical issue of sustainable assessment, which entails and affects the environment and socio-economic aspects (Klinglmair et al., 2013; Van Oers and Guinée, 2016). Bio-economy is the term of utilizing renewable biological resources and/or nature-based resources for needs-producer sectors (agriculture, forestry, fisheries, food industries, etc.) (Pelli et al., 2018) with the basis of sustainability conceptualizations. Bosman and Rotmans (2016) explained that the concept of bio-economy was introduced due to the co-evolution of economic, social, technological, institutional, and ecological developments.

The bio-economy concept application might vary, depending on the livelihoods, social community, and the (natural) resources. Indonesia's biodiversity and net primary production are touted among the highest on earth (Agus, 2018, 2019), although it is more supported by a fast organic cycle than decayed acid soil with low fertility. Excessive exploitation of natural resources in tropical ecosystems has caused significant problems in economic, environmental, and socio-cultural aspects (Agus, 2019; Agus et al., 2019d, 2019e, 2020a). Therefore, those abundant resources will end up as a story to the future generation without a proper management system.

The grand challenge of climate change, population growth, food security, Covid-19 pandemic, and economic crisis caused bio-economy conceptualization is an urgent need for the highest-income-contributing sector such as agriculture and forestry sector. The covid-19 pandemic impacted Indonesia's socio-economy in all industrial sectors, including the agriculture sector. Dwia et al. (2020) stated that the agricultural commodities continued to decline due to the prohibition for farmers to leave their homes so that the crops became poorly maintained and less controlled. The Covid-19 pandemic has a severe impact on the lives and livelihoods of family farmers in Southeast Asia and resulted in increased poverty and hunger in the region. During the pandemic, the volume of food production was reduced by 3.11% or 17.03 million tons due to the decline of agricultural farm labor, which affected 100.77 million people.

The concept of bio-economy might be differently implemented among nations. In Tanzania, where the social community is the key driving force to the bio-economy conceptualization, community-based natural resources management (CBNRM) was introduced (Shanker, 2004). Meanwhile, in Indonesia, Integrated Bio-cycle Management or IBM was introduced as a new paradigm, from extraction to empowerment, from red and green economic concept to blue economic concept (Agus et al., 2020b). The red economy concept refers to the exploitation of natural resources, resulted in environmental and life damage. The green economy concept considers more about environment and health values; however, this concept was less efficient, effective, fund-consuming, and dangerous. The blue economy was developed by Gunter Pauli from ZERI Foundation in 2009 (Agus et al., 2016) through the acceleration of the natural cycle process by empowering land resources (land, water, and minerals), biology (plants, animals, and humans) with economic added value, environment, social culture, technology, sustainable management (Agus et al., 2016). The blue economy concept is thought of as an improvement concept that offers efficient investment, increased innovation, increased funding, job creation, social capital development, and entrepreneurial stimulation (Agus, 2018). It is carried out by utilizing waste and abandoned goods into food, energy, and employment, thereby transforming poverty into sustainable development, and scarcity becomes available. The blue economy has provided new creative and innovative opportunities (Agus et al., 2020b).

Adopting the concept of blue economy and the basis sustainable conceptualization of bio-economy, the Integrated Bio-cycle Farming System (IBFS) is introduced as an alternative system that harmoniously combines the agricultural sector non-agricultural aspects in the management of landscape ecology. IBFS is expected to provide additional benefits for farmers with small, medium, and abundant capital, through recycling organic waste into renewable resources to produce high-value production, such as organic fertilizer (liquid and solid), animal feed, and gas energy (Agus, 2018). As an integrated system, IBFS is developed through several systems, named ICM

(Integrated Crop Management), INM (Integrated Nutrient Management), IMM (Integrated Moisture Management), and Integrated Pest Management (IPM). The whole system requires a synergic collaboration and network between four important pillars: Academic, Business, Community, Government, Media, and Funding (ABCGMF), featuring the economic, environmental, and sociocultural approaches for sustainable development (Agus et al., 2017, 2019e, 2020c). Collaboration between stakeholders is expected to create an innovative and sustainable agro-prenuer atmosphere that positively impacts developing local economies (Agus et al., 2020b).

5. Agricultural supply chain management

Amidst the global health emergency caused by a coronavirus, it is evident that particular attention must be given to the human and animal relationship. The predicament leads to the one health approach in which humans and animals' health and microbes are connected under a shared socio-ecosystem and should be coordinated in a multidisciplinary manner (Ahmad and Hui, 2020; Häsler et al., 2014). Minimizing ecosystems degradation and fragmentations improve the protection and integrity of the links between natural resources management and human well-being. The continuous conversion of natural habitat through unsustainable practices and land-use changes contributes to the loss of animals' natural habitats (Dale et al., 2000).

Agricultural supply chain management consists of multifaceted processes that transform agricultural produce from their raw stage to the consumption phase (Kumar et al., 2017). The supply chain's structure and organization are distinctive depending on commodities and crops; they are also reliant on the demand. The decline in human activity during the global pandemic, including in the agricultural supply chain arena, particularly fishing and mobility, a sharp reduction of carbon emissions is monitored (Liu et al., 2020), and other improving parameters of environmental quality (Somani et al., 2020). These allow for the ecosystems to find their way to their equilibrium state.

The limited anthropogenic activities because of the lockdown restriction might not last in perpetuity. To actively respond to the caution of living in tropical ecosystems, another significant way to establish a lasting agricultural supply chain management environment is to promote deforestation-free commodities. The location of farms and plantations as sources of agricultural commodities directly influences the integrity and health of ecosystems, especially the integrity of tropical forest ecosystems. (Curtis et al., 2018. Public authorities (Pirard et al., 2015), to effectively create a healthy scene for the economy and human well-being to thrive post-pandemic, should orchestrate a global connectedness of green agricultural supply chains.

6. Precision and smart agriculture

Recently the global attention has risen on precision agriculture, a farming practice which has been introduced in the 1980s, along with the pressure of food and sustainability needs and the augmentation of intelligent farm technologies (Wolfert et al., 2017). Precision agriculture is defined as an approach where inputs are used in precise amounts to obtain higher average yields, improve efficiency and profitability while minimizing unintended impacts in the long run (Whelan and Taylor, 2013). It utilizes advanced technologies such as Global Positioning Systems (GPS), Geographic Information Systems (GIS), yield monitors, Variable Rate Technology (VRT), Unmanned Aerial Vehicle (UAV), automatic guidance, sensors, and other spatial management technologies (Lambert et al., 2004). However, precision agriculture has been significantly adopted in developed countries, with few adoptions or related studies in developing countries. Whereas the benefits of precision agriculture can be summarized as (i) increase efficient use of inputs usage; (ii) reduce risk of environmental hazard due to excessive application of inputs, i.e., chemical fertilizer; and (iii) improve process control (Cook and Bramley, 1998).

Precision agriculture has been distinguished to be successful in making precise management decisions in different cropping systems and enhancing crop production while minimizing environmental impact throughout countries (Koch and Khosla, 2003; Cisternas et al., 2020). In conventional agriculture, crops are normally managed based on soil, nutrients, moisture, weeds, and insect conditions, which resulted in numerous over-application or under-applications of pesticides, irrigation, fertilizers, and other treatments. On the harvesting aspect, monitoring and mapping yields enable farmers to obtain helpful information on spatial and temporal variability in crop yields, helping farmers in the decision-making process (Data Dragon EU, 2020).

Precision agriculture in the husbandry sector or known as precision livestock farming, has also awakened global attention. Noting that the world demand for meat will increase by at least 40% by 2030 (FAO, 2009), Berckmans (2017) proposed precision livestock farming to be less use of the feeder, less manure higher productivity. Banhazi et al. (2012) recorded the adoption of precision livestock farming, including precision feeding, automated livestock monitoring, weight management via camera sensor, and virtual fencing, is increasing in several countries.

Meanwhile, precision agriculture in fisheries, yet still limited, but is constantly growing (Christiani et al., 2019). The future development of precision fisheries will utilize sensors on satellites that can detect sea temperature, turbidity, topography, winds, and other variables; radar can collect information even in worse weather; airborne and floating drones can provide additional information about oceanographic conditions. The use of these IoT in fisheries could address uncertainty around catches and help manage risks.

Forestry has lagged behind most other agriculture sectors in adopting precision agriculture because it is not simply the adoption of digital technologies. Forestry also involves a paradigm shift, from a highly manual and analog system with broad-brush management prescriptions to a system with digital data capture and planning, granular management prescriptions, and tight operational control. The goal of precision forestry is to deploy high-resolution data to support site-specific tactical and operational decision-making, to allow for highly repeatable measurements, actions, and processes to grow and harvest trees, as well as to protect and enhance biodiversity, wildlife habitat, and other environmental resources. The current implementation of precision forestry but still limited is the uses of UAV, laser scanning (lidar), and soil sensors (Holopainen et al., 2014). These technologies have proven the potential to significantly improve forest-management results, including higher forest productivity, lower operational costs, and improved planning effectiveness.

Despite high-profile reports above, there is a perception that the adoption of precision agriculture has been slow. The adoptions and developments of precision agriculture were predominantly instigated in developed countries. The US is a leading country in adopting precision agriculture for crop production, with about 90% of the crop yield monitors in the world have been operating in the US. Furthermore, the adoption rate of automatic guidance technology in crop farms in the US reached about 60–80%. In Europe, the adoption of precision agriculture for crop production reached 70%, with greater utilization in fertilizing and spraying (Armagan, 2016). Europe also has been dramatically developing precision livestock farming, with examples of commercial adoption include the use of robotics in dairying, measurement of water usage, egg counting, bird weighing, computerized feed systems, automated disease detection, growth measurement, and real-time production site data capture (Guarino et al., 2008).

In Asia Pacific countries, Mordor Intelligence (2016) reported that China, Japan, India, and Australia lead in the adoption of precision agriculture technologies. For instance, an international agribusiness forum, Agribusiness Global, reported that in 2020, more than drones (UAV) sprayed 80% of rice grown in the Northeast part of China and has already repressed tractors and manual sprayers. Another adoption is in India, which FAO (2012) recorded the declining water table and degrading soil health have been distinguished as their primary concerns. One of the precision agriculture adoption under a program "Laser Land Leveling" has indicated a wide range of benefits such as saved water to the tune of 25–30% and time by 30%, also improved the productivity by 10–15% in the country (ADB, 2012). Meanwhile, in Australia, the range of precision agriculture adoptions has increased year by year, predominantly in crop, livestock farming, and fisheries (Bramley and Trengove, 2013).

Despite the rising global attention on precision agriculture, unfortunately, Indonesia is mainly still undergoing partial research. An irony since Indonesia is known as the emerald of the equator, which has high natural resource values and has excellent potential (Agus, 2019). To date, Indonesian farmers still use conventional farming systems and rarely put the environment at risk. Therefore, Agus et al. (2020) emphasized an immense challenge that needs to be addressed by Indonesia in developing a sustainable farming system that covers all aspects: economy, environment, and socioculture.

FAO (2018) has considered bio-economy plays an essential role in developing countries, particularly those in the tropics (Thurston, 1969), given the importance of their agricultural sector and the use of biomass to produce energy, medicinal plants, and as a building material, in addition to food, feed, and fiber. More specifically, because tropical agriculture management is still relatively traditional, less effective, and efficient, causing the productivity is lower than in temperate regions.

Precision agriculture has become one of the leading solutions to combat food insecurity issues due to the Covid-19 pandemic. The World Bank (2020) notes that as the Covid-19 crisis unfolded, it hits food production, disrupts food supply chains, creates a loss of income, and causes food insecurity in many countries. With current circumstances and regulation under the Covid-19 situation, precision agriculture is remarkably possible to monitor farms while not being physically present through automation, minimizing the need to contact other people, which is crucial during these times. The Covid-19 crisis is set to accelerate the agricultural sector's push for utilizing IoT technologies more widely. Farmers have had their regular business routines wholly disrupted, and digital technologies have proven to be a lifeline during this crisis, making agriculture and food security more resilient than ever.

The emerging technologies of the Internet of Things (IoT) are considered to play a significant role in various applications of the agriculture sector, including providing solutions of many traditional farming issues, i.e., drought response, yield optimization, land suitability, irrigation, and pest control. The IoT application to agriculture is often referred to as Smart Agriculture; hereafter will be mentioned as SA (Bacco et al., 2019). SA is the concept of innovative farming using the latest technologies to increase agricultural products' quantity and quality. Extensive research has been conducted on the current SA implementation, with one of the most widespread adoptions is precision agriculture (Benke and Tomkins, 2017; Shamshiri et al., 2018).

Precision agriculture is one of the sustainable agriculture practices and one of the pathways to achieve the bio-economy. To implement and use these innovations, not only do related stakeholders need to make financial investments, but they also need to invest in the development of new skills, especially for farmers. The sophistication of technology and the compatibility of emerging technologies with current processes and existing equipment are further considerations in the decision-making process. The challenge of adopting a new system like the bio-economy is great (Agus et al., 2020), but it is also clear. It is new wisdom of managing the living resources economically and sustainably, even during the worst crisis.

7. Conclusions

Tropical ecosystems have a high and optimal temperature, sunlight intensity, rainfall, and organic cycles throughout the year, so they have the world's highest growth rate and biodiversity. The Covid-19 pandemic has caused a widespread, tragic, human, and life disaster and healing the earth's environment. This healing is far more dramatic than any systematic anthropogenic effort to reduce the impact of emissions. An integrated organic-cycle cultivation system based on the universe's nature could have superior bio-economic prospects in managing tropical natural resources cured by Covid-19.

References

Achard, F., 2002. Determination of deforestation rates of the world's humid tropical forests. Science 297 (5583), 999–1002. https://doi.org/10.1126/science.1070656.

ADB, 2012. Developments in Precision Agriculture Use in Asia. Asian Development Bank (ADB) Knowledge Event Repository. https://events.development.asia/materials/20120412/developments-precision-agriculture-use-asia.

Aguilar, A., Twardowski, T., Wohlgemuth, R., 2019. Bioeconomy for sustainable development. Biotechnol. J. 14 (8), 1800638. https://doi.org/10.1002/biot.201800638.

Agus, C., 2018. Development of blue revolution through integrated bio-cycles system on tropical natural resources management. In: Leal Filho, W., Pociovalis, T.D., Borges de Brito, P., Borges de Lima, I. (Eds.), Towards a Sustainable Bioeconomy: Principles, Challenges and Perspectives, World Sustainability Series. Springer, Cham, pp. 155–172.

Agus, C., 2019. Integrated bio-cycle system for sustainable and productive tropical natural resource management in Indonesia. In: Singh, H.B. (Ed.), Bioeconomy for Sustainable Development. Springer-Nature, Singapore, pp. 201–216, https://doi.org/10.1007/978-981-13-9431-7.

Agus, C., Putra, P.B., Faridah, E., Wulandari, D., Napitupulu, R.R.P., 2016. Organic carbon stock and their dynamics in rehabilitation ecosystem areas of post open coal mining at tropical region. Procedia Eng. 159, 329–337.

Agus, C., Wulandari, D., Primananda, E., Hendryan, A., Harianja, V., 2017. The role of soil amendment on tropical post tin mining area in Bangka Island Indonesia for dignified and sustainable environment and life. IOP Conf. Ser. Earth Environ. Sci. 83, 012030.

Agus, C., Ilfana, R., Azmi, F.F., Fahreza, A.I., Widiyatno, Wulandari, D., Purwanto, Harun, M.K., Yuwati, T.W., 2019d. The effect of tropical peat land-use changes on plant diversity and soil properties. Int. J. Environ. Sci. Technol. https://doi.org/10.1007/s13762-019-02579-x.

Agus, C., Wulandari, D., Cahyanti, P.A.B., Bantara, I., Hutahaean, B.P., Lestari, T., 2019e. Environmental site engineering and integrated bio-cycles management for rehabilitation of degraded tin mining land in tropical ecosystem. In: IOP Conf. Series: Earth and Environmental Science. 2009 Earth and Environment Sciences, p. 012013.

Agus, C., Primananda, E., Nufus, M., 2020a. Integrated bio-cycle system for rehabilitation of open-pit coal mining areas in tropical ecosystems. In: Filho, W.L., de Brito, P.R.B., Frankenberger, F. (Eds.), World Sustainability Series International Business, Trade and Institutional Sustainability. Springer Nature Switzerland AG, pp. 515–528, https://doi.org/10.1007/978-3-030-26759-9_33. Retrieved from.

Agus, C., Meilania, N., Ambar, P., Margaretha, A.W., NurAini, I.H., Catur, S., Primananda, E., 2020b. Tropical biological natural resource management through integrated bio-cycles farming system. In: Venkatramanan, V., et al. (Eds.), Sustainable Bioeconomy., https://doi.org/10.1007/978-981-15-7321-7_10.

Agus, C., Cahyanti, P.A.B., Widodo, B., Yulia, Y., Rochmiyati, S., 2020c. Cultural-based education of Tamansiswa as a locomotive of Indonesian education system. In: Leal Filho, W., et al. (Eds.), Universities as Living Labs for Sustainable Development, World Sustainability Series. Springer, Cham, pp. 471–486. https://www.springer.com/gp/book/9783030156039.

Agus, C., Nugraheni, M., Pertiwiningrum, A., Wuri, M.A., Hasanah, N.A.I., Sugiyanto, C., Primananda, E., 2021. Tropical biological natural resource management through integrated bio-cycles farming system. In: Venkatramanan, V., Shachi, S., Ram, P. (Eds.), Sustainable Bioeconomy Pathways to Sustainable Development Goals. Springer Nature Singapore, pp. 209–238, https://doi.org/10.1007/978-981-15-7321-7. Retrieved from.

Ahmad, T., Hui, J., 2020. One health approach and coronavirus disease 2019. Hum. Vaccines Immunotherap. 16 (4), 931. https://doi.org/10.1080/21645515.2020.1732168.

Ajayi, O.F., Akinbobola, T.O., 2019. Links Among Bioeconomy, Food Security, and Sustainable Development Goals in Nigeria. Encyclopedia of the UN Sustainable Development Goals, pp. 1–11, https://doi.org/10.1007/978-3-319-69626-3_106-1.

Armagan, Z.E., 2016. Global trends in agriculture and technological solutions. In: Fifth World Summit on Agriculture Machinery, 21 January 2016, Istanbul, Turkey.

Ayaz, M., Ammad-Uddin, M., Sharif, Z., Mansour, A., Aggoune, E.-H.M., 2019. Internet-of-things (IoT)-based smart agriculture: toward making the fields talk. IEEE Access 7, 129551–129583. https://doi.org/10.1109/access.2019.2932609.

Bacco, M., Barsocchi, P., Ferro, E., Gotta, A., Ruggeri, M., 2019. The digitisation of agriculture: a survey of research activities on smart farming. Array 3–4, 100009. https://doi.org/10.1016/j.array.2019.100009.

Banhazi, T.M., Lehr, H., Black, J.L., Crabtree, H., Schofield, P., Tscharke, M., 2012. Precision livestock farming: an international review of scientific and commercial aspects. Int. J. Agric. Biol. Eng. 5 (3), 1. https://doi.orh/10.3965/j.ijabe.20120503.00.

Bates, A.E., Primack, R.B., Moraga, P., Duarte, C.M., 2020. COVID-19 pandemic and associated lockdown as a "Global Human Confinement Experiment" to investigate biodiversity conservation. Biol. Conserv. 248, 108665. Retrieved from https://doi.org/10.1016/j.biocon.2020.108665.

Benke, K., Tomkins, B., 2017. Future food-production systems: vertical farming and controlled-environment agriculture. Sustain. Sci. Practice Policy 13 (1), 13–26. https://doi.org/10.1080/15487733.2017.1394054.

Berckmans, D., 2017. General introduction to precision livestock farming. Anim. Front. 7 (1), 6–11. https://doi.org/10.2527/af.2017.0102.

Bidang data & Satgas Penanganan Covid-19, 2021. Analisis Data Covid-19 Indonesia. Retrieved from Indonesia: https://covid19.go.id/p/berita/analisis-data-covid-19-indonesia-update-28-maret-2021.

Bosman, R., Rotmans, J., 2016. Transition governance towards a bioeconomy: a comparison of Finland and The Netherlands. Sustainability 8, 1017.

Bramley, R., Trengove, S., 2013. Precision agriculture in Australia: present status and recent developments. Engenharia Agrícola 33 (3), 575–588. https://doi.org/10.1590/s0100-69162013000300014.

Christiani, P., Claes, J., Sandnes, E., Stevens, A., 2019. Precision Fisheries: Navigating a Sea of Troubles With Advanced Analytics. https://www.mckinsey.com/industries/agriculture/our-insights/precision-fisheries-navigating-a-sea-of-troubles-with-advanced-analytics.

Cisternas, I., Velásquez, I., Caro, A., Rodríguez, A., 2020. Systematic literature review of implementations of precision agriculture. Comput. Electron. Agric. 176, 105626. https://doi.org/10.1016/j.compag.2020.105626.

Coll, M., 2020. Environmental effects of the COVID-19 pandemic from a (marine) ecological perspective. Ethics Sci. Environ. Pol. 20, 42–55. Retrieved from https://doi.org/10.3354/ESEP00192.

Cook, S.E., Bramley, R.G.V., 1998. Precision agriculture—opportunities, benefits and pitfalls of site-specific crop management in Australia. Aust. J. Exp. Agric. 38 (7), 753. https://doi.org/10.1071/ea97156.

Curtis, P.G., Slay, C.M., Harris, N.L., Tyukavina, A., Hansen, M.C., 2018. Classifying drivers of global forest loss. Science 361 (6407), 1108–1111. https://doi.org/10.1126/science.aau3445.

Dale, V.H., Brown, S., Haeuser, R.A., Hobbs, N.T., Huntly, N., Naiman, R.J., Valone, T.J., 2000. Ecological principles and guidelines for managing the use of land sup> 1. Ecol. Appl. 10 (3), 639–670. https://doi.org/10.1890/1051-0761(2000)010[0639:EPAGFM]2.0.CO;2.

Data Dragon EU, 2020. The Benefits of Yield Monitoring and Mapping. Data Dragon EU. https://datadragon.eu/2020/08/03/the-benefits-of-yield-monitoring-and-mapping/.

Dohong, A., Abdul Aziz, A., Dargusch, P., 2018a. A review of techniques for effective tropical peatland restoration. Wetlands 38 (2), 275–292. Retrieved from https://doi.org/10.1007/s13157-018-1017-6.

Dohong, A., Aziz, A.A., Dargusch, P., 2018b. Carbon emissions from oil palm development on deep peat soil in central Kalimantan Indonesia. Anthropocene 22, 31–39. Retrieved from https://doi.org/10.1016/j.ancene.2018.04.004.

Dwia, A.T.P., Andi, A.U., Suwandi, S., Sudarmo, Seniwati, 2020. The economic impact of COVID-19 outbreak on the agriculture sector. Int. J. Agric. Syst. 8 (1). https://doi.org/10.20956/ijas.v8i1.2337.

Ebed de Rosary, 2020. Begini Dampak Pandemi Bagi Masyarakat di Kawasan Konservasi Perairan Indonesia Timur. Retrieved from https://www.mongabay.co.id/2020/05/07/begini-dampak-pandemi-bagi-masyarakat-di-kawasan-konservasi-perairan-indonesia-timur/.

FAO, 2009. How to Feed the World in 2050. FAO, Rome. www.fao.org/fileadmin/templates/wsfs/docs/expert_paper/How_to_Feed_the_World_in_2050.pdf.

FAO, 2012. The State of the World's Land and Water Resources for Food and Agriculture. http://www.fao.org/3/i1688e/i1688e05.pdf.

FAO, 2018. Assessing the Contribution of Bio-economy To Countries' Economy. http://www.fao.org/3/I9580EN/i9580en.pdf.

Guarino, M., Jans, P., Costa, A., Aerts, J.-M., Berckmans, D., 2008. Field test of algorithm for automatic cough detection in pig houses. Comput. Electron. Agric. 62 (1), 22–28. https://doi.org/10.1016/j.compag.2007.08.016.

Halimatussadiah, A., Widyasanti, A.A., Damayanti, A., Verico, K., Qibthiyyah, R.M., Kurniawan, R., et al., 2020. In: Halimatussadiah, A., Widyasanti, A.A., Damayanti, A., Verico, K., Qibthiyyah, R.M., Kurniawan, R., et al. (Eds.), Thinking Ahead : Indonesia' s Agenda on Sustainable Recovery From COVID-19 Pandemic. Institute for Economic and Social Research Faculty of Economics and Business, Universitas Indonesia (LPEM FEB UI) and Ministry of National Development Planning/National Development Planning Agency (BAPPENAS), Jakarta.

Hao, R., Yu, D., Liu, Y., Liu, Y., Qiao, J., Wang, X., Du, J., 2017. Impacts of changes in climate and landscape pattern on ecosystem services. Sci. Total Environ. 579, 718–728. https://doi.org/10.1016/j.scitotenv.2016.11.036.

Harrison, M.E., Ottay, J.B., D'Arcy, L.J., Cheyne, S.M., Anggodo, Belcher, C., et al., 2020a. Tropical forest and peatland conservation in Indonesia: challenges and directions. People Nat. 2 (1), 4–28. Retrieved from https://doi.org/10.1002/pan3.10060.

Häsler, B., Cornelsen, L., Bennani, H., Rushton, J., 2014. A review of the metrics for One Health benefits. Rev. Scient. Tech. Off. Int. Des. Epizooties 33 (2), 453–464.

Holopainen, M., Vastaranta, M., Hyyppä, J., 2014. Outlook for the next generation's precision forestry in Finland. Forests 5 (7), 1682–1694. https://doi.org/10.3390/f5071682.

Iyyanki, V., Muralikrishna, Manickam, V., 2017. Chapter Three—Natural Resource Management and Biodiversity Conservation., https://doi.org/10.1016/B978-0-12-811989-1.00003-8.

Keesing, F., Belden, L.K., Daszak, P., Dobson, A., Harvell, C.D., Holt, R.D., et al., 2010. Impacts of biodiversity on the emergence and transmission of infectious diseases. Nature 468 (7324), 647–652. Retrieved from https://doi.org/10.1038/nature09575.

Klinglmair, M., Sala, S., Brandão, M., 2013. Assessing resource depletion in LCA: a review of methods and methodological issues. Int. J. Life Cycle Assess., 1–13. https://doi.org/10.1007/s11367-013-0650-9.

Koch, B., Khosla, R., 2003. The role of precision agriculture in cropping systems. J. Crop. Prod. 9 (1–2), 361–381. https://doi.org/10.1300/j144v09n01_02.

Kristinsson, H.G., Jörundsdóttir, H.Ó., 2019. Food in the bioeconomy. Trends Food Sci. Technol. 84, 4–6. https://doi.org/10.1016/j.tifs.2018.10.011.

Kumar, C., Murugaiyan, P., Madanmohan, G., 2017. Agri-food supply chain management: literature review. Intell. Inf. Manag. 9, 68–96. https://doi.org/10.2139/ssrn.3093241.

Lambert, D.M., Lowenberg-DeBoer, J., Griffin, T.W., Peone, J., Payne, T., Daberkow, S.-G., 2004. Adoption, Profitability, and Making Better Use of Precision Farming Data. RePEc. https://ideas.repec.org/cgi-bin/htsearch?q=ADOPTION%2C+PROFITABILITY%2C+AND+MAKING+BETTER+USE+OF+PRECISION+FARMING+DATA.

Laurance, W.F., Sayer, J., Cassman, K.G., 2014. Agricultural expansion and its impacts on tropical nature. Trends Ecol. Evol. 29 (2), 107–116. https://doi.org/10.1016/j.tree.2013.12.001.

Liu, Z., Ciais, P., Deng, Z., Lei, R., Davis, S.J., Feng, S., et al., Schellnhuber, H.J., 2020. Near-real-time monitoring of global CO_2 emissions reveals the effects of the COVID-19 pandemic. Nat. Commun. 11 (1), 1–12. https://doi.org/10.1038/s41467-020-18922-7.

Loh, E.H., Zambrana-Torrelio, C., Olival, K.J., Bogich, T.L., Johnson, C.K., Mazet, J.A.K., et al., 2015. Targeting transmission pathways for emerging zoonotic disease surveillance and control. Vector-Borne Zoonotic Dis. 15 (7), 432–437. Retrieved from https://doi.org/10.1089/vbz.2013.1563.

Loures, L., Chamizo, A., Ferreira, P., Loures, A., Castanho, R., Panagopoulos, T., 2020. Assessing the effectiveness of precision agriculture management systems in mediterranean small farms. Sustainability 12 (9), 3765. https://doi.org/10.3390/su12093765.

Mongabay, 2014. Human Population Growth. https://rainforests.mongabay.com/08population.html.

Noor, T., 2020. Pandemic Covid-19 and its effects that limiting air pollution: a review. J. Res. Lepidoptera 51 (2), 963–974. Retrieved from https://doi.org/10.36872/lepi/v51i2/301150.

Page, S.E., & Hooijer, A. (2016). In the line of fire: the peatlands of Southeast Asia. Philos. Trans. R. Soc. B: Biol. Sci., 371(1696). Retrieved from https://doi.org/10.1098/rstb.2015.0176.

Pelli, P., Kangas, J., Pykäläinen, J., 2018. Service-based bioeconomy-multilevel perspective to assess the evolving bioeconomy with a service lens. In: Towards a Sustainable Bioeconomy: Principles, Challenges and Perspectives, World Sustainability Series.

Phillips, C.A., Caldas, A., Cleetus, R., Dahl, K.A., Declet-Barreto, J., Licker, R., et al., 2020. Compound climate risks in the COVID-19 pandemic. Nat. Clim. Change 10 (7), 586–588. Retrieved from https://doi.org/10.1038/s41558-020-0804-2.

Pigford, A.A.E., Hickey, G.M., Klerkx, L., 2018. Beyond agricultural innovation systems? Exploring an agricultural innovation ecosystems approach for niche design and development in sustainability transitions. Agric. Syst. 164, 116–121. https://doi.org/10.1016/j.agsy.2018.04.007.

Pirard, R., Fishman, A., Gnych, S., Obidzinski, K., Pacheco, P., 2015. Deforestation-Free Commitments: The Challenge of Implementation–An Application to Indonesia. vol. 181 CIFOR, https://doi.org/10.17528/cifor/005572.

Pongsiri, M.J., Roman, J., Ezenwa, V.O., Goldberg, T.L., Koren, H.S., Newbold, S.C., et al., 2009. Biodiversity loss affects global disease ecology. Bioscience 59 (11), 945–954. Retrieved from https://doi.org/10.1525/bio.2009.59.11.6.

Quigley, M.C., Attanayake, J., King, A., Prideaux, F., 2020. A multi-hazards earth science perspective on the COVID-19 pandemic: the potential for concurrent and cascading crises. Environ. Syst. Decis. 40 (2), 199–215. Retrieved from https://doi.org/10.1007/s10669-020-09772-1.

Rupani, P.F., Nilashi, M., Abumalloh, R.A., Asadi, S., Samad, S., Wang, S., 2020. Coronavirus pandemic (COVID-19) and its natural environmental impacts. Int. J. Environ. Sci. Technol. 17 (11), 4655–4666. Retrieved from https://doi.org/10.1007/s13762-020-02910-x.

Shamshiri, R., Kalantari, F., Ting, K.C., Thorp, K.R., Hameed, I.A., Weltzien, C., Ahmad, D., Shad, Z.M., 2018. Advances in greenhouse automation and controlled

environment agriculture: a transition to plant factories and urban agriculture. Int. J. Agric. Biol. Eng. 11 (1), 1–22. https://doi.org/10.25165/j.ijabe.20181101.3210.
Shanker, R.J., 2004. Community-based natural resources management—experiences from Africa. Short notes. In: Conference: National Workshop on Watershed Management. https://www.researchgate.net/publication/245031316.
Shareef, A., Hashmi, D., 2020. Impacts of COVID-19 pandemic on Air Quality Index (AQI) during partial lockdown in Karachi Pakistan. J. Health Environ. Res. 6 (3), 93. Retrieved from https://doi.org/10.11648/j.jher.20200603.17.
Sharma, R.S., Panthari, D., Semwal, S., Uniyal, T., 2021. Aftermath of industrial pollution, post COVID-19 quarantine on environment. In: Chakraborty, C., Roy, S., Sharma, S., Tran, T.A. (Eds.), The Impact of the COVID-19 Pandemic on Green Societies (The Impact). Springer International Publishing, pp. 141–167. Retrieved from https://doi.org/10.1007/978-3-030-66490-9.
Simonovic, S.P., Kundzewicz, Z.W., Wright, N., 2021. Floods and the COVID-19 pandemic—a new double hazard problem. Wiley Interdiscip. Rev. Water 8 (2), 1–18. Retrieved from https://doi.org/10.1002/wat2.1509.
Somani, M., Srivastava, A.N., Gummadivalli, S.K., Sharma, A., 2020. Indirect implications of COVID-19 towards sustainable environment: an investigation in Indian context. Bioresour. Technol. Rep. 11, 100491. https://doi.org/10.1016/j.biteb.2020.100491.
Supriatin, L.S., Martono, M., 2016. Impacts of climate change (El Nino, La Nina, and Sea Level) on the coastal area of cilacap regency. Forum Geografi 30 (2), 106. Retrieved from https://doi.org/10.23917/forgeo.v30i2.2449.
Susilawati, S., Falefi, R., Purwoko, A., 2020. Impact of COVID-19's pandemic on the economy of Indonesia. Budapest Int. Res. Critics Inst. (BIRCI-J.) Human. Soc. Sci. 3 (2), 1147–1156. Retrieved from https://doi.org/10.33258/birci.v3i2.954.
Syaharani, S., Tavares, M.A., 2020. Nasib Target Emisi Indonesia: Pelemahan Instrumen Lingkungan Hidup di Era Pemulihan Ekonomi Nasional. J. Hukum Lingkungan Indonesia 7 (1), 1–27. Retrieved from https://doi.org/10.38011/jhli.v7i1.212.
Thurston, H., 1969. Tropical agriculture, a key to the world food crises. Bioscience 19 (1), 29–34. https://doi.org/10.2307/1294626.
United Nation (UN), 2017. World Population Prospects. https://www.un.org/development/desa/en/news/population/world-population-prospects-2017.html.
United Nations Environment Programme—UNEP, 2016. UNEP Frontiers 2016 Report. Emerging Issues of Environmental Concern. United Nations Environment Programme. Retrieved from https://web.unep.org/frontiers/sites/unep.org.frontiers/files/documents/unep_frontiers_2016.pdf.
Van Oers, L., Guinée, J., 2016. The abiotic depletion potential: background, updates, and future. Resources 5, 1–12. https://doi.org/10.3390/resources5010016.
Voeks, R.A., 1996. Tropical forest healers and habitat preference. Econ. Bot. 50 (4), 381–400. Retrieved from https://doi.org/10.1007/BF02866520.
Wang, M., Liu, F., Zheng, M., 2020. Air quality improvement from COVID-19 lockdown: evidence from China. Air Qual. Atmos. Health 2. Retrieved from https://doi.org/10.1007/s11869-020-00963-y.
Weisse, M., Goldman, E., 2021. Kerusakan Hutan Hujan Primer Meningkat Sebesar 12% dari Tahun 2019 hingga Tahun 2020. Retrieved from https://wri-indonesia.org/id/blog/kerusakan-hutan-hujan-primer-meningkat-sebesar-12-dari-tahun-2019-hingga-tahun-2020.
Wesseler, 2015. Agriculture in the Bioeconomy Economics and Policies. Wageningen University. https://www.researchgate.net/profile/Justus_Wesseler/publication/280723637_Agriculture_in_the_Bioeconomy/links/55c3158f08aebc967defecc9/Agriculture-in-the-Bioeconomy.pdf.

Whelan, B., Taylor, J., 2013. Precision Agriculture for Grain Production Systems (Illustrated ed.). CSIRO Publishing.
Wisuda, A., 2020. Sisi Positif Wabah Corona Bagi Terumbu Karang Indonesia. Retrieved from https://www.mongabay.co.id/2020/06/01/sisi-positif-wabah-corona-bagi-terumbu-karang-indonesia/.
Wolfert, S., Ge, L., Verdouw, C., Bogaardt, M.-J., 2017. Big data in smart farming—a review. Agric. Syst. 153, 69–80. https://doi.org/10.1016/j.agsy.2017.01.023.
World Bank (2020). Food Security and COVID-19. https://www.worldbank.org/en/topic/agriculture/brief/food-security-and-covid-19\.
Wu, X., Nethery, R.C., Sabath, M.B., Braun, D., Dominici, F., 2020. Exposure to air pollution and COVID-19 mortality in the United States: a nationwide cross-sectional study Xiao. MedRxiv 165, 1–13.
Zahawi, R.A., Reid, J.L., Fagan, M.E., 2020. Potential impacts of COVID-19 on tropical forest recovery. Biotropica 52 (5), 803–807. Retrieved from https://doi.org/10.1111/btp.12851.

Further reading

Agus, C., Azmi, F.F., Ilfana, Z.R., Widiyatno, Wulandari, D., Harun, M.K., Yuwati, T.W., 2018. The impact of forest fire on their biodiversity and soil characteristic on tropical peatland. In: Filho, W.L. (Ed.), Climate Change Management Book Series, Handbook of Climate Change and Biodiversity. Springer, Zug, Switzerland, pp. 287–303, https://doi.org/10.1007/978-3-319-98681-4_18.
Agus, C., Primananda, E., Faridah, E., Wulandari, D., Lestari, T., 2019a. Role of arbuscular mycorrhizal fungi and *Pongamia pinnata* for revegetation of tropical open-pit coal mining soils. Int. J. Environ. Sci. Technol. 15 (11), 1–11. https://doi.org/10.1007/s13762-018-1983-5.
Agus, C., Anggari, P.E., Faridah, E., Wulandari, D., Suginingsih, Ratnaningrum, Y.W.N., Winarni, W.W., Lestari, T., Sunaryo, Y., Pertiwingrum, A., Napitupulu, R.P., Primananda, E., 2019b. Role of watering interval and organic pot on the growth of exotic fast-growing species on coal mining media. IOP Conf. Ser. Earth Environ. Sci. 308, 012055. https://iopscience.iop.org/article/10.1088/1755-1315/308/1/012055.
Agus, C., Hendryan, A., Harianja, V., Faridah, E., Atmanto, W.D., Cahyanti, P.A.B., Wulandari, D., Pertiwiningrum, A., Suhartanto, B., Bantara, I., Hutahaean, B.P., Suparto, B., Lestari, T., 2019c. The role of soil organic amendment of humus paramagnetic and compost for remediation of post tin mining tailing media and their growth of *Reutealis trisperma* seedling. Int. J. Smart Grid Clean Energy 8 (5), 556–561. https://doi.org/10.12720/sgce.8.5.556-561.
Harrison, M.E., Wijedasa, L.S., Cole, L.E.S., Cheyne, S.M., Choiruzzad, S.A.B., Chua, L., et al., 2020b. Tropical peatlands and their conservation are important in the context of COVID-19 and potential future (zoonotic) disease pandemics. PeerJ 8, 1–43. Retrieved from https://doi.org/10.7717/peerj.10283.
Meijer, M., Rizwanul Fattah, I.M., Saiful Islam, A.B.M., Uddin, M.N., Ashrafur Rahman, S.M., Chowdhury, M.A., et al., 2020. Relationship between weather variables and new daily covid-19 cases in Dhaka, Bangladesh. Sustain. (Switzerl.) 12 (20), 1–10. Retrieved from https://doi.org/10.3390/su12208319.

CHAPTER 11

Towards the fifth innovation era in agricultural innovation

Per Frankelius and Matilda von Rosen
Agtech 2030, Department of Management and Engineering, Linköping University, Linköping, Sweden

1. Introduction

Times are always changing and so is agriculture. In the history of agriculture, we have witnessed many turning points due to social, technical or natural factors or processes. Present time is not an exception. In this chapter we will argue that we are right now in the beginning of a new era—the fifth innovation era.

In 2020 the Covid-19 pandemic hit the world and hindered agricultural workers to travel across country borders. But this pandemic is only one kind of challenge, among many others, that the agricultural industry have had to deal with. Other challenges, as will be described more in detail further down in this chapter, is climate change, changing of food habits and lack of infrastructure for coping with new technological paradigms.

One can observe an interplay between these new challenges, new technology and new needs for change of social behavior related to farming industry as well as consumers—or even society as a whole. Innovation is probably one of the best ways to cope with the challenges of today. However, it is easy to think only about technology when talking about innovation. Instead, we will here define innovation in another way and end up in the conclusion that technology is often necessary but not sufficient for realizing innovation.

2. Challenges of world agriculture

Agriculture is connected to society as a whole and has always been so through the history. Agriculture stands for an essential part of all the food production in the world. Moreover, the productivity expansion in agriculture during the 19th century paved the way for both workforce supply to the upcoming manufacturing industry and for expansion of the demand of products from the factories. Agriculture of today is intertwined with other parts

Agricultural Bioeconomy
https://doi.org/10.1016/B978-0-323-90569-5.00003-2

of society in many ways. So, both agriculture and society as a whole share the need for understanding and coping with the agricultural-related challenges. During our years of working in the agriculture context we have notified quite many challenges and we will here present our view on "the biggest challenges of the 21st century."

2.1 Challenge 1: Increasing population

During the last decades the population has been increasing very much. The global population is now estimated to be around 7–8 billion people. The population growth varies among different parts of the world. In some parts it is even estimated to decrease. However, as a whole we shall be prepared to feed more people in the future. Some scenarios point at 9–10 billion people by 2050, and the UN Department of Economics and Social Affairs estimated 10–12 billion by the end of the 21st century (United Nations, 2019).

2.2 Challenge 2: New food habits

This challenge has probably even more impact than the growing population and it is about new food habits—alongside higher income levels—in many parts of the world, which to a large part mean more interest in eating more meat and more "crop-expensive" meet. This will lead to an increased demand for grain etc. used as animal feed. Because more crops are needed in order to produce meat, compared to vegetable food, this means increased need for production output. Beef cattle, for example, might need up to 10 kg of feed to produce 1 kg of meat (Silva, 2018).

2.3 Challenge 3: Protectionism and domestic preparation

In many countries there is an increasing level of domestic preparation and protectionism. The Russian invasion of Ukraine in 2022 reinforced this trend. This leads to difficulties in the global trade of food products, and might hinder the export of certain food, which means the importing countries have to start produce the food for themselves, which in turn can be difficult due to climate zones, machinery infrastructure, skills profile, etc.

2.4 Challenge 4: Protecting biodiversity

During the last decades there has, in some regions in the world, been an increased desire to protect biodiversity. This affects the regulations of use of chemicals. In Europe, for example, there has been an extensive debate around glyphosate and other chemicals. This, however means goals conflict.

In 2020, the US Department of Agriculture sounded the alarm that the EU's new agricultural policy jeopardizes the access to food. According to their analysis, that policy will "*reduce agricultural production by 7% to 12%*" (Beckman et al. 2020, abstract). According to them it will drive up global food prices by up to 89%. Indirectly, this would increase the number of starving people by up to 185 million—all according to their analysis.

2.5 Challenge 5: Fossil-free energy and products

This challenge has to do with the need to reduce the use of fossil fuels and products derived from fossil oil. For the farmers this means several things. One important part for the farmer is to be able to run the farm on fossil-free machines in a feasible way, both field machines, such as tractors, but also powering the grain drying facilities. Another important implication of this challenge is the fact that the farmer's arable land will most likely be used to meet the increasing demands growing crops for fossil-free biofuels and fossil-free fibers. The latter is used more and more as replacement for plastics. Using the arable land for other purposes than food production, means that there are conflicts between these challenges. A significant portion of the grain is already being utilized for biofuel, not least ethanol and biodiesel. Silva (2018) estimated that 40% of the U.S. corn crop was used for producing ethanol.

2.6 Challenge 6: More energy independence

This challenge is related to the one above but here it is not the climate concern that is the main driver. Instead the driver for many countries, as well as for individual businesses, is to become more energy independent. They want to be less reliant on suppliers and countries whose political agenda or systems are not considered ethically sound. Besides diesel and natural this can also include mineral fertilizer, which is connected with energy.

2.7 Challenge 7: Arable land sealed for future agricultural use

This challenge concerns the fact that arable land become used for other purposes than agriculture. The seriousness of this is mirrored in the report by Mission Board for Soil Health and Food (2020) pinpointing the large soil sealing in Europe. Every year 2.5% of agricultural land is sealed due to expansion of cities and logistics centers, etc. Because of the fact that large parts of the productive arable land is already in use it is not easy to expand it. We need more food and therefore cropland in the future but, according to

FAO (2017), "*The global expansion of agricultural land has stabilized over the last 20 years at around 4.9 billion hectares (ha)*" (p. 32).

2.8 Challenge 8: Lack of trust in the food safety

There is a growing public interest in food safety. Consumers want to know about the food they are eating, and policy-makers want to secure that food on societal level. This put pressure on the whole food production chain or system. There will be more demand on, for example, reliable food traceability. This will lead to higher costs in the production chain. Will these extra costs be possible to charge the consumer in the end?

Another issue related to this challenge is the growing trend among consumers that do not trust the large-scale farming systems and instead require small-scale production, local production and organic production. Small-scale food production is not as efficient as large-scale production and this trend therefore can decrease production volume. Going away from large scale farming can have many good effects, but one cannot ignore that it might be a trade-off between this demand and production volume.

2.9 Challenge 9: Big preharvest losses: pests, weeds and insects

This challenge concerns different issues that threatens the crop while still growing, such as crop pests, invasive species of weeds or insects (sometimes promoted by climate change). In a study published 1994 Orke et al. (1994) studied eight crops that together represented half the world's crop areas. They found that problems with pests, weeds and insects accounted for preharvest losses of 42% during the years 1988–90 (insects 15%, weeds 13%, and pests 13%). According to the same report there were postharvest losses of about 10%, related to factors such as drying and storage problems. Regarding pests Bradshaw et al., (2016) confirmed the Orke picture and estimated annual crop losses to plant pests to 20%–40% of global production. The economic value of the losses from plant diseases was by Agrios, (2005) estimated at US $220 billion annually.

2.10 Challenge 10: Climate changes that causes new prerequisites for crops

This challenge is due to the climate effects. The arable land has already started to be affected by factors such as droughts and flooding. Extreme storms can also have direct effects on plant cultivation. There is also shortage, or risk of shortage, of fresh water in several parts of the world that endangers crop irrigation. Major river basins face water scarcity due to low supply and high demands from not only agriculture but also industry and cities.

2.11 Challenge 11: Farm animal diseases

This challenge refers to diseases affecting farm animals that in turn also can affect humans. We are here dealing with both old and new kinds of diseases but most of all we have to deal with new spreading logics. The world has, during recent years, suffered from several pandemic animal diseases, such as bovine spongiform encephalopathy and avian influenza. Some of these are so-called transboundary also called zoonotic diseases. One historic example was the mad cow disease during the 1990s. One recent example is the spread of African swine fever virus (ASFV). The reason why this is a bigger challenge today than before is well described by FAO (2017):

> *The risk of serious outbreaks is increasing as more people, animals, plants and agricultural products move across international borders, and as animal production systems become more intensive. Because they can have serious socio-economic consequences, including the disruption of regional and international livestock markets and trade, transboundary animal diseases are a constant threat to the livelihoods of livestock keepers, especially in low- and middle-income countries. They undermine food security and prevent livestock sectors from achieving their full economic potential (p. 56).*

Regarding the pandemic of Coronavirus disease (Covid-19) caused by SARS-CoV-2 it has been classified as a zoonotic disease. However, as Haider et al. (2020) conclude, no animal reservoir has yet been found, so the real background of this pandemic is still not clarified.

2.12 Challenge 12: Lack of manpower

This challenge is about dealing with lack of manpower for agricultural activities. Despite all automation many activities in agriculture are still in the need of manual work. Not least is this true for horticulture. But also milking robots or modern machinery in fields need skilled and dedicated people. The salary levels are lower in agriculture in relation to many other jobs and the work environment is not always so good. When the day is over the green worker might smell ammonia and need careful cleaning. Workplace accidents are still common. In order to enter the workplace, on the farm, a car is often needed, since there oftentimes is no existing possibility to commute.

2.13 Challenge 13: Finite resource of phosphorus

The important resource phosphorus, seems to be finite. Rosemarin et al. (2011) wrote: "*A conservative estimate of longevity of the resource shows that at a 1% exponential increase for the next 50 years followed by zero increase, the global reserves would last 235 years. If one uses the UN global population growth rate to determine future demand, with a stabilization by 2100, the current global reserves would last 172 years.*" *(abstract)*. Despite efforts to get phosphorus by new means (like extracting it from residual waste from mines or municipal treatment plants) this challenge is going to be hard to cope with in the future.

2.14 Challenge 14: Security issues

Farming faces security issues of various kinds. The modern agriculture is very dependent upon digital systems and advanced machinery. There are many risks related to this. In Sweden for example there are lots of thefts of precision agriculture equipment. Some observers are also afraid of digital cyberattacks that for example can stop modern tractors on distance.

2.15 Challenge 15: Lack of infrastructure for communication

This challenge is about lack of infrastructure for communication. Part of modern agriculture relies on digital information systems for management of both animals and crop production. Precision farming can solve many goal-conflicts between production volume and environmental care. However, digital data needs communication infrastructure and in many countries, Germany for example, the infrastructure does not suffice on the countryside.

2.16 Challenge 16: Soil fertility

In many regions soils are degraded due to modern management methods and other factors. In order to understand this challenge some facts from the European Union and the Mission Board for Soil Health and Food (2020) are stated here. In the EU are 2.8 million potentially contaminated sites. Out of all soils within the EU 83% has residual pesticides, and 21% of has cadmium levels above the limit for drinking water. Soils with heavy metal content, potentially unsafe for food production, is 6% of the EU soils. Moreover, cropland soils lose carbon at a rate of 0.5% per year and 23% of land has compacted subsoil. One indicator of soil health is the amount of humus (organic matter) and another is the number of earthworms.

2.17 Challenge 17: Food waste

Food waste is already substantial. According to *The Guardian* about 30% of the world's agricultural land area is used to produce food that will be wasted: "*Each year 1.3 billion tonnes of food, about a third of all that is produced, is wasted, including about 45% of all fruit and vegetables, 35% of fish and seafood, 30% of cereals, 20% of dairy products and 20% of meat*" (Lyons et al., 2015). More distances between production and consumption might lead to longer and more complex cold chains, for example, and waste as a result of disturbances is to be expected (Silva, 2018).

2.18 Challenge 18: Animal welfare

Recent year's sustainability discussion is partly rooted in the Brundtland report published in 1987 (Brundtland, 1987). In this report ecological diversity was central, but not how individual animals are treated. Animals were not either included in the meaning of "social sustainability." Also when the Sustainable Development Goals were decided by UN member countries in 2015 no specific mention of Animal Welfare was included. However, animal welfare in the agricultural sector is considered by many as a substantial challenge. Consumers are becoming more enlightened and demanding regarding this, but also farmers take this challenge seriously, and for example is looking at the possibility to reduce animal suffering by discovering sick animals in early states.

2.19 Challenge 19: The need to decrease the use of antibiotics in the animal production

Antibiotic resistance is one of the biggest threats to global health, food security, and development today. A growing number of infections—such as pneumonia, tuberculosis, gonorrhoea, and salmonellosis—are becoming harder to treat as the antibiotics used to treat them become less effective (WHO, 2021). Many researchers are already warning us about the fact that we are heading towards a new era where we will need to deal with antibiotic resistance. Already today 500,000 people die annually, due to multiresistant bacteria. The prediction is that this number will increase dramatically if we do not take forceful actions. This is a global risk and international collaboration is important.

In order to reduce the need for antibiotics in animal production the best way is to prevent the diseases from arising. This is done by better hygiene and care for the animals, as well as finding relevant vaccines. Better breeding programs which will produce healthier animals is also important, together with only giving antibiotics to animals under veterinary supervision, and not to use antibiotics for growth promotion or to prevent diseases in healthy animals.

2.20 Challenge 20: Coping with disruptive X factors

Our last challenge is coping with "X factors" such as global epidemic spreading (with the Covid-19 in mind) or supranational political events like new guidelines from the United Nations. The Russian invasion of Ukraine in 2022 was also an unexpected disruptive event. Such “X factors” are not new phenomenon but due to the more interconnected society agriculture and the rest of society must be prepared for more frequent and high-impact X factor events.

3. Challenges as drivers of innovation

In our view there are three main pathways to cope with the agricultural challenges: evolution, imitation and innovation.

Evolution can be defined as a slow and gradual process of change driven by internal and external forces. From a specific actor’s perspective, evolution can be about doing what this actor has done for a long time, albeit better and better, step by step. Compare this with the kaizen concept, to make continuous improvements of existing processes.

Imitation means copying and occurs when someone starts doing something that someone else is already doing or has done before. An advantage of imitation is that it can involve the dissemination of “best practice” and “best available technology” (BAT).

Innovation is often incorrectly equated with new technology or new products. However, in an etymological study (Frankelius, 2009) was found that the original meaning of the term innovation is a phenomenon that occurs when something principally new—technology or not—develops and gains ground among users, at a certain point in time. Besides the technology-or-not-dimension this definition also places ground-gaining as important as the “the innovative thing in itself.”

Because of the magnitude of the challenges facing agriculture, innovation probably is the best way to cope with them. One positive thing with the challenges connects with the old saying that "necessity is the mother of the invention."

Let us illustrate innovation in relation to the challenges that Covid-19 caused the fruit and vegetable industry in Australia 2020. The harvest period, which begins in October, loomed for Manbulloo Limited with over 1000 hectares of mango trees. Robots had not yet been able to replace human hands in harvesting mangoes, so Manbulloo and others in the industry were dependent on human labor. Major and Fitzgerald (2020) described the matter in ABC Rural: "*The $14.4 billion industry relies heavily on seasonal workers from the Pacific Islands and Timor Leste, as well as tens of thousands of backpackers on working holiday visas.*"

Manbulloo and other Australian companies, like the rest of the world, were now shocked by an external factor of a kind that is not common in business economics theory: the Covid-19. It put an end to the influx of foreign labor that is so important for the mango farmers. Not least, as Major and Fitzgerald wrote, it's about backpackers. They combine adventure with jobs in the green sector. Mango farmers in Australia are dependent upon 40,000 guest workers during harvest time, but in June 2020 the forecast was that only 8000 would be available.

So, what happened to the mango harvest in Australia? The harvest was conducted from October 2020 to February 2021 and most businesses were severely understaffed. However, the situation with a shortage of seasonal staff accelerated the development of two things. Firstly, the mango farmers started to strive for growing different varieties of mangos, that ripens on different times. In this way, the harvest would be more prolonged and the farmers would not get as high work peaks, but could cope with lower seasonal staffing. Secondly, the innovative process of developing an automatic mango harvester was accelerated. A prototype was tested during 2021 year harvest on several different farms, and it was believed that a mango harvester would be on the market within a couple of years. Maybe the mango farmers' problem with finding seasonal workers will be just a memory in the future? No matter the answer, the mango case illustrated that challenges are connected to innovation. In the next section we will discuss different innovation eras in the history of agriculture (Fig. 1).

Fig. 1 The mango picker robot. Photo kindly provided by Kerry Walsh, University of Australia.

4. Historic reflections: The first four innovation eras

Agriculture has historically undergone periods of great change in relation not only to problems and challenges but also to certain innovations. We refer here to these periods as innovation eras. We will summarize these periods of radical change to provide the background to our main point: The entering of the fifth innovation era today.

4.1 The first innovation era (approximately 10,000 B.C. to 1750 A.D.)

This first innovation era started with the Neolithic Revolution. People at the Indus Valley, along the Yangtze River (China) and the Middle East, gave up the nomadic life, settled down and started to grow the cereals barley and wheat, about 10,000 BC. They did also invent the first tools for farming. So writes Percy Blandford: *"A Babylonian drawing of 1316 B.C shows an ox-drawn plough with a funnel, or hopper device, into which seeds are being dropped as the plough progresses"* (Blandford, 1976, p. 77). Alongside crop cultivation also animal domestication came along as well as livestock and poultry.

4.2 The second innovation era (approximately 1750–1900)

The second innovation era of great change was the expansion during the mid of 18th Century through land reforms, drainage and crop rotation. The horse took place in agriculture and was faster and more flexible than oxen. With the horse came horse-drawn iron ploughs, reapers, seed drills and other implements. Jethro Tull, born 1674 in Berkshire, England, was one of the inventors. He early embraced the scientific method of his innovation business and invented a seed drill machine around 1733. Before this the fields were usually sown simply by sprinkling seeds over the ground. With Tull's machine, small rows were created in which the seeds were dropped. Horses drew the machine. The result became that a larger portion of seeds grew, and hence the crops increased significantly. In fact Tull himself connected his precision machine with row hoeing. His book from 1731 had the main title Horse-Hoeing Husbandry. By means of his sowing machine it was possible to hoe effectively due to the seeding in precise straight lines.

At the end of this era steam engines developed. The world's first high pressure steam engine was invented by the Englishman Richard Trevithick in 1802. This invention made its way into society by Hayle Foundry that built it in 1811 for the farmer Christopher Hawkins. He wanted it to run a threshing machine on his farm in Probus, Cornwall. The engine came into operation in 1812.

4.3 The third innovation era (approximately 1900–2000)

The third innovation era was fueled by the mechanization in the wake of electricity and the tractor which heralded, among other things, milking machines, combine harvesters and grain elevators. The milking machine made life much easier for livestock farmers. Fertilizer spreading had been done as early as 1886, but the combination of centrifugal spreaders and better granules improved spreading much with start in the 1930s and gaining momentum in about 1945 (Fig. 2).

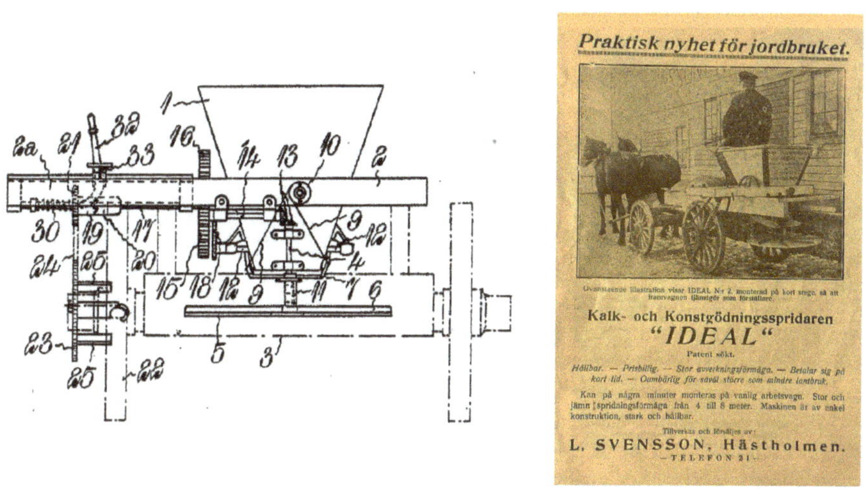

Fig. 2 A patent from L. Svensson company 1930 alongside a flyer of the centrifugal spreader. From Crister Stark collection.

Many other innovations made its breakthrough during this era. Electric fence, hydraulics, silage methods, reversible plows and four wheel drive tractors are examples. We also have to mention pesticides. Roundup (glyfosate) came to market 1976 and became instrumental for many parts of the world's crop production.

4.4 The fourth innovation era (approximately 2000–today)

The fourth innovation era, according to our interpretation, took off partly by animal handling robotics, partly by satellites. One main innovation here was robotics in milking. The breakthrough occurred around the year 2000 but movements took off 20 years earlier. In 1992 Lely installed its Astronaut on a farm in the Netherlands. DeLaval launched the Voluntary Milking System (VMS) 1998. Despite many problems with the robot systems it all turned out to be a success story. The diffusion became fast after the year 2000. In 2017 estimations pointed at about 35,000 robots installed worldwide (Salfer et al., 2017) (Fig. 3).

Fig. 3 What then is the motivation to invest in a robot from the customer's point of view? Ewa Lööw, at the farm Storgården in Sweden commented: "We put in our first Lely A2 in the autumn of 2005. That was after I had problems with pain in my arms and hands and had problems milking an entire work shift. … We had three choices. Stop with cows, hire someone or invest in a robot. … I do not regret the selection. It became a completely different job and our children became interested in the cows when a robot and a computer came in." Lööw pointed at Physical working environment, staff and attraction among young people. There are also animal welfare aspects, for example increased freedom for the cows and that the cows become milked more times per day. *(Photo: Per Frankelius.)*

One trigger of the second main innovation forming the fourth era was the Navstar GPS-prototype 1978 as the first step towards the GPS system. As soon as the U.S. government cancelled the "selective availability" function in the year 2000 farmers got centimeter precision. The RTK technology and StarFire (John Deere) improved the precision to about just 3–6 cm. Satellites paved the way for assisted steering and later auto steering. Notably, this was a time far earlier than the foundation of the Tesla company and self-driving cars. To sum up, satellites made precision farming possible.

Some precision farming pioneers began to embrace and develop combinations of yield monitoring and GPS. Already in September 1992, one year after the outbreak of the Kuwait war, an article in the *Seattle Times* was written about the farmer John Ruth at Amana Farms. By using a sensor on the combine harvester that read the crop yield every second as the combine harvester progressed—in combination with map software, a database and "the same satellite system that navigated troops through the Kuwaiti desert" as he himself said—he could see exactly how much oats a certain part of the fields gave.

5. The new fifth innovation era starting now

We argue that we currently are in the beginning of the new fifth era. There are some themes we derive from the challenges mentioned earlier in this chapter, themes that we think will form the new fifth innovation era for agriculture. The themes are grouped into five different areas, these are: new crop and cultivation concepts, field robotics, fossil-free energy and products, smart connected systems and animal welfare technology systems. We will discuss each and one of these five areas below.

5.1 New crop and cultivation concepts

In the fifth era new crop and cultivation concepts are forming. Among driving forces are challenges of today that are partly related to real climate change—not least drought and extreme weather but also social movements in the wake of climate care concerns. Moreover fears of chemicals have changed the political frames as well as consumer's preferences. Above this many farmers face the need to boost their economic situation. At the same time technology and new methods are developing.

Efforts are made in plant breeding to get improved crops. Here phenotyping and DNA-technology plays an important role in combination with simulation and novel computing. One key concept is high speed crop selection. For example KeyGene is developing fossil-free rubber by crossing the

Russian dandelion with other species. Plant breeding targets include crops with more salt or draught tolerance. Drought tolerance is related to deeper root systems. Researchers have even tried to improve the photosynthesis. In 2012 started the international research project Realizing Increased Photosynthetic Efficiency (RIPE) funded by the Bill and Melinda Gates Foundation. The aim partly was to make crops more productive by improving the photosynthesis through DNA-technology. As a result, the amount of biomass in the experimental plants improved by about 15% (Kromdijk et al., 2016).

The fifth era also includes new cultivation concepts. Precision is taken to higher heights through new sensors and artificial intelligence. One illustration of this is manure spreading. By means of a NIR sensor (near infra-red) the farmer can measure the amount of different nutrition's going through the a fertilizer spreader pipe in real-time. Dinamica Generale patented one kind of such NIR technology in 2004.

Another example is roe hoeing. In 2008 System Cameleon was presented by the company Gothia Redskap (now in the Lyckegård company). It was a system of tools for seeding, roe-hoeing and milling fertilizers. The machine could cut the weeds as close as just a centimeter from the crop—at a speed of 15 km/h. The concept was made possible by means of the combination of stable mechanical linkage and a camera that in real-time identified the plants and automatically controlled the hoes so that they ran exactly between the sown lines. Later this system used RTK-GPS alongside angle-sensors connected with wheel steering.

Still one example is SeedEye from the Väderstad company. The core of it is an infrared light sensor placed on each seed hose, that can read each wheat seed that flows through the hose. By this the farmer can set exactly how many seeds he or she wants per square meter (Fig. 4).

Among concepts being alternatives to both chemical and mechanical weed control the company Zasso is one example that presented the Electroherb in 2019. In front of the tractor is a unit with hanging high voltage electrodes. When they are moved over the weeds, an electric current is generated down through the weeds and their roots, which causes the weeds' vascular system to be damaged, and the weeds die. The power is generated by a power take-off generator at the rear of the tractor. Other related concepts are treatment with hot steam, microwaves or infrared radiation.

Various forms of biological control should also be mentioned. For example, the company M3 and USDA collaborated in 2015 in an attempt to

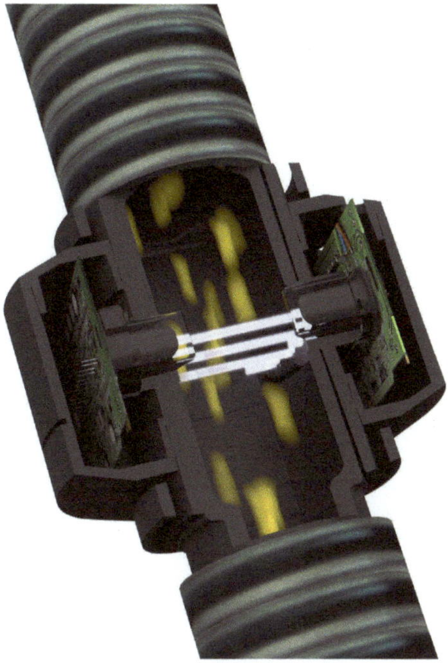

Fig. 4 SeedEye sensor kit. *(Illustration: Väderstad.)*

bomb cotton fields with sterile butterfly larvae as an alternative to chemical plant protection products.

There are many reasons to expand the vegetation on the field. Bare soil does not for example bind as much CO_2 as soil covered by vegetation. Carbon capturing in soil is one of the main issues in the fifth innovation era. Therefore concepts develop about combinations of annual and perennial crops including main and cover crops. Some farmers have started to sow new crops directly in growing crops. Core concepts are cover or and companion cropping. To succeed is needed new machinery, and development is ongoing at many companies, not least seeder, planter and row hoeing producers (Fig. 5).

Still other targets in the fifth innovation era are soil health and reducing soil compaction. New technologies such as tire construction, automatic tire inflation systems, and lightweight material can decrease soil compaction. Using crops with different ripening times is still one method. By such crop combination the farmer can have a wider time window to harvest even with a smaller and lighter combine. Better timing of activities in relation to soil conditions is vital. The Swedish innovation platform Agtech 2030 invented

Fig. 5 Companion crop (*Medicago lupulina*) in wheat. *(Photo: Martin Krokstorp.)*

the Compaction Prevention System (CPS) that tells the farmer the compaction risk every meter on the go and also provides maps of compaction risk covering the whole farm. This system got a silver medal in Agritechnica Innovation Award 2020.

Last, but not least, we shall mention the diffusion of new irrigation concepts such as subsurface drip irrigation. In 2015, Netafim, in collaboration with the De Jager Farms dairy farm in Chowchilla (California), conducted a test that involved spreading liquid manure through an underground drip irrigation system.

5.2 Field robotics

There are several driving forces behind the desire to use field robots. Through some of these, the farmer can reduce the weight of the field vehicle and thereby reduce soil compaction and energy consumption. Furthermore, robots can be a solution to the problem of finding labor.

There has long been talks about robots in agriculture and many attempts and experiments have been made. But going from experiment to robust and functional concept has not been as easy. However, a trend-break has taken place. Some have the vision of smaller robots working in swarms on the fields. One example is the Oz robot from Naïo Technologies, France. The first robot prototype from that company was presented in 2012. In 2013 the first sale of the small robot "Oz" started, and in 2014 ten robots

Fig. 6 The Oz robot from Naïo Technologies. *(Photo: Per Frankelius.)*

were on test for horticultural missions around France. The Oz robot is now adapted for row hoeing of maize and similar crops (Fig. 6).

Some companies, like Fendt, have visions about robot swarms working on the fields. One special aspect of this is that the farm is not that fragile if one unit is out of order. The other can still function.

Field robots also have other shapes, not least ordinary tractors that can be set in "robot mode". John Deere, for example, launched a market-ready autonomous 8R tractor in Las Vegas in January 2022. Hybrid concepts such one driver managing many tractors on the same field at the same time, is also part of the fifth innovation era. Moreover we can mention the concept in which a robot is following a manned master unit (master-servant concepts). Fendt, for example, showed the Guide Connect in 2011 but did not reach the market. The U.S. company Smart HP in 2018 presented a grain cart tractor that autonomously follows the combine on the field. In Sweden is invented the Fieldgofer concept that can handle both grain and fertilizers. Raven Applied Technology shall also be mentioned.

Field robotics can be said to include drones, despite the fact they are above ground. Technology and new legislation pave the way for new concepts. American Robotics, for example, in 2017 revealed the Scout concept. A drone is placed in a cassette and can take off and land for itself at given time points. The farmer gets overviews of crop fields and grasslands once every day. The fifth innovation era includes drone-aided agricultural concepts.

5.3 Fossil-free energy and products

The agricultural industry will need to start using fossil-free fuels and energy. One example is the project The Carbon-Free Farm driven by the MIT and NASA engineer Jay Schmuecker at Pinehurst Farm in Iowa, USA. They built a hydrogen tractor already in 2010 by installing a Ford 460 V8 otto engine into a John Deere 7819 tractor and mounting tanks on the roof. In 2011, they started to produce hydrogen by means of solar power. In 2015, he could present the world's first tractor to run on NH_3.

Within the agricultural industry the carbon footprint needs to be reduced also when it comes to its necessities. The Yara company is working on developing carbon-free fertilizer. The aim is to use it on farms from 2024. This carbon-free fertilizer probably will be between two and five times more expensive than regular fertilizers.

The arable lands can be used to produce many of the fossil-free products and fuels (like ethanol or biodiesel) of the future. Already there is a lot of product development going on within the car industry, using natural fibers, both from the forest and the fields. One product used for this is flax. An advantage of the new materials is the reduced weight, because lighter cars can drive further with the same amount of electrical energy. "We can make car interiors 50% lighter. We can reduce the amount of plastic by 80%," says co-founder and CEO Christian Fischer of Bcomp (Fig. 7).

Fig. 7 In 2018, an electric and competition-adapted Tesla was presented with a body and wing made of the ampliTex material based on the agricultural product flax. The result was construction parts that are light, high-strength and with up to 30% better cost-effectiveness than carbon fibers. Furthermore, the crash behavior was better because there are not as sharp phrases as when using carbon fiber. Manufacturer of the material is the Swiss company Bcomp. *(Photo: © Bcomp/Jaime de Diego.)*

The arable land can also be used as platform to create wind and sun energy. The Agrosolary concept, for example, combine solar trackers with vegetation lines on the fields. Inventors behind this are Kurt Hansson and Landberg Solutions.

5.4 Smart connected systems

Animals, vehicles, crop fields but also silo facilities are more and more equipped with sophisticated sensors in combination with advanced computing such as artificial intelligence. Moreover, these systems are connected by means of wireless communication. The connectedness also includes new opportunities regarding remote support and control of machinery. This path of development has got a boost during the Covid-19 epidemic. At the same time sensors and communication systems pave the way for traceability that can be of interest for both suppliers and consumers in the agri-product systems.

Part of this is predictive maintenance. One example is Expert Alerts, presented by Deere & Company in 2017—actually before aircraft engine companies. Through this concept the farmer receives warnings about deficiencies that can lead to operational disruptions. It is thus a forecasting tool. The key is large amounts of information about many vehicles that are collected in John Deere's computers in combination with special algorithms to predict problems before they occur, with major financial damage as a result.

Remote support is another possibility of the connected farming world. This includes both digital and personal concepts. A milking robot manager or a tractor driver can ask for help and get it instantly thanks to the fact that an expert geographically far away can connect to the machine at the same time having personal video communication in turn including options such as augmented reality. In the U.K. was in 2019 launched the South West Dairy Development Centre in Somerset. One of their projects is about connecting the farmer with veterinaries by means of augmented reality glasses. Remote support is also important for field machinery.

All vital systems in modern agriculture are becoming connected in the fifth era. Besides animals, stable machinery and field machinery we mentioned grain handling. The Grain Cloud concept from the company Skandia Elevator was presented in 2018. The user interface is an app that collects information from the grain plant's control system in turn collected by a local computer (including a programmable logic controller, PLC) located at the silo and drying facility. The app shows which crop is in which silo, which conveyers, fans, heaters etc. are running and estimated drying times. The app alerts and sends warnings in the event of a grain flow jam, machine failure and any need for service. This means freedom for the farmer to be able to have the office on the mobile phone where alarms and warnings can be received as soon as something important happens (Fig. 8).

Connected systems pave the way for traceability and added value to agri products. Through the use of blockchains, for example, food sellers can give customers evidence and information regarding were the food come from. They just have to scan the QR code and will see where and when, for example, the grain in the bread was grown. These new concepts replaces the paradigm of bulk handling of products for batch handling and may lead to a lot of new method in the handling of grain also after harvests. Small batch-divided silo systems is but one possible future scenario.

Fig. 8 Grain cloud contributes to a calmer life for the farmer through better control of the equipment. Here the project manager Malin Heder tests a prototype of that system. Photos: Skandia Elevator and Linköping University. *(Photo, left and right: Skandia Elevator. Photo, top: Thor Balkhed/Linköping University.)*

5.5 Animal welfare technology systems

In the fifth era the farmers will get information about its animals 24/7. Already there are some systems in the market, but there are still a lot of possible developments to be done.

With the combination of camera surveillance and artificial intelligence, the farmer can monitor all kinds of animals. One area is the monitoring of calves around the clock in the calf huts where calves are often kept for their first 2 weeks. This system flags signs of illness via an app several days earlier than the farmer would otherwise see symptoms with the naked eye. The most common calf disease is diarrhea and by taking measures earlier, the farmer reduces calf mortality. Early detection of diarrhea reduces the need for prescription drugs, but above all increases animal welfare. The company behind this, launched in 2021, is Futuro Farming in Germany.

The Swedish company Videquus started, in the Spring 2021, promoting a camera surveillance, that via AI will alert the horse owner when the horse is about to foal.

Another important animal welfare example can be found in the egg industry. Roosters cannot lay eggs, and are ill-suited to food, so newly hatched roosters are killed and ground down or burned. It is estimated that seven billion roosters worldwide are killed immediately after hatching. Therefore, efforts have been made to detect the sex of eggs. The German company Respeggt GmbH has developed a system where a small sample is taken out of the egg in order to sex the egg. In Sweden a team led by the Linköping university professor Anita Lloyd Spetz and the Dansic company has developed another method for detecting rooster eggs. Either way, this will increase animal welfare, as well as reduce the food waste, since the rooster eggs can be used within the food industry.

The new technology will also speed up the breeding development. Healthier and stronger animals, better suited for its purpose, will be bred. The Swedish company Viking Genetics have speeded up their breeding program via genomic selection, via a combined index of different genetic traits that are heritable through mating bulls with cows. The purpose, of course, is to develop the cattle's "genetic capital" to achieve higher profitability and functionality of the herd by breeding new generations of cows with higher capability, e.g. for milk production and resistance to diseases.

6. Enablers and disablers in the fifth innovation era

At the beginning of this chapter we stated that technology is often necessary but not sufficient for realizing innovation. The above mentioned five area's development are dependent on a number of other parameters, we call them potential enablers and disablers, we will discuss them in the text below.

For many farmers potential technological systems in rural areas are dependent on communication infrastructure. There needs to be a fast and robust working internet, not least since a lot of the data is stored in the cloud.

The authorities have an important role in encouraging and making new innovations financially viable. Who is going to pay for the more expensive carbon-free fertilizer and food traceability?

Furthermore, the authorities have an important role when it comes to global handling of the need for less use of antibiotics.

It is also important that the agricultural industry as a whole start to work with getting new needed common standards in place. The faster the better.

Another important enabler is that the consumers use their power to choose the "right" food. The more the consumers ask for meat produced with less antibiotics or bread with a stated carbon footprint the faster these changes will come through. Food is not just calories for the body, it is also a statement of who you are, and what you as a consumer think is important.

7. Conclusion

Our analysis took off with a definition of 20 challenges facing agriculture. Because of the magnitude of these challenges, as well as conflicting goals that can be derived from these, we argued that innovation is a more promising path forward than evolution or imitation. Then we defined four innovation eras in the history of agriculture and argued we right now are approaching a fifth, summarized in Fig. 9.

Our definition of innovation eras was based on an assumption of an expanding change rate, so the last eras were much shorter in time compared with the first eras. The first era in our analysis lasted 11,750 years, the second era 150 years, the third era 100 years and the fourth era 20 years. We are not alone in this kind of interpretation of acceleration. Consider for example Ray Kurzweil's book *The Singularity Is Near: When Humans Transcend Biology*

Fig. 9 The signum of the fifth innovation era in agriculture.

(Kurzweil, 2005). However while Kurzweil treat technology we have tried to describe interaction of innovation and challenge-related forces like social reaction to climate change, and we have argued that "X factors" such as the Covid-19 epidemic can stimulate innovation.

The synthesis of our analysis is that agriculture is both coping with and driving dramatic change and already now approaching a fifth era, only 20 years after the entering of the fourth. The length of the fifth era cannot be known beforehand but there are reasons to believe it will be quite short, maybe only 12 years or so because of "the acceleration of acceleration."

The bottomline of our foresight is that the agricultural industry will play an important role, when it comes to handling several of the current global challenges.

References

Agrios, G.N., 2005. Plant Pathology, fifth ed. Elsevier Academic Press, New York.

Beckman, J., Ivanic, M., Jelliffe, J.L., Baquedano, F.G., Scott, S.G., 2020. Economic and Food Security Impacts of Agricultural Input Reduction Under the European Union Green Deal's Farm to Fork and Biodiversity Strategies, EB-30. U.S. Department of Agriculture, Economic Research Service. https://www.ers.usda.gov/publications/pub-details/?pubid=99740&fbclid=IwAR1R7QnRWoAEMLbn9-v2kBk7CAgqVVOf3QgQRcVRCjsKIqkN4R3VhsmUXQ0.

Blandford, P.W., 1976. Old Farm Tools and Machinery: An Illustrated History. Gale Research Company, FL.

Bradshaw, C.J., Leroy, B., Bellard, C., Roiz, D., Albert, C., Fournier, A., Barbet-Massin, M., Salles, J.M., Simard, F., Courchamp, F., 2016. Massive yet grossly underestimated global costs of invasive insects. Nat. Commun. 7, 12986. https://doi.org/10.1038/ncomms12986.

Brundtland, G.H., 1987. Our Common Future. Oxford University Press, Oxford.

FAO, 2017. The Future of Food and Agriculture—Trends and Challenges. Food and Agriculture Organization of the United Nations, Rome.

Frankelius, P., 2009. Questioning two myths in innovation literature. J. High Technol. Manag. Res. 20 (1), 40–51. https://doi.org/10.1016/j.hitech.2009.02.002.

Haider, N., Rothman-Ostrow, P., Osman, A.Y., Arruda, L.B., Macfarlane-Berry, L., Elton, L., Thomason, M.J., Yeboah-Manu, D., Ansumana, R., Kapata, N., Mboera, L., Rushton, J., McHugh, T.D., Heymann, D.L., Zumla, A., Kock, R.A., 2020. Covid-19-Zoonosis or Emerging Infectious Disease? Front. Publ. Health 8 (596944), 1–8. https://doi.org/10.3389/fpubh.2020.596944.

Kromdijk, J., Glowacka, K., Leonelli, L., Gabilly, S.T., Iwai, M., Niyogi, K.K., Long, S.P., 2016. Improving photosynthesis and crop productivity by accelerating recovery from photoprotection. Science 354 (6314), 857–861.

Kurzweil, R., 2005. The Singularity Is Near: When Humans Transcend Biology. Viking, New York.

Lyons, K., Swann, G., Levett, C., 2015. Produced But Never Eaten: A Visual Guide to Food Waste. The Guardian, 12 August 2015 https://www.theguardian.com/environment/ng-interactive/2015/aug/12/produced-but-never-eaten-a-visual-guide-to-food-waste.

Major, T., Fitzgerald, D., June 2020. Backpacker Stream Dries Up Causing Looming Rural Labour Shortage. ABC Rural. https://www.abc.net.au/news/rural/2020-06-23/farmers-warn-no-seasonal-workers-will-leave-fruit-to-rot/12374016.

Mission Board for Soil Health and Food, 2020. Caring for Soil is Caring for Life—Ensure 75% of Soils are Healthy by 2030 for Food, People, Nature and Climate. European Commission, Brussels.

Orke, E.C., Dehne, H.W., Schönbeck, F., Weber, A., 1994. Crop Production and Crop Protection: Estimated Losses in Major Food and Cash Crops. Elsevier, Amsterdam.

Rosemarin, A., Schroder, J.J., Dagerskog, L., Cordell, D., Smit, A.L., 2011. Future supply of phosphorus in agriculture and the need to maximise efficiency of use and reuse. Int. Fertil. Soc. Proc. 685, 1–28.

Salfer, J.A., Minegishi, K., Lazarus, W., Berning, E., Endres, M.I., 2017. Finances and returns for robotic dairies. J. Dairy Sci. 100, 7739–7749. https://doi.org/10.3168/jds.2016-11976.

Silva, G., 2018. Feeding the World in 2050 and Beyond. Part 1. Productivity Challenges. Michigan State University Extension. https://www.canr.msu.edu/news/feeding-the-world-in-2050-and-beyond-part-1.

United Nations, 2019. World Population Prospects 2019. Department of Economic and Social Affairs. Online Edition. Rev. 1. https://population.un.org/wpp/Download/Standard/Population/.

WHO, 2021. Antibiotic Resistance. July 31, 2020 https://www.who.int/news-room/fact-sheets/detail/antibiotic-resistance?fbclid=IwAR2Keu7l42jFq_zjDNUNlcHffyBD5HaT2KVWCItHsGdw7WnzHBghIiWZJiQ.

Further reading

Grigg, D., 1992. The Transformation of Agriculture in the West. Blackwell, London.

Ordolff, D., 1980. Melkzeuge automatisch Ansetzen-Demnächst Wirklichkeit? Landtechnik 35 (5), 222–224.

Roche, J.R., Berry, D.P., Bryant, A.M., Burke, C.R., Butler, S.T., Dillon, P.G., Donaghy, D.J., Horan, B., Macdonald, K.A., Macmillan, K.L., 2017. A 100-Year Review: a century of change in temperate grazing dairy systems. J. Dairy Sci. 100 (12), 10189–10233. https://doi.org/10.3168/jds.2017-13182.

CHAPTER 12

The role of collaboration and entrepreneurship in strengthening the participation of primary producers in the bioeconomy

Nunzia Gabriella Fasolino, Matteo Zavalloni, and Davide Viaggi
Department of Agricultural and Food Sciences, University of Bologna, Bologna, Italy

1. Introduction and overview

The main goal of the last Updated Bioeconomy Strategy (2018) is to foster the transition to a sustainable bioeconomy in the upcoming years. In this respect, the European strategy has, at its core, some pressing global socio-economic and environmental challenges, such as an efficient use of the natural resources, food security and climate change mitigation. Starting from recent years, the European Union has promoted an increasing number of initiatives in the context of bioeconomy, and many past and future policies are linked to the theme (Fund et al., 2018). The ambitious goal of the Bioeconomy Strategy (2018) is the full integration of different and multi-output value chains with an optimization of the waste and residues as feedstock for other value chains (Carus and Dammer, 2018). Moreover, it promises to have a high impact also on the economic growth, rural development, and job creation.

In this context, to provide a grounded and contextual approach, a combination of often unrelated and highly interdisciplinary knowledge in different subsectors is needed (Viaggi, 2018). Bio-based businesses are inherently interconnected because they require a high degree of interactions, communication, and coordination within complex and interdependent areas (Antikainen and Valkokari, 2016). Moreover, new solutions are asked to facilitate the exchange of best practices and experiences to acquire and disseminate knowledge.

Agricultural Bioeconomy
https://doi.org/10.1016/B978-0-323-90569-5.00013-5

Primary production has been finally recognized for its important potential in contributing to the ambitious objectives of the European bioeconomy strategy (2018). Target stakeholders such as primary producers, cooperatives processors and producer association are asked not only to provide raw materials but also to contribute to the process towards the sustainable transition. Furthermore, the bioeconomy is seen as a good opportunity for increasing their profitability and to strengthen their competitiveness in the longer term. Finally, the recent emphasis of the bioeconomy on natural resources and ecosystem services matches with the growing attention at the role of agriculture as a provider of public goods (Viaggi, 2015). However, the involvement of farmers in bio-based initiatives is still limited and their position is still confined into the area of the primary sector. The adoption of biobased solutions appears still hampered in Europe due to two main challenges faced by primary producers: (i) poor collaboration and (ii) low level of awareness and knowledge creation. Consequently, due to those starting conditions, low level of interactions exists, and few new business relations are established (Borge and Bröring, 2020).

However, the entrepreneurial activity of primary producers has a key role in contributing to the sustainable bioeconomy, even if their role has been neglected for long time in the entrepreneurship literature. Entrepreneurial orientation is important in stimulating collaborative behavior and in establishing new relations among producers, in the context of bio-based economy. Furthermore, given the complexity of emerging areas such as bioeconomy, usually those interactions need a specific knowledge on optimization of processes, high degree of investments in innovation and technologies. However, little attention has been paid until now to the key role of person-based relationships and to the importance of supportive environment and networks in the knowledge transfer process. Despite the increasing interest in this matter, research in the field of collaboration networks in the bioeconomy itself is still composed of fragmented and unconnected themes without a contextual focus. This is even more true when it comes to new strategies to enhance the involvement of farmers in the bioeconomy going beyond the primary production.

This book chapter attempts a state-of-art review on collaborations and networks in the field of the bioeconomy. Even if collaboration networks already exist implicitly in the bioeconomy, the subject does not have evidence in the literature. In this sense, the study aims at filling the current research gap. First, the study starts focusing on the evidence reported in the literature reviewing the role of interactions and entrepreneurship into the bioeconomy, using a systematic approach. Then, we focused on practical business case operating successfully in the European bioeconomy. The chapter is organized as follows. In Section 2, it is provided a description of the methodology used for the research and we illustrate the outcome of the search in the literature. In Section 3, the results are discussed. Furthermore, we extend our attention to new initiatives in the European bioeconomy. Finally, conclusions are pointed out in Section 4.

2. Methodology

2.1 Overview

The study uses the method of the systematic literature analysis following the Preferred Reporting Items for Systematic Reviews (PRISMA) guidelines to analyze the contributions from the literature. We systematically researched from the recognized database "Scopus" a defined research query; the term "bioeconomy" was used as the first keyword for the research of the topic. Then, the outcome was searched based on the presence of other selected keywords, namely "entrepreneurship" and "collaboration" within the results shown. We intentionally decided to use the operator "AND" to limit the results to contributions having both terms in their abstract, title or keywords. The field was later further restricted to peer-reviewed studies of journal articles and reviews written only in English.

Moreover, additional relevant articles identified when reading the full-text articles or checking the reference lists of the selected contribution (e.g., the "snowball" search strategy) were screened using the same methodological criteria setting the boundaries that delimited the scope of our study. Later the results were added to the main list of references; all the results were examined in more detail when they were considered potentially relevant. The eligibility criteria to obtain the final sample were established according to the research question. The first round of screening was the elimination of all the duplicates. The second selection was carried out based on a careful reading of the title and abstract of the papers. We eliminated studies when the contributions concerned other fields which were considered not related according to the purpose of our research. Next, the articles were analyzed after reading the full manuscript when it was available. The entire process is illustrated in the flow diagram in Fig. 1. The final selection of publications was then subjected to a qualitative review by the authors.

The results of literature analysis are discussed and enriched with empirical cases from the European bioeconomy. In doing so, we try to link what the

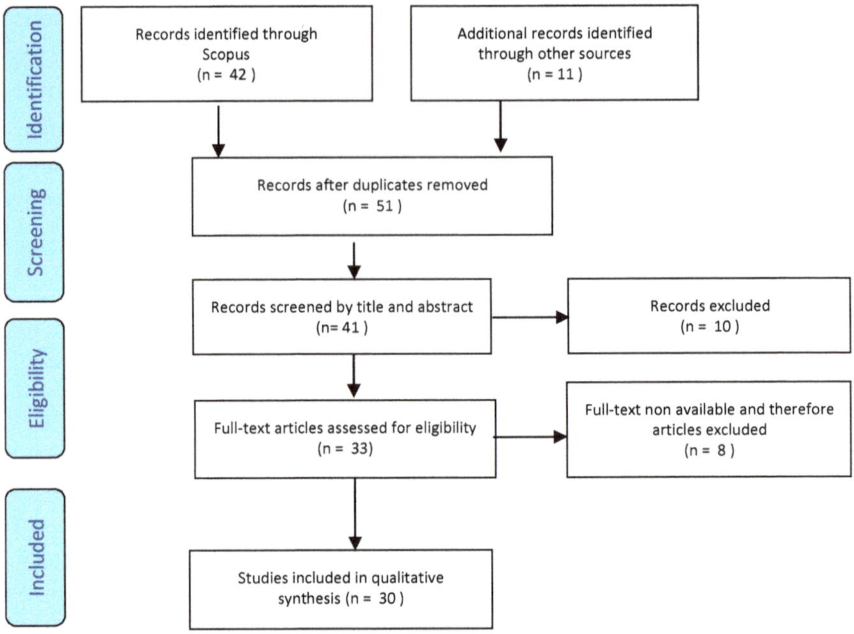

Fig. 1 Reporting items for systematic review adapted from the "Preferred reporting item for systematic reviews" PRISMA.

literature describes in the context of entrepreneurship and network to practical business case already operating in the European bioeconomy. The main purpose is to provide empirical examples to make the readers understand what bioeconomy is about.

The approach of the research is developed in the context of the H2020 EU research project entitled Coopid (COOPeration of bioeconomy clusters for bio-based knowledge transfer via Innovative Dissemination techniques in the primary production sector) where visits to the presented business in the European bioeconomy were organized with the aim of a wider knowledge dissemination among stakeholders.

2.2 Constitution of the Corpus of analysis and descriptive analysis

The research is based on the identification of the string presented in Table 1 in November 2021 searching from different sources.

From the main database (Scopus), the search yielded 42 results of peer reviewed papers. The snowballing method added 11 more articles, resulting in 53 articles being screened by title and abstract. Duplicates ($n = 2$) were

Table 1 Research question.

Source	String	Results
Scopus Database Search	(TITLE-ABS-KEY (bioeconomy)) AND ("entrepreneur*" AND "collaboration") AND (LIMIT-TO (DOCTYPE, "ar") OR LIMIT-TO (DOCTYPE, "re"))	42
Snowball Search Strategy	Keyword "entrepreneurship" or "bioeconomy" or "network"	11

eliminated. The listed studies were organized by year of publication and in alphabetically order of the first author's name. After the material collection, sources were selected based on inclusion and exclusion criteria already described in the previous section. However, the selection procedure is illustrated in more detail in the following figure. The final outcome based on the inclusion criteria yielded a total of 30 articles.

3. Results

3.1 Results of the literature review

We examined 30 full-text publications according to our search. The year with more studies that entered our search is 2021 as illustrated in Fig. 2. The results show a significant increase in the number of publications starting from 2018.

Regarding the methodology, we have found a predominance of qualitative studies in the form of case studies in European countries. The selected contributions reviewing collaboration and entrepreneurship in the

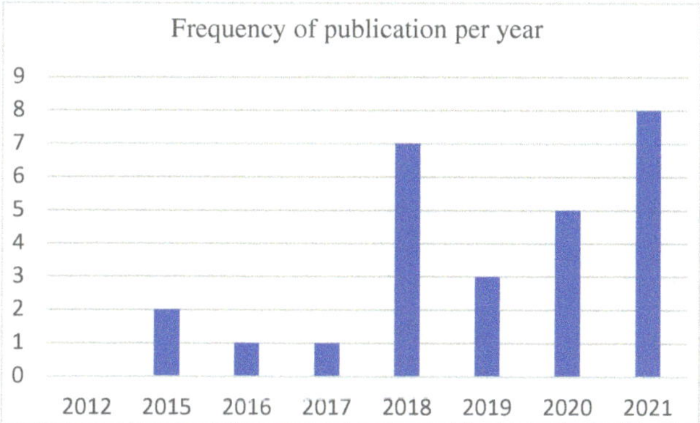

Fig. 2 Frequency of publication per year.

bioeconomy are published in 20 different journals. The results highlight a high degree of heterogeneity among the subject area of the journals meaning that the issue has been analyzed from different perspective in the literature. More than the half of the articles accounts only for one article per journal, while the other journals ($n = 3$) have at most 2 contributions each.

Among the results, a wide range of research focuses on specific technology-based processes in the bioeconomy: these are strictly related to the developing of technologies aimed at reducing the dependence on non-renewable resources. In this regard, energy is the main recurrent topic. This is reflected also by thematic journals and from the words "biomass" or "biofuels" appearing within the title, abstract or the keywords. This high number of publications reflects in part the first goal of the bioeconomy which is to merely link in finding alternative solutions to fossil fuels based on biological resources. Finally, the area concerning the innovation from different perspective (product, process, and organizational) is also recurrent in the outcome obtained.

European bioeconomy has emerged as a new paradigm with the aim of minimizing the impact of economic activities on natural recourses while strengthening the competitiveness of firms and supporting local systems (Fritsche et al., 2020). This is the main reason why scientific research have shown growing interest on bioeconomy issues in the latest years. Studies on bioeconomy embed different fields of interest mainly related to research and innovation referring to various stakeholders such as producers, organizations and users. While the interest in the bioeconomy is growing and it also reflected by the numerous Bioeconomy strategies worldwide, a comprehensive economic analysis is still missing (Viaggi et al., 2021). In relation to this, Viaggi et al. (2021) provide a holistic framework reviewing the recent literature on the Bioeconomy focusing on the four main aspects of supply, demand, market, and business models identifying also pathways for future economic research.

Focusing on primary production in the bioeconomy, in the context of entrepreneurial studies, agriculture has been neglected for long time as a research field. The main reason is because in the past it has been considered a traditional sector with low tech investments and low innovation rate (Condor, 2020). However, the European Commission (2018) in the first *Bioeconomy Strategy* already had identified the role of agricultural entrepreneurs and private actors. In this regard, the primary production—especially, agriculture and forestry—has a great potential to directly contribute to the five objectives of the strategy. Agricultural activities do not only impact farm management but also affect the utilization of natural resources, the environment, and the land conservation, implying indirectly a consequence for society (Kuckertz et al., 2020). According to some scholars, enterprises in the bioeconomy are, by definition, oriented towards the concept of sustainable entrepreneurship where the competitive position is measured not only by economic advantages but also on the creation of social and environmental value (Horisch, 2015). The process of creating commercial opportunities from natural science is defined as bioentrepreneurship by Wulandari (2021). While entrepreneurship inherently presupposes a level of innovation activities, and, hence, risk, due to the nature of biological resources, bioentrepreneurship presents even more risks. Indeed, knowledge-based areas such as the bioeconomy intrinsically include a risk component; for example, limited information, low level of knowledge transfer and consumer acceptance could put the actors more at risks. Part of the literature stresses the importance of consumers acceptance which is also connected to the issue of the access to information. As a matter of fact, media and communication campaigns play a pivotal role in defining new trends among the population (Viaggi et al., 2021). For instance, stimulating the demand for bio-based products could be archived through a coherent policy strategy.

From the business perspective, due to the complexity of bioeconomy, new processes rely on a high degree of investments. Sołtysik et al. (2019) define this type of innovations as eco-innovation which are not determined by demand and supply but depend on many intercorrelated factors as laws, market position and legal standards (e.g., environmental ones). Furthermore, eco-innovative activities rely more on external source of knowledge compared to general innovations; for example, they result from collaborations with research centers as universities (Del Rio et al., 2015). This condition requires that farmers show an extra willingness to innovate, and to, again, accommodate a risk-taking behavior. Given these conditions, the adoption of innovative bio-based businesses is still highly compromised because of the existence of high costs, specific knowledge and high market uncertainty (Borge and Bröring, 2020).

Beside this, an additional issue is how to reduce risks related to external factors for complex and multi-layered systems which are intrinsically part of the bioeconomy. For instance, the outbreak of Covid-19 emergency has exposed the global systems to a higher vulnerability using pressure on supply, disrupting food chains and increasing the level of food insecurity. The pandemic has sharpened the awareness of potential future and unpredictable crises. Under all these conditions, the need to adapt societies and economies to be more resilient to shocks is clear. An interesting point highlighted by Viaggi and Zavalloni (2021) concerns the evolution of business relations from value chain to value web. In this regard, the structure of interactions among the actors involved is not defined linearly, as it is in a standard value chain, but rather as a web of relationships. Therefore, the higher number of relations should ensure a higher degree of flexibility. This new concept of value web promises a better resilience of the whole economic systems (Viaggi et al., 2021). Even though, the interdependence could also have the practical consequence of increasing the complexity of the processes and relations.

However, the bioeconomy promises to bring new business opportunities especially in the primary sector where extra revenues can be generated beyond the mere agricultural production. Furthermore, it is fundamental to stimulate the upgrading of the simple businesses present to more innovative ones and to build integrated value chains embedding all the actors involved starting from primary producers. The bioeconomy itself, by definition, interconnects different business sectors even with surprising links in terms of raw materials and feedstock.

To overcome the often-limited economic size of firms, a large degree of cooperation between different agents is required to develop complex knowledge (Stegmann et al., 2019). Historically, there is a lack of tradition among primary producers to cooperate with other sectors of relevant industries (e.g., energy, pharmaceutical, textile, etc.). The collaborative behavior poses a number of opportunities for primary producers: a significant bargaining power as a consequence of higher production volumes when economies of scales exist; stronger competitiveness and better access to market; portfolio diversification in production while reducing risks and creating new economic benefits, better investment capacity due to pooling resources and control over the whole production; facilitated access to information and to new knowledge. Networking and collaboration emerge as a new organizational approach among business actors. Moreover, the collaborative behavior has also been defined in the literature as a collaborative advangtage supporting lon-term parthnerships between firms (Kanter, 1994). Thereby, in several contributions, the establishment of new relations is even presented as an actual innovation (Viaggi et al., 2021). However, once again, uncertainty could also increase given the growing degree of interdependence.

On the other side, poor cooperation implies a lack of practical knowledge on bioeconomy issues by stakeholders. In general, stakeholders as innovation brokers, advisers or public authorities have showed to have a limited knowledge about what bioeconomy is (Wilde and Hermans, 2021). In particular, primary producers, being unaware of potential benefits of alternative market possibilities for their by-products, show little interest in participation in the bioeconomy and in seeking support for developing new business relationships.

Another issue refers to the limited access to information in the bioeconomy. The available data about biobased innovations are mostly confidential or even protected by patents or intellectual property rights (Del Rio et al., 2015). Accessible data on the successes or failures could boost the involvement of other agents in the bioeconomy. However, the information is often not easily accessible to the public. For this reason, new initiatives are supporting alternative learning processes. In this regard, there is an extra need to stimulate and promote the sharing of knowledge and expertise via alternative knowledge dissemination or sharing techniques in order to create new networks among different entities already involved in the European bioeconomy (Borge and Bröring, 2020).

In addition, consistently with this, entrepreneurial orientation has been identified as a factor positively correlated to knowledge transfer (Lancker et al., 2016). These aspects are also defined by the concept of *entrepreneurial culture*, characterized by openness, positive attitude, and creativity (Kuckertz et al., 2020). However, time and efforts are required to establish relationships and to initiate new contacts especially when it comes to cross sectoral cooperation. Enhancing awareness and communication among entrepreneurs is a necessary condition for the collaboration among themselves and their integration in the bioeconomy. This aspect is reflected by Kuckertz et al., (2020) as they have investigated the influence of informal collaboration among entrepreneurs in terms of knowledge creation in networks dynamics. Moreover, according to Morrison and Rabellotti (2009), relationships promote the sharing of knowledge especially within the established cluster. However, little attention has been paid to the key role of person-based relationships in the knowledge dissemination process. In this sense, the establishment of networks could be beneficial in emerging knowledge-based sectors such as the bioeconomy, given the interactions and the knowledge transfer through new dissemination strategies and activities.

While collaboration has been largely identified as a precondition for the development of the bioeconomy, actions to stimulate new interactions are often lacking especially when considering private initiatives. On the other side, entrepreneur capacity is not considered the decisive bottleneck in blocking bioeconomy opportunities. The stakeholders ask that the collaborative initiatives should be supported by institutions at national and international level (Wilde and Hermans, 2021). To tackle the issue already mentioned, on primary producers' participation, European institutions have increasingly promoted new cooperation opportunities through funded projects involving the member states. On many occasions, a multi-actor approach is proposed to stimulate cooperation not only among private actors but also involving public authorities at different levels. Empirical application of the European bioeconomy will be discussed in the next section.

3.2 Initiatives in the European bioeconomy and case studies of new business opportunities through cooperation

Barriers still exists in the development of sustainable business model especially in the primary production. As a matter of fact, the bioeconomy and the circularity of the processes appear still as theoretical concepts for farmers. The EU is struggling to identify new strategies in order to stimulate the implementation of the bioeconomy, starting from primary sector, and to encourage a meaningful participation of farmers. In this regard, a growing number of initiatives are focusing in how to successfully engage farmers in the process of strengthening the European bioeconomy. Among them, the H2020 EU research project entitled Coopid provides a recent example of collaborative approach valorizing existing leading initiatives. The main goal of the research project is to stimulate the adoption of biobased solution in the primary sector. In doing so, project partners are trying to establish a network of European bioeconomy clusters starting from 10 European countries in the primary sector. The project enhances the development of the bioeconomy through connecting successful business model with the active participation of primary producers. The project has managed to mobilize target actors as that is primary producers' associations and cooperatives, other industry & business, academia & researchers as well as regional/local public sector bodies, having a wide geographical scope. The novel aspect of the research is focused on the transformative role of stakeholders in boosting collaboration in bioeconomy cluster. As a matter of fact, the knowledge dissemination is based on the organization of visits to selected successful biobased business where primary producers participate actively.

In the context of the research projects, different success stories were selected in the European scenario. In the winery sector, an example is represented by the Italian cooperative Caviro. The company developed a virtuous circular business model where a range of different bio-based solutions are adopted reaching social, environmental, and economic sustainability. In addition, the valorization of the residues from the wine production is facilitated due to the low transaction and transport costs of the byproducts. The cooperative adopts a high technology based on innovative biorefinery process. As in this case, the sustainability aspects of the circular bioeconomy have contributed to rural development in the areas while creating new job opportunities. The role of primary producers is fundamental: the group relies on the important function of its members which at the same time own a smaller business and are co-owners of the cooperatives. In this sense, primary producers provide grapes as raw materials and, at the end they obtain the residues of the entire production process as natural fertilizer. To promote a long-term relationship between Caviro and its members, the cooperative pays attention in the dissemination of information to guarantee a transparent decision-making process.

Entrepreneurs are attracted by new business opportunity in the bioeconomy where residues are considered as source of extra profit. It is the case of *Fiusis*, an Italian company collecting pruning residues from olive tree and then converting them to bioenergy. The residues from olive tree are usually costly to dismiss by farmers and even dangerous when burnt in open fields. In fact, valorized byproducts as this one can offer extra economic benefits when companies buy them to produce pellets for energy purposes, adding high value to this byproduct. Moreover, the valorization of the residues creates a new commercial opportunity for the farmers based on the innovative products obtained. On the other hand, the process stimulates the improvement of local infrastructures, the enhancement of collaborations with consortia and companies in the agro-food and other sectors while contributing to the promotion of a wider application of the bioeconomy in rural areas.

4. Discussion and conclusions

The purpose of this book chapter is to increase the understanding of the challenges and opportunities inherent to the role of agriculture and primary producers in the development of the European bioeconomy. Moreover, the research contributes at the development of a more comprehensive framework discussing the key role of collaborative behavior in supporting the development of bioeconomy and exploring the active role of stakeholders in the process of knowledge dissemination. Despite the potential advantages, the adoption of innovative bio-based solutions in the European primary production sector is still limited due to two main reasons: (i) poor cooperation and (ii) low awareness and knowledge. Supporting at the same time the research for the required knowledge, co-developing skills, expertise and competence, is essential in creating an innovation-supporting environment. The collaborative behavior is also influenced by external factors that shape the stakeholder's attitude. The main condition influencing the engagement of primary producers has been identified as a join effect of factors, not only acting at individual level but also in relation to the social, institutional contexts and markets where the actors operate.

In essence, individuals with entrepreneurial drive are more likely to undertake new initiatives. However, we suggest that the impact would be higher if the entire process could be supported by top-down initiatives such as European projects. In this regard, a better integration of primary producers in the bio-based value chains ensures a fairer distribution of value, which is directly associated to societal, environmental, and economic benefits. Finally, we have investigated how informal collaboration among entrepreneurs positively affects the knowledge dissemination. The promotion of cooperation and networking among different stakeholders contributes to the progress of the bioeconomy development, going beyond the mere spheres of the primary production. Finally, the most recent Covid-19 pandemic has shown that global systems are vulnerable when influenced by significant shocks. Under those conditions, the market becomes unpredictable. It is therefore important to learn from the lessons of past crises increasing the flexibility and the resilience of all the actors involved in the bioeconomy.

The main limitations of this study are represented by the limited number of contributions reviewed, compared to similar works in other fields. The reason can be linked to the very specific research topic on collaboration issues in the bioeconomy, that counts at the moment a limited number of papers. This suggests that further research is needed to analyze the impact of relationships and entrepreneurship in the bioeconomy. More qualitative and case-study-based research may be useful in this direction. However, with the growth of the bioeconomy sector, it may be expected (and sought) a growing number of quantitative studies and hence the possibility of more structured cross-cutting reviews.

The Covid-19 pandemic is still causing disruptions at key bottlenecks in the global value chain causing increasing instability: even in the future, the potential for shocks and instability will remain a key factor affecting society's organization. Hence, future research will need to include more explicitly contributions on organizational adaptation to ensure flexibility and resilience as well as in the adoption of creative solutions and in the prediction of unexpected shock.

Acknowledgments

We acknowledge support of this work by the research project H2020 COOPID "COOPeration of bioeconomy clusters for bio-based knowledge transfer via Innovative Dissemination techniques in the primary production sector" funded by the European Commission (reference number for the project 101000519).

References

Antikainen, M., Valkokari, K., 2016. A framework for sustainable circular business model innovation. Technol. Innov. Manage. Rev. 6 (7).

Borge, L., Bröring, S., 2020. What affects technology transfer in emerging knowledge areas? A multi-stakeholder concept mapping study in the bioeconomy. J. Technol. Transf. 45, 430–460.

Carus, M., Dammer, L., 2018. The "Circular Bioeconomy"—Concepts, Opportunities and Limitations. Nova-Institut, Hürth (Germany).

Condor, R., 2020. Entrepreneurship in agriculture: a literature review. Int. J. Entrepreneurship Small Bus.

Del Rio, P., Penasco, C., Desiderio, R.-J., 2015. What drives eco-innovators? a critical review of the empirical literature based on econometric methods. J. Clean. Prod.

European Commission, 2018. A Sustainable Bioeconomy for Europe: Strengthening the Connection Between Economy, Society and the Environment Updated Bioeconomy Strategy. Bruxelles.

Fritsche, U., Brunori, G., Chiaramonti, D., Galanakis, C., Hellweg, S., Matthews, R., Panoutsou, C., 2020. Future transitions for the Bioeconomy towards Sustainable Development and a Climate-Neutral Economy—Knowledge Synthesis Final Report. Publications Office of the European Union.

Fund, C., El-Chichakli, B., Patermann, C., 2018. Bioeconomy Policy (Part III): Update Report of National Strategies Around the World. Berlin.

Horisch, J., 2015. The Role of Sustainable Entrepreneurship in Sustainability Transitions: A Conceptual Synthesis against the Background of the Multi-Level Perspective. Administrative Sciences.

Kanter, R.M., 1994. Collaborative Advantage: The Art of Alliances. Harvard Business Review.

Kuckertz, A., Berger, E.S., Brändle, L., 2020. Entrepreneurship and the Sustainable Bioeconomy Transformation. vol. 37 Environmental Innovation and Societal Transitions, pp. 332–344.

Lancker, J., Wauters, E., Huylenbroeck, G.V., 2016. Managing innovation in the bioeconomy: an open innovation perspective. Biomass Bioenergy, 60–69.

Morrison, A., Rabellotti, R., 2009. Knowledge and Information Networks in an Italian Wine Cluster. Vol. 17(7).

Sołtysik, M., Urbaniec, M., Wojnarowska, M., 2019. Innovation for Sustainable Entrepreneurship: Empirical Evidence from the Bioeconomy Sector in Poland. Administrative Science.

Stegmann, P., Londo, M., Junginger, M., 2019. The circular bioeconomy: its elements and role in European bioeconomy clusters. Resour. Conserv. Recycl.

Viaggi, D., 2015. Research and Innovation in Agriculture: Beyond Productivity? Bio-based and Applied Economics.

Viaggi, D., 2018. The Bioeconomy. CABI.

Viaggi, D., Zavalloni, M., 2021. Bioeconomy and Circular Economy: Implications for Economic Evaluation in the Post-COVID Era. Circular Economy and Sustainability.

Viaggi, D., Bartolini, F., Raggi, M., 2021. The Bioeconomy in Economic Literature: Looking Back, Looking Ahead. Bio-based and Applied Economics.

Wilde, K., Hermans, F., 2021. Innovation in the bioeconomy: perspectives of entrepreneurs on relevant framework conditions. J. Clean. Prod.

Wulandari, S., 2021. Bioentrepreneurship approach as a pillar to accelerate the integrated farming system implementation. In: OP Conference Series: Earth and Environmental Science.

CHAPTER 13

Biofuel co-products for livestock feed

Ugo De Corato and Egidio Viola
Department of Bioenergy, Biorefinery and Green Chemistry (TERIN-BBC-BIC), Italian National Agency for New Technologies, Energy and Sustainable Economic Development (ENEA), Bari, Italy

Abbreviations

C	carbon
CDO	corn distillers oil
CGF	corn gluten feed
CGM	corn gluten meal
CO_2	carbon dioxide
DDG	dried distillers grain
DDGS	dried distillers grain with soluble
DLUC	direct land-use change
EC	European Commission
EU	European Union
FAO	Food and Agriculture Organization
GHG	greenhouse gases
ha	hectare
IEA	International Energy Agency
ILUC	indirect land-use change
LUC	land-use change
N	nitrogen
P	phosphorus
RED	renewable energy directive
SRF	short rotation forestry
US	United States
USDA	United States Department of Agriculture
WBA	World Bioenergy Association
WDG	wet distillers grain
WDGS	wet distillers grain with soluble

1. Introduction

The increasing price and environmental impact derived from fossil fuels have increased the production of biofuels to reach unprecedented volumes over the last 15 years. The biofuel industry has evolved rapidly over the last decades developing new processing techniques and expanding the biofuel

Agricultural Bioeconomy
https://doi.org/10.1016/B978-0-323-90569-5.00010-X

crops and agri-based commodities as an important feedstock. Biofuels can be produced domestically leading to low fossil fuel importations, but the application of biofuel co-products is still far to support their production costs. The standardization of co-products quality both within an individual crop or between several crops to search new feedstocks with more safety standards are needed since much researches concerned the potential co-products feeding value. The production of biofuels which do not compete with crop food land and the valorization of agro-waste are believed to be the main challenges for making sustainable biofuels supply chains. The sustainable use of residues, litters, and wastes for biofuels which do not require arable lands with limited or zero environmental risks needs to be globally encouraged and promoted by using agri-based feedstocks including surplus forest growth, as well as establishing energy crop plantations on unused land that may be less expensive than creating large-scale energy plantations on fertile land (Ullah et al., 2015). The assessment of the impacts that extensive biofuel production may cause to food supply chains and to ecosystem services has become considerably important given the increasing requirement of land for biofuel. At the same time, there are growing opportunities and requirements to use biomass for providing energy for heat, power, fuel, and bio-based materials in consideration of increasing population growth and change of their consumption patterns, dietary preferences, and fruit/vegetable losses due to phytosanitary concerns. Although biomass from energy crops is expected to play substantial roles in the future energy systems, nevertheless the bioenergy worldwide potential is still too limited because lands are still too multi-functional, being needed for food, feed, timber, fiber production, biodiversity conservation, and climate change protection. Furthermore, the bioenergy implications for climate change mitigation remains often unclear due to large uncertainties about the crop yield improvement and land availability for biomass plantations. Large-scale cultivation of energy crops is likely to affect bioenergy potentialities, global food prices, and water scarcity in arid and semi-arid soils. In the last 35 years the global energy supplies have nearly doubled, but the relative contribution from renewables has increased from 13% to 19% by including 9.3% of traditional biomass and 9.7% of new renewable source (Fig. 1).

Although the contribution of modern renewables is still marginal when compared to total global renewable energy supply, nevertheless the integrated policies for energy, land use, and water management are continuously needed (Popp et al., 2014). Major emphasis for development of bioenergy systems has been given by alternatives to fossil fuels. For instance, the

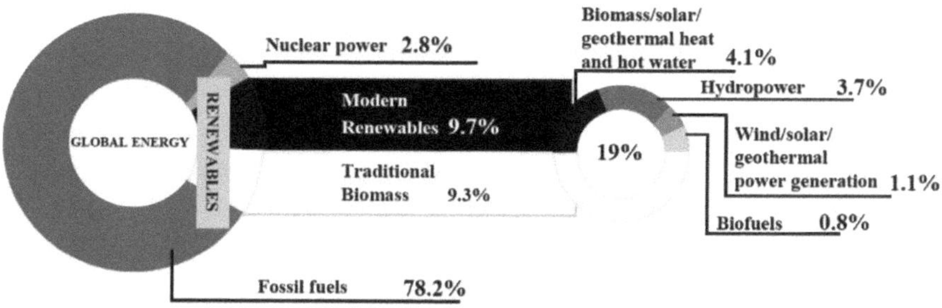

Fig. 1 Estimated renewable energy share of global final energy consumption in 2011. *(Taken from Popp, J., Harangi-Rákos, M., Gabnai, Z., Balogh, P., Antal, G., Bai, A., 2016. Biofuels and their co-products as livestock feed: global economic and environmental implications. Molecules 21, 285 by alignment with the copyright rights of the CCBY license of the MDPI journals.)*

transportation sector is responsible for about 20% of total energy worldwide. Among them, biofuel is currently the fastest growing bioenergy sector for the transport, even though they represent around 3–4% of total road transport fuel and only 5% of total bioenergy consumption. Biofuels have received the most attention, although just a small fraction of biomass is still used globally for their production. Many countries have incentived production of liquid biofuels for transportation to enhance domestic energy security, to develop spur economic, and to reduce emissions of Greenhouse Gases (GHG) and other pollutants (Popp et al., 2016). The installed production capacity values showed that the global biofuel production capacity increased from 1.58 billion liters in 2010 to 4.21 billion in 2013 although the actual production volumes were not still available (WBA, 2014). Global expansion of biofuel production is accelerated during the next decade, although at slower pace than over the last half decade. The two primary biofuels produced worldwide are ethanol and biodiesel, being predominantly derived from agricultural commodities, while considerable researches and developments are in progress to commercialize new feedstock. Currently, around 80% of the global production of liquid biofuels is under the form of ethanol, as global fuel ethanol production reached 108 billion liters and those of biodiesel amounted to 28 billion liters in 2012–14 (Popp et al., 2016). The two tops ethanol producers, the United States (US) and Brazil, accounted for around 75% of total production, while biodiesel is far less concentrated than ethanol around the world. The European Union (EU) remains the core business of global biodiesel production with 12 billion liters, such representing about 77.6% of total output. Ethanol production in the US is expected to be relatively flat over the next decade due to decline of gasoline. Nevertheless, most of the additional ethanol production is expected to take place in Brazil (USDA, 2015). Moreover, Indonesia will surpass the US and Brazil in the latter years of the outlook period to become the second largest biodiesel producer behind the EU. Global ethanol and biodiesel production are both expected to expand in order to reach, respectively, almost 135 and 39 billion liters by 2024 (FAO, 2015). In 2040, the share of biofuels in road transport fuels would range from 5% to 18% globally, whose from 11% to 31% in the EU, and from 11% to 29% in the US (IEA, 2015a). The bulk grains transformed worldwide into fuel ethanol overall includes maize, both in the US and China. However, an increase in the offtake of wheat for fuel ethanol has also been observed in Canada and the EU. For these reasons, the increased biofuel production has led to criticisms and concerns about food availability, while it is feared that rising demand for agricultural land will lead to deforestation, grassland conversion, and increased GHG emissions. The main issue is based on the expected impacts of biofuel production due to introduction of dedicated targets and policies (Elobeid et al., 2012; Khanna and Zilberman, 2012). Most models were developed to evaluate agricultural-environmental policies that were later adapted in order to incorporate biofuel production (Witzke et al., 2008). The biofuel supply chains produce both fuel and other important co-products, whose the type and quantity of them strongly depends on the chain. Earlier models did not included co-products from the second generation biofuel technology (Golub and Hertel, 2012). Today, central to the debate on the impact of biofuel production is how the increased biomass requirement could fit with the area expansion, yield improvement, and increased cropping intensity. However, such increases were not sufficient to fully satisfy biofuel policy objectives in the US and the EU. Further expansion of biofuel production is also expected in Brazil, China, and Argentina. The economic impacts of biofuel policies should be considered with care since a strong expansion in biofuel production has been observed by changing the biofuel policies.

Co-products generation in early biofuel impact assessments was ignored leading to an overestimation of land requirements and GHG emissions. The outputs of such co-products are relatively high in the US, EU, and China due to the large share of grains for ethanol production with high feed yields, whereas, it is quite low in Brazil where ethanol production is dominated by sugarcane which generates no feed co-products. Likely, co-products yield is low for rapeseed and soybean in biodiesel industry. The estimated impacts of biofuel production often have used models with limited ability to incorporate economic and environmental aspects, thereby overlooking the generation of biofuel co-products. Although a significant development to commercialize next-generation biofuels is still under consideration, these are unlikely to be produced at high quantity in the short term. Furthermore, several co-products are not likely to have any application in the animal feed market (Makkar, 2012).

2. Structure of the chapter

The present chapter first describes the feedstock suitable to produce both traditional or advanced biofuels by comparing them into the fuel market framework. Afterwards, it focuses biofuel co-products suitable for livestock feed. Finally, it analyzes the economic and environmental impacts related with the increasing expansion of biofuel market worldwide due to the increased production of co-products from the bioethanol and biodiesel supply chains. This chapter has outlined, in the last section, how the replacement of cereals and oleaginous crops with biofuel co-products into livestock feed rations has determined the net cropland reduction for cultivation of energy crops, thereby opening the way to food crops. Finally, it was concluded from this contribution that the back-flow of biofuel co-products into livestock market has rationalized the cropland use by reducing GHG emissions as well as in developing and developed countries.

3. Biofuel for a sustainable market

The biofuel expansion between 2000 and 2010 in Brazil, US, Indonesia, Malaysia, China, Mozambique, South Africa, and the EU was investigated by Langeveld et al. (2014). About 30–35 million hectares (ha) of feedstock was needed for biofuel in 2010, while the global cropland ratio for biofuels is around 2% with wider differences among countries and regions. In 2010, the aforementioned countries produced 86 billion liters of ethanol and 15 billion liters of biodiesel representing 95% of global biofuel production. Land dedicated to biofuel was calculated of 32 million ha in 2010 with an increase of 25 million ha when compared to 2000. Land required to produce biofuels grown from 30 million ha in 2010 to 100 million ha by 2050 (IEA, 2011).

First generation biofuels are produced basing on well-known processes and feedstocks (Ullah et al., 2015). Apart from sugarcane ethanol, the traditional biofuels will hardly be sustainable in the future as large-scale production would take away feedstock and land from food production and forestry.

In addition, biofuels are rather expensive and offer only limited reduction in terms of GHG emissions if compared to fossil fuels. Biofuels from lignocellulosic matter may be sourced from crop residues or other agro-wastes by planting and harvesting fast-growing trees or grasses. A direct competition with food is still likely because such crops grow best and are most easily harvested on fertile lands already dedicated to food crops. Using cropland to grow trees and grasses rather than food crops for biofuels, will probably not reduce, let alone eliminate, competition for cropland. Trees and grasses will have a hard time for producing more biofuels per hectare than food crop-based biofuels. Alternatively, cellulosic crops might rely on harvesting existing forests or producing fast-growing trees or grasses or woody savannas. Some researchers point to abandoned farmland as a good candidate for biomass production that thus avoids competition for land. But, abandoned farmlands typically regenerate into forests, woodlands, or grasslands if left alone which also provide climate benefits that are already assumed and counted in climate change assessments. These benefits would be sacrificed by using these land for biofuels. Although there is capacity to increase plant production per hectare by increasing yield or by enhancing use of lands that are degraded or marginal, nevertheless this capacity is already needed to meet rising demands for food and wood products by preserving ecosystems. There are some feedstocks that avoid the competition for land, such as agro-wastes, agricultural and forestry residues, but, such residual biomass might contribute only modestly in replacing fossil fuels in the long time. Phasing out the dedicated use of land to generate biofuels, would reduce the food gap and keep it from greatly expanding. Searchinger and Heimlich (2015) defined the land for bioenergy as the production of bioenergy that sacrifices alternative outputs from land. For instance, some feedstocks for biofuel production would not fit this definition because the land produces firstly protein (animal meal) and secondarily cooking oil, biofuel feedstocks, and other biofuel co-products. These authors certainly did not promote bioenergy production from such feedstock, but croplands can be very different in terms of suitability and productivity for different vegetation types. Searchinger and Heimlich (2015) correctly pointed out that redirecting wastes like crop residues and municipal solid waste into biofuel production is a good use of these resources. Land that may not be suitable for crops could be suitable for growth of other vegetation types because of differences in nutrients, water, climate, and other requirements. Further, Wang and Dunn (2015) treated marginal/degraded lands as not existing. However, Lynd et al. (2009) raised concerns on dedicated cellulosic crops that are not a promising option for a biofuel feedstocks because they require land that do not give high yield sufficiently. Reports indicated that significant areas of marginal land still exist worldwide that should be used to produce cellulosic crops that are still currently under-utilized (Cai et al., 2011). Many studies have identified land where cellulosic biomass can be grown to avoid competition between cellulosic biomass and food production (Perlack et al., 2011; Werling et al., 2014).

The development of second and third generation biofuels for diversification and de-carbonization of biofuel markets is needed in longer time (Ullah et al., 2015). The conversion of renewable resources into biofuel requires transition processes that are currently under commercial demonstration or under development, with small plants in operation and large plants under construction or planned all over the world. Since 2013 advanced biofuels have followed good progress trend with nine commercial-scale plants commissioned in the EU, whose seven of them build from 2014 to 2015. Policies that mandate blending levels and provide capital incentives along with the development of secure local feedstock supply chains, have been fundamental because there is a significant bioenergy potential able to reduce the costs, but a sustained long-term policy commitment would be needed which may face risks from the lower oil price environment (IEA, 2015b). Further reductions in processing costs will be necessary to achieve a product that will be competitive with first generation biofuel. A combination of feedstock for first and second generation biofuel (e.g., corn cobs together with corn stover) can eliminate some bottlenecks and lead to major competitiveness (Paulova et al., 2013). IEA reports higher water consumption in the production of advanced biofuels when compared with conventional biofuels, referring specifically to lignin-cellulosic fuel ethanol (IEA, 2010). Local impacts on water quality and availability should be carefully monitored because the major part of water consumed would be for feedstock cultivation, so that biofuels from wastes and residues should have a reduced water footprint (Kretschmer et al., 2013). The processing of feedstock into biofuels by steam explosion requires high energy inputs. However, other environmental impacts resulting from the processing of biomass through advanced conversion technologies should be investigated and adequately addressed, such as water consumption (Ullah et al., 2015). In Europe, the policy requires that every member state obtain 10% of its transport fuel from renewable sources by 2020, however, new legislation limit the contribution of biofuels derived from sugar, starch, and oil crops due to sustainability concerns. This, in combination with adequate national legislation to biofuel policies, has raised producers to possibility of developing advanced biofuels that has lagged due to the lack of EU-wide policy support, although some member states started to enact national policies. Italy, for example, has announced an increasing of 0.6% next-generation biofuel by 2018, such promising to be more sustainable with higher GHG emission reductions. These estimated values are based either on biofuel feedstock or on land not used for food production and farming since forestry and agricultural residues, and other organic wastes (including municipal solid waste) would provide between 50 and 150 EJ/year, while the remainder would come from energy crops, surplus forest growth, and increased agricultural productivity (Fig. 2).

3.1 Feedstock for the first generation biofuel: Multifunctional (annual) crops

Biofuel can be obtained by the following technological processes (Naik et al., 2010; De Corato et al., 2018): (1) oil trans-esterification for the

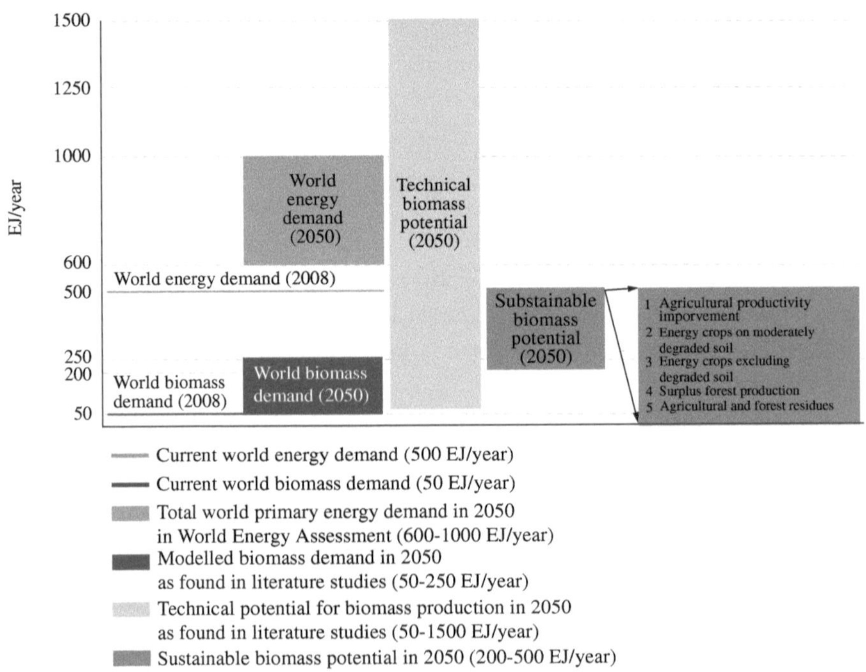

Fig. 2 Global bioenergy sources. *(Taken from Popp, J., Harangi-Rákos, M., Gabnai, Z., Balogh, P., Antal, G., Bai, A., 2016. Biofuels and their co-products as livestock feed: global economic and environmental implications. Molecules 21, 285 by alignment with the copyright rights of the CCBY license of the MDPI journals.)*

production of biodiesel from oilseeds, (2) steam explosion of fresh and/or air-dried plant biomass for the production of fuel ethanol from lignin-cellulosic sources, and (3) pyrolysis of plant biomass for the production of charcoal (solid), bio-oils (liquid) and gaseous fuels. In addition, landfill is a very old technique in which all municipal waste and refuse accumulate in large ditches, valleys, or specially excavated holes in the land where bacteria decompose waste to produce gaseous biofuel, especially methane. Landfill methane is used not only for electricity generation but also for firing brick kilns and producing steam in industry. The same purpose is achieved, according to the latest technologies, through the anaerobic biological digestion of organic solid waste and plant biomass into bioreactors for concomitantly producing biogas and digestate as a co-product. The biogas is then tapped and used either in houses or in power stations, while digestate could also be composted and then used as organic input in agriculture. Biofuel is not the only product made from energy crops such as rapeseed, wheat, and soybean. Rapeseed oil is used for biodiesel production (0.35 kg/kg of raw rapeseed) by trans-esterification (with methanol) with the main by-product glycerin (0.035 kg/kg). Rapeseed meal (0.5 kg/kg) is a protein-rich by-product that can be used as feed or for process energy. Other possible by-products are honey and straw. Starch from wheat is the main resource for ethanol production (0.3 kg/kg of wheat grains). Dried distillers grain with soluble (DDGS) is a protein-rich by-product of ethanol production (0.3 kg/kg) suitable for feed or process energy. Other by-products are bran and straw (used in stables). Soybean meal is the main product from soybean (0.76 kg/kg of soybeans). From the soy oil about 0.16 kg/kg of biodiesel can be produced with 0.016 kg/kg of glycerin as a by-product.

Fuel ethanol is mostly made from maize and wheat in the US and China. Sugarcane is the predominant feedstock for ethanol production in Brazil and South Africa. In 2014, most fuel ethanol was produced from sugar crops (61%) with the remainder aliquot from cereals grain (39%). Fuel ethanol feedstocks vary significantly depending on the country or region. For example, ethanol production in the US is based largely on maize, Brazil relies primarily on sugar crops, whereas China on maize, sweet sorghum, cassava, and other non-grain crops. The fuel ethanol share demand of corn, wheat, and other coarse grains is about 6% of global grains production. The fuel ethanol sector accounts for 13% of global maize consumption and 20% of global sugarcane production. An estimated 143 million tons of grain is used globally for fuel ethanol. The US is the global leader in grain ethanol production by accounting for roughly 90% of total grain use, followed by the EU, China, and Canada. The European and Canadian producers principally use wheat and maize for ethanol. About one-third of the volume of cereals grain processed for ethanol was used to produce livestock feed, whereas the equivalent of two-thirds of the grain used to produce only ethanol.

On the other hand, soybean and rapeseed oil has a 70% share of the total feedstocks used in biodiesel production worldwide. Rapeseed, canola, and soybean represent the most common source of oilseeds for biodiesel. An estimated 6 million tons of rapeseed oil and 7 million tons of soybean oil is used globally for biodiesel production, representing roughly 70% of the total feedstocks used in global biodiesel production (Licht, 2013). The biodiesel share demand of rapeseed, soybean, and oil palm is around 11% of global vegetable oil production. The share of waste biodiesel feedstocks, such as animal fats and used cooking oils, increased to 15% in the total biodiesel outputs. Global biodiesel production is based largely on vegetable oil, mostly from rapeseed (Europe) and soybean (US, Brazil, Argentina). The highest biodiesel yields were observed for oil palm (Indonesia and Malaysia), but there is a feedstock of minor importance, such as castor bean in Brazil, sunflower in the EU, and *Jatropha curcas* and coconut in Mozambique. Biodiesel production also includes industrial by-products such as exhausted cooking oils (the main feedstock in China) and animal fats. In Europe, the relative share of cooking oils and tallow in biodiesel production is increased as the EU policy has allowed to such feedstock to be double-counted in transportation targets. Continuously growing demand for protein meal has been the main driver in recent years behind the expansion of oilseed production. This has increased the share of protein meal in the value of oilseeds and favored soybean over other oleaginous crops. When compared with coarse grains and other feed ingredients, protein meal prices have stayed relatively high where global oilseeds production have reached 530 million tons (soybean 320 and rapeseed 70 million tons) in the 2014/2015 marketing year. At the same time, soybean production increased faster than rapeseed and sunflower. Vegetable oil production increased to 180 million tons (out of this 60 million tons palm oil) since vegetable oil for biofuel production accounted for about 20 million tons per year. Demand growth has slowed due to stagnating biodiesel production from vegetable oils in developed countries. Production of rapeseed in Canada and the EU is expected to grow much slower than in the previous decade as high oil-containing oilseeds, like rapeseed are more affected by the slower growth in vegetable oil prices.

Among the best feedstock for biofuel, such as maize and sugarcane for producing fuel ethanol, and soybean, rape, sunflower, and canola for biodiesel, only sugarcane appeared to have a real future although from maize will depend the yield improvement rate that could be achieved. The partition rate between the increase in demand and the increase in production suggested that maize will lose long-term economic viability, unless the already high rate of yield per hectare improvement can be increased yet further. The yields of soybean, rape, sunflower, and canola are still too low to contribute significantly to future energy supply with the only exception of oil palm (*Elaeis guineensis*) because soybean and rapeseed produce far too little biofuel per hectare to be competitive without mandates and subsidies.

3.2 Feedstock for the second generation biofuel: Perennial grasses and agricultural residues

Plant biomass is among the most available renewable resources that provide liquid, gaseous, and solid biofuels as it is organic matter made of hemicelluloses, cellulose, lignin, and minor amounts of extractives that is directly or indirectly related to plant growth. Plant biomass generally includes energy crops, agricultural trees, wood, firewood plantations, forestry residues, plant and animal waste, agro-industrial and municipal waste (De Corato et al., 2018). Biomass currently contributes to approximately 12% of the total energy supply worldwide in develop countries, while its contribution ranges from 40% to 50% in developing countries (Garg and Datta, 1998). The importance of plant biomass as a natural feedstock regards the biofuel chains that could increase in the near future because biomass is a promising source of raw material. New feedstocks suitable for producing second generation biofuel are: (1) herbaceous/woody perennial crops (grasses), and (2) agricultural/forestry residues.

3.2.1 Perennial grass

Biofuel from lignocelluloses energy crops cultivated on marginal land is likely to be cost-efficient contribution. However, extreme territorial conditions and climatic change of marginal land together to resistant species or varieties are required for rendering biofuels most competitive. Herbaceous perennial crops (*Miscanthus sinensis*, *Arundo donax*, *Panicum maximum*, *Sorghum bicolour*, and *Cynara cardunculus*) and woody species that grow quickly as short rotation forestry (SRF) (*Populus tremula*, *Eucalyptus* spp., *Salix purpurea*, and *Robinia pseudoacacia*) may be economically sustainable and environmentally eco-friendly because their long-term cultivation increases the carbon dioxide (CO_2) levels captured from the atmosphere into the biomass and the soil. These crops mobilize mineral nutrients from the stems and leaves into rhizomes and roots at the end of growing season, such reducing the fertilizer need for crops. These species can rehabilitate the quality of marginal land by restoring the soil organic matter content (Alshaal et al., 2013; El-Ramady et al., 2015; Jones et al., 2015; Pandey et al., 2016). Further requirement for such crops that growth rapidly are high biomass yielded without irrigation (high water-use efficiency) associated with low dose of pesticides (Ge et al., 2016). On the other hand, some biomass plants have invasive potential in the riparian areas. Particularly, shoot and rhizome fragments of *A. donax* (giant reed) can disperse and quickly populate new habitats and ecosystems during the floods along the rivers. Non-food energy crops such as woody plants (poplar, willow, *Eucalyptus* spp.) and/or herbaceous perennial grasses (*Miscanthus* sp., switchgrass, giant reed) can be used as feedstock to produce biogas as well in anaerobic co-digestion with animal manure and slurry (Stolarski et al., 2015; Allwright and Taylor, 2016). In addition, fast-rotation of non-food crops grown on marginal and non-arable lands and the solid organic fraction of urban bio-waste are considered as well.

Giant reed is nowadays considered among the best bioenergy/biorefinery crops according to very rich literature (Giessow et al., 2011; Nackley and Kim, 2015; Maceda-Veiga et al., 2016). Giant reed is a perennial and rhizomatous species which has been previously introduced as an ornamental/crop plant (Fig. 3) as it is a sterile plant without germinating seeds that can be easily

Fig. 3 Giant reed (*Arundo donax*) cultivation. *(Picture taken from Wikipedia.)*

and quickly propagated through rhizome or stem. Propagules can be produced also by hydroponic or in vitro cultures.

Giant reed produces high biomass yield per hectare and can be adapted to different types of soil and quality of water (Triana et al., 2015) as it has lower agronomic input requirement than food crops (Testa et al., 2016), and can be a good candidate to recover soil organic matter by contrasting soil decline (Alshaal et al., 2014a,b) having ability to bioremediate the contaminated soils and wastewaters from heavy metals (overall hexavalent chromium) (Elhawat et al., 2014, 2015). Despite there are limited data available about the production of biogas or bio-ethanol from giant reed, there is nevertheless increasing demand for the production of this perennial grass since low agronomic and energetic inputs are required for cultivation. From the second half of the 1990s, giant reed is regarded as one of the most promising source for biomass industry due to high biomass production per hectare. For example, a 12-year field trial in Central Italy without irrigation has produced annually on average 38 tons dry matter per hectare (Searchinger and Heimlich, 2015) and 20 tons dry matter on sandy soil without fertilization (Wang and Dunn, 2015). Due to its high biomass yielded per hectare and chemical composition, large amounts fuel ethanol can be produced (11.000 L/ha in case of 45 tons/ha of harvested biomass) (Franscisco et al., 2010; Shatalov and Pereira, 2012; Rabemanolontsoa and Saka, 2013; Corno et al., 2014; Komolwanich et al., 2014; Silva et al., 2015). In literature is reported that from 12.960 to 15.228 L/ha ethanol could be produced (Corno et al., 2014), which is higher than *M. sinensis* (Ge et al., 2016). Despite increasing interest for giant reed, propagation and production systems should be well optimized by reducing costs of bioenergy and bio-products production to enhance economic feasibility. It is necessary to improve sugar yield and reduce lignin content by optimizing the pretreatments process of giant reed for increasing ethanol production. Energy input and efficiency of converting giant reed into bioenergy and bio-products are both significantly influenced by the content of minerals, cellulose, hemicelluloses, lignin, moisture, and several recalcitrant compounds. Although giant reed can be easily cultivated almost in all climatic zones, nevertheless cold climate seems to be a strong limiting factor that has convinced researchers to select cold-resistant giant reed varieties (Cai et al., 2011; Kretschmer et al., 2013). But, giant reed has invasive potential in Mediterranean areas or warmer climates; so that, giant reed cultivation should be avoided in case of establishment of giant reed plantations surrounding water conservation areas in order to reduce potential hazards for agroecosystems. Finally, due to the resulting recalcitrant materials in giant reed, adequate pre-treatments of raw material is a crucial step to hydrolyze cellulose and hemicelluloses by removing or rearranging lignin structure and converting cellulose chain for enzymatic hydrolysis (Viola et al., 2016). Therefore, lignin-cellulose fractionating is more complex and overall more expensive for giant reed than sugary and starchy biomass used for first generation fuel ethanol.

3.2.2 *Agricultural/forestry residues*

A wide range of agricultural residues and plant-waste largely produced (corn stalks, rice husks, corn stover, wheat straw, cropping residues, unmarketable products, vegetable processing leftovers, etc.) can be used for producing second generation biofuel (Table 1).

Moreover, several agro-industrial wastes (spent coffee-ground, defatted olive marc, sugar beet bagasse, tomato-waste, cardoon-waste, vegetable-waste, etc.) are considered the most important organic inputs (De Corato et al., 2018). Such feedstock is considered sustainable as becomes waste material from food crop production that do not compete with food crops for land. Two main types of forestry resources are used for producing second generation biofuel: (1) residues from harvesting of wood (straws, woodchips, wood shavings, aspen chips, olive brash residues, branches, foliage, roots, tubers, etc.), and (2) complementary felling which describes the difference between the maximum sustainable harvest level and the actual harvest needed to satisfy the wood demand (Ullah et al., 2015).

3.3 Feedstock for the third generation biofuel: Microalgae

Marine, river and lake microalgae can be useful alternative source of biomass used to produce overall biodiesel and biogas and more else high-value products in a marine biorefinery model (Fig. 4).

Primarily, growth of microalgae for the production of oil and energy conversion has focused on species which have a great potential as feedstock for third generation biofuels. Algae have been cultivated commercially since the 1950s, mainly for the pharmaceutical industry, but have only recently received special attention as a source of bioenergy. Numerous studies of algae and other water-based biomass has been performed to identify stronger candidates for biomass accumulation rates, as well as lipid content for biodiesel production. Depending on their composition, different algae species may be suitable for a range of biofuels. Algae allows to reach a potentially high productivity per hectare, since they can grow on non-arable land, can utilize a wide variety of water sources (fresh, brackish, saline and

Table 1 Classification of residual biomass suitable to be recycled in an agri-based circular economy system for producing second generation biofuel (De Corato et al., 2018).

Residual biomass typology	Feedstock for producing fuel ethanol, biodiesel, and biogas
Crop residues	Straws, stover, husks, cobs, press cakes from rape and soybean seeds, olive marc and lees, grapes, olives and other fruit, bagasse, sugar beet pulps, palm kernel meals, palm oil mill effluents, empty fruit bunches and nutshells.
Forestry residues	Treetops, branches, stumps, leaves, sawdust, cutter shavings, scrap wood, wood pulp and wood chip.
Food processing residues	Oil-less seed cakes/meals, molasses, defatted olive marc, spent coffee-grounds, tea bags, exhausted beet pulps, tall oil pitch and animal fats.
Seaweeds and aquaculture fisheries residues	Brown, red, and green seaweeds, fish scales, viscera and scraps.
Municipal bio-wastes	Used cooking oils, exhausted engine oils, biomass fraction of mixed Municipal solid waste, animal manure, biomass fraction of industrial waste and sewage sludge.
Wet Municipal solid waste (European Waste Code: 200108)	Un-marketable products as vegetable and fruit by-products; food wastes as meat, fish, bread, eggs, cheeses, pastries, coffee grounds, tea bags; cropping residues as grass, leaves, flowers, twigs, extinguished wood ashes.
Green Municipal solid waste (European Waste Code: 200201)	Vegetable processing leftovers, large pruning, cuttings, and garden waste.

wastewater) and can potentially recycle CO_2 and other nutrient wastes. However, algae cultivation faces several challenges related to availability of locations with sufficient sunshine and water, required nutrient inputs, and oil extraction (Darzins et al., 2010). Algae feedstock for alternative biofuels are still not economically competitive with fossil fuels at the present time. Furthermore, it has not yet been demonstrated that algae production systems offer improved sustainability characteristics. Like fossil fuel, algae

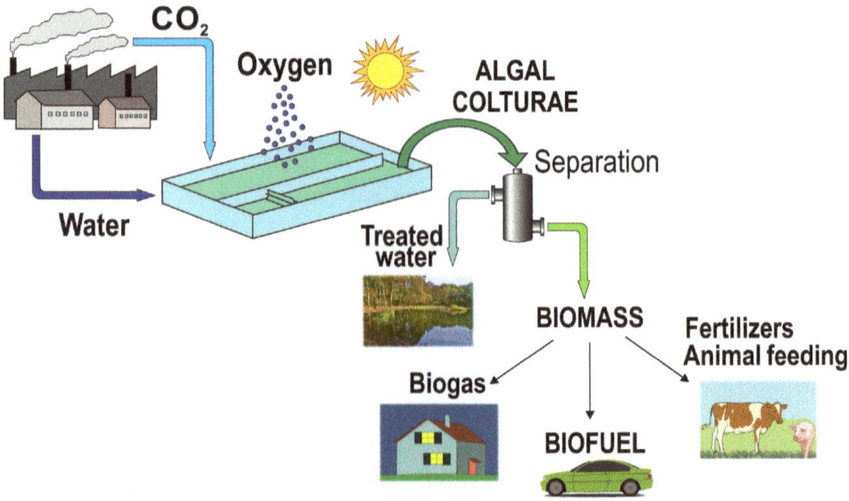

Fig. 4 Model of a microalgal biorefinery. *(Adapted from IEA, Task-42 by the second author.)*

fuel releases CO_2 when burnt, but unlikely fossil fuel, algae fuel release CO_2 that is removed from the atmosphere via photosynthesis as the algae or plant grew. Additionally, algal biomass productivity per hectare could be higher than energy crops. Based on high level of innovation demonstrated within the algal industry for biofuels in just the past decade, it is likely that new technologies will continue to be introduced in the future. In fact, the introduction of technological innovations will be critical if the sector is recently going to achieve a commercial success (Sikes et al., 2011).

Growth of marine, river and lake biomass from microalgae is not without challenges because the maximum growth rate of microorganisms typically occurs under very specific conditions. Furthermore, open pond algal systems are susceptible to contamination from various airborne microorganisms that can decrease overall algal productivity. But, the main concern is the ability for producer to separate algae from water. This issue could be turned into a profitable opportunity as it is feasible to use algae as a means for tertiary wastewater treatment in order to reuse trace nutrients, such as phosphorous-nitrogen-containing compounds, or to use industrial processes as a way to absorb CO_2 by entraining algal cultures to gaseous exhaust streams (Diltz and Pullammanappallil, 2013). Photosynthetic microalgae and cyanobacteria could serve as an attractive feedstock because they have higher growth rates that require much less land area if compared to plants and they can grow in areas that cannot usually support mainstream agriculture. Algae can be cultivated at the sea level or on non-arable land, so there is no competition with food (Ndimba et al., 2013). Despite significant progress, reliable and cost-effective production of lipid-protein-rich algal biomass have not been demonstrated at scales more than 10 m^2. Productivity and cost still remain the two fundamental barriers to algae commercialization (Huntley et al., 2015) since commercially viable production of biofuel from microalgae will depend on effective strategies to generate high-volume of biomass and low-value biofuel along with high value-added co-products. Several companies and government agencies are funding efforts to reduce capital and operating costs and make algae fuel production most commercially viable. Commercial cellulosic biofuel production began in the US in 2013, while biofuels from algae are not yet commercially produced.

Microalgae have emerged as a profitable biomass source that plays the dual role of wastewaters bioremediation and generation of biogas and biodiesel. Growing microalgae on different types of wastewaters has been studied over the past decades whenever the success of such studies depends on performance of the selected microalgae strains (Zhou et al., 2014). Moreover, the use of wastewaters treatment becomes an economically attractive alternative (Kligerman and Bouwer, 2015) as various types of algal biofilm reactors were integrated with wastewaters treatment plants to overcome current limitation of algal biodiesel production. So far, the integration of algal bioreactor and wastewater has been limited to municipal wastewaters, while only few agricultural wastewater amounts have been reused for algal bioreactor. However, algal biofilm reactors integrated with wastewater would have high potential for high productivity of algal biomass for biofuel and efficient wastewater treatment only if conditions are optimized (Hoh et al., 2016). Wastewater purification-based algae production for generating biogas with a closed technology is one of the innovative technologies (Bai et al., 2012). Even, though biogas and algae production for wastewater purification are suitable also individually for disposal, nonetheless they are much more effective when used in combination among them. The explanation is that anaerobic microorganisms are primarily able to decompose the organic matter in wastewater, while algae are more suitable for utilizing inorganic nutrients; therefore, they complement each other from wastewater purification aspects. At the same time, algae can be used for biogas production, and therefore they can also be used for making completely closed purification systems, including biogas plants without using arable land. Biogas and algae plants both produce renewable energy of any type individually (heat, electricity, fuel). In comparison with other procedures algae-based wastewater purification, this is more expensive, but it provide an environmental friendly and long-term solution for purification of fermented wastewater rich of inorganic matter that had previously been purified of all organic matter by the biogas plant. The biogas plant plays a key role in providing feedstock (fermented, separated wastewater), yield increase (waste heat and CO_2) and the use of the end-product (gasification of the algae biomass). For this reason, algae-based wastewater purification makes it possible to utilize and dispose the harmful end products of the biogas plant which would otherwise go to waste entirely in the biogas plant. The biogas plant mineralizes nitrogen (N) and phosphorus (P) provided by algae addition, and it also produces bio-methane which is significant from the point of view of CO_2 emission and its profitability (Sialve et al., 2009). Wastewater produced in the biogas plants is not suitable for further gasification, but it contains a significant amount of N, P, and potassium, as well as fermented organic matter which cannot be drained directly to living waters. Algae can be effectively used for the purification of fermented and non-fermented wastewaters. Nowadays, the conventional process of oxidation needs of significant amount of mechanical energy which can be generated with solar energy in the case of algae. The amount of biogas produced from algae largely depends on the species and the pretreatment technology. The gasification of residues after oil extraction may have a significant role in the sustainability of algae oil methyl-ester production (Sialve et al., 2009). Whereas, Becker (2007) performed theoretical calculations with regard to the bio-methane and ammonium production ability of various species: the theoretical biogas yield of the algae species were between 470 and 800 Nm^3/ton organic matter which can be considered significant in comparison with conventional crops. In order to optimize the C/N ratio (from 20 to 35), and since microalgae produces 55–110 times as much N as rapeseed, it is not advised to ferment algae with other high N content feedstock (e.g., food industry waste) (Hughes et al., 2012). The blending of algae with glycerin increases the methane content of biogas to be obtained from glycerin by around 10% (from 48–58% to 59–69%), therefore it could significantly increase the efficiency of utilization of waste-like polluted glycerin originating from biodiesel production for energy production purposes as it cannot be utilized in any other way (Salerno et al., 2009). The results of experimental trials on biogas yields from various algae species showed that the C/N ratio of algae is not ideal for biogas production in itself and there are rather significant differences between the algae yields. The biogas yield of algae in water is typically lower than those in wastewater (Ward et al., 2014). Algae which contain less than 40% oil can be used for biogas production directly, while those which contain more than 40% employed for biogas after extraction of oil (Ward et al., 2014). Although the economic indexes may be significantly more favorable in this case, there is still the risk factor of the special difficulties of extracting algae oil. The significance of the high oleic acid content in *Chlorella* spp. is that it makes its composition similar to rapeseed oil, which is the most suitable feedstock for oil production and its quality level that is in line with the standard. The most significant fatty acid of rapeseed oil is the oleic acid, while that of sunflower and soybean oil is the linoleic acid. At the same time, if the oil extracted from *Chlorella* is used as biofuel, its engine-related characteristics are even better than those of rapeseed oil in terms of cold start.

3.4 General overview of biofuel production from biomass and related value added co-products

The hierarchical distribution of biomass types, chemical building blocks, and biofuel co-products is showed in Fig. 5.

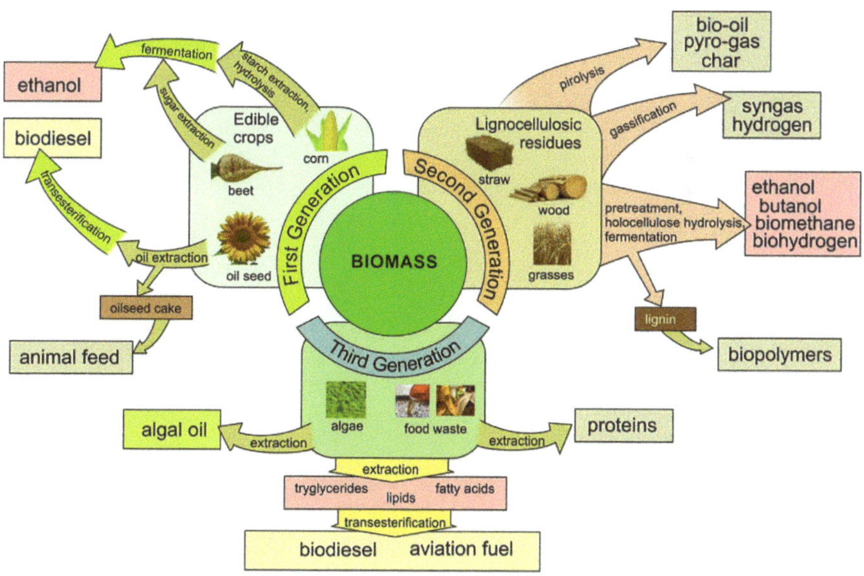

Fig. 5 Hierarchical distribution of biomass types, chemical building blocks, and biofuel co-products. The production of crude glycerin, as the main by-product of the biodiesel supply chain, was not shown for reasons of graphical feasibility. *(Original elaboration of the second author.)*

4. Biofuel co-products for livestock feed

4.1 Background

The biofuel production industry based on first generation feedstock as wheat, maize, rapeseed, and soybean generates a range of valuable protein-rich co-products, such as oil-less seed meal and DDGS, which can be used as animal feed in the livestock industry (Fig. 6).

Because protein-rich crops require a relatively large amount of arable and fertile lands for a given output when compared with cereal crops, the use of co-products can therefore reduce the net land-use. The co-products here considered overall include DDGS and soybean or rapeseed oil-less meal, while palm kernel cake and sugar beet pulp were produced in negligible amounts though are also an important component of feed concentrates. The following key definitions should be clarified before further description of the following sections of the chapter. Corn ethanol: Ethanol produced from corn (maize) through a fermentation process. Dry milling: A corn fractionation process where corn germ and corn bran are removed from the corn kernel and the remaining portion of corn kernel is further fermented to yield

Fig. 6 Feedstuffs for livestock feed from biofuel chain co-products and related food crops. *(Pictures of the authors.)*

ethanol. Distillers' co-products: Co-products from distillation of cereal grains for ethanol production. They can be a substitute for conventional animal feed (e.g., corn, soybean meal, and urea). Distillers' co-products displacement ratio: Ratio of the amount of distillers' co-products to the amount of conventional animal feed that is displaced by the distillers' co-product for livestock.

The two main industrial processes widely employed overall in the US to produce co-products from the fuel ethanol chain, that are mainly used in livestock feed, are the wet and dry milling (Prieler and Fisher, 2009). Wet milling generates as value-added co-products vegetable oil as corn distillers oil (CDO), corn gluten meal (CGM), and fiber. Dry milling instead provides wet distillers grain with soluble (WDGS) or wet distillers grain (WDG) and, after drying, DDGS or dried distillers grain (DDG), respectively (Fig. 7).

Very many papers were found in literature about the market of CDO, CGM, WDGS, WDG, DDGS, and DDG. The opportunities and challenges of these bioethanol co-products as a livestock feed have been detailed in a governmental handbook (FAO, 2012), the estimated displaced products and ratios of distillers' co-products from corn ethanol plants under the

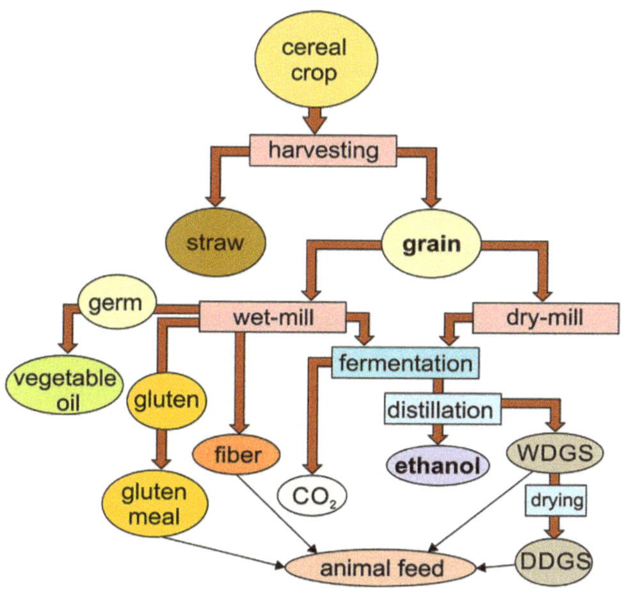

Fig. 7 Process flow and related by-products of ethanol production from starchy crops. *(Original elaboration of the second author.)*

implication of a lifecycle analysis assessment have been critically evaluated in the US (Arora et al., 2010), the global economic and environmental implications of bioethanol co-products employed as livestock feed have been analyzed in the biofuel industry (Popp et al., 2016), whereas utility and potential employment of DDG as a fertilizer source for corn was investigated (Nelson et al., 2009).

4.2 Livestock feed potential

Fuel ethanol chain from sugary crops (overall sugar beet and sugarcane) produces by-products which include either beet-top used as green fodder or beet-pulp and filter cake in cattle feed. Over 90% of the ethanol produced comes from the dry milling process and 10% by wet milling. Both milling processes utilize only the starch fraction of the corn kernel for ethanol production. The remaining fraction of proteins, fats, fibers, and other nutritional components remains available for animal feed. In distillers, DDGS and DDG remain nutritional components from corn kernel that, being concentrated by a factor of three, represent typical distiller co-products which have at least three times as much protein and fat as an equivalent amount of corn. If distiller grains are being fed to livestock in close proximity to the ethanol plant, the drying step can be avoided and related co-product is the WDG. An estimated 85% of existing dry mills have the capability to extract corn oil, which is then sold as an individual feed ingredient or as a feedstock for biodiesel production. In the wet milling process, corn oil from the germ is either extracted on-site or sold to crushers who extract the corn oil. The remaining fiber, gluten, and starch components are further separated and sold as corn gluten feed (CGF) or CGM, while starch is processed in one of the following ways: fermented into ethanol, dried and modified corn starch, or processed into corn syrup. One-third of grains that feed the ethanol process is returned into the animal feed market, most often in the form of distillers grains, CGF, and/or CGM. These nutrient-dense co-products are fed to beef cattle, dairy cows, swine, poultry, and fish. In the 2013/2014 marketing years, the US ethanol industry produced an estimated 39 million tons, the EU 4 tons, and China 2 million tons of high-quality feed, namely 41 million tons of distillers' grains and around 4 million tons of CGF and CGM. Estimated average potential in the US is around 62 million tons per year for DDGS feed consumption, an amount much higher than the current DDGS supply. CGF and meal production remained relatively constant in the last decade, as the majority of growth in ethanol production has come from dry mills. As for the wet milling co-products, CGF is primarily fed to dairy and beef cattle, while CGM is fed to a wide range of species including poultry and fish. CGM, which is a very high concentrated source of protein, is used as an ingredient in pet food as well. Much more studies are under way to increase the co-product inclusion levels to best take advantage of their nutritional and economic benefits. While ethanol co-products are used by the domestic markets in the EU and China, the US exported around 10 million tons of distillers' grains and approximately 2 million tons of CGF and CGM. These exports were volumetrically equivalent to about 24% of total 2013/2014 corn grain exports (Wisner, 2015). Today, more than 85% of existing dry mills have capability to extract CDO in the US, which is then sold as an individual feed ingredient or as a feedstock for biodiesel production. In many cases, co-products are a more effective source of energy and protein than compounds they are replacing in the animal diet. Distillers' grains provide approximately 130–150% of the energy of an equivalent amount of corn when fed to beef. By economically displacing traditional feed ingredients, ethanol co-products effectively reduced the livestock and poultry industry's demand for maize and protein meal. The back-flow of biofuel co-products into feed markets has highly increased agricultural land use as well, where at least one-third of every hectare dedicated to ethanol output could be reused for producing feed as alternative to biofuel. Distillers' grains have generally sold at a significant discount to competing feed ingredients, thereby meaning they are a good added-value for livestock and poultry feeders. Moreover, as prices for soybean meal and other protein feeds spiked in 2013 due to the 2012 drought, DDGS was increasingly used as a replacement for protein meal. Thus, DDGS prices have averaged 109% the price of corn and 40% the price of soybean meal since the beginning of 2013 until mid-2014. A ton of DDGS contains approximately 270 kg of protein, while a ton of high-protein soybean meal contains 436 kg of protein. Thus, DDGS served as a lower-cost means of meeting protein needs during the period in which the 2012 drought affected global feed supplies. Although the feed co-product output is expected to grow more slowly in the coming years, nevertheless a number of emerging technologies may change the composition by further improving the quality of feed co-products. About 85% of dry mills extract from the stillage on the back end of the process where classified as CDO that can be sold into the feed market (particularly for poultry) or used for biodiesel output. CDO displaces higher value fat and vegetable oil that are typically used in food or feed applications. The feed co-products resulting from oil extraction typically have lower fat and higher protein content than conventional distillers' grains. These co-products are characterized by reduced fat content and can be useful for certain species. Dry fractionation is another technology that may emerge more broadly in the dry mill ethanol industry. This practice allows dry mill ethanol producers to separate the corn germ and other components from the starch at the beginning of the ethanol process. The germ can be processed and sold as feed or feedstock in further processing for other uses. Further, some dry mill facilities are being examining the potential of converting corn kernel fiber to cellulosic ethanol. This process would reduce the fiber content of DDGS which may make the product more palatable for monogastric animals. Additionally, a number of ethanol processes are likely to improve the nutritional quality and utility of ethanol co-products. The success of new technologies will depend on the demands of the ethanol industry's customers in the animal feed market.

The main feedstock used to make biodiesel are soybean or rapeseed oil by using an alcohol source (methanol) for trans-esterification. Seed meal co-products from soybean and canola are recycled after oil extraction. An estimated 80% of soybean seed and 60% of rapeseed is left as oil-less seed meal/cake/pellet after the extraction process, thereby generating a significant quantity of them. These co-products can be used immediately as an animal feed without further treatment. About 7 million tons of soybean oil and 6 million tons of rapeseed oil is used in biodiesel production contributing to almost 3.6 million tons of rapeseed meal and 5.6 million tons of soybean meal output. Taking into account that 210 million tons of soymeal and 40 million tons of rapeseed meal is produced a year globally, the co-products of biodiesel production have a moderate impact on the feed market. The main direct by-product of biodiesel production is crude glycerin that constitutes about 10% of the co-products of such biofuel chain. There is a limited demand for glycerin suitable for producing food, beverage, personal care, and oral products as well as pharmaceutical and industrial uses. Crude glycerin can be used effectively in livestock rations to replace fossil-based glycerin, which can also be reutilized as a feed ingredient for livestock rations as crude glycerin contains similar energy to that of maize for pigs.

Utilization of microalgae biomass into livestock feed industry is another growing market (Becker, 2007). Algal biomass is a valuable feed supplement or substitute for conventional protein sources because many studies shown the suitability of algae as a potential animal feed and replacement for protein sourced from soybean and fish meal. Unfortunately, the actual trend is to avoid of using live algae due to their actual high cost of production. The use of microalgae as feedstuff is more recent than other feedstock where the addition of algae into poultry rations at 51% rate offers the most promising perspectives for their commercial use in animal feeding (Spolaore et al., 2006). Also, microalgae may serve as almost the sole protein source in laying hens and several countries they are officially approved as chicken feed (Gouveia et al., 2008). In pigs ration has assumed the algae addition of about 33% without concerns (Yap et al., 1982). The feed intake of the pigs on the algal meal diets was not affected until the third period of the experiment when added at 10% rate, thereby it has been suggested that the acceptance from a palatability point of view would not be a major issue, although the manipulation of the gut microbiome with algae meal is a possible source of concern, although a larger number of piglets raised in commercial environments would be required to better define this as an issue (Henman and Australia, 2012). However, it should be expected that the most suitable for feeding with algae are ruminants as they are able to digest even an un-processed algae cell wall (Gouveia et al., 2008). Manufacturers of these organisms worldwide recognized different potentials and focused just on the food and feed industry (Abo et al., 2019). Although microalgae are eaten as a human food in China and Chad (Africa) being considered as a solution to the world's food shortage, nevertheless their use on a global scale appeared to be still too limited to health food and food supplements (Milledge, 2011). The life cycle assessment induces to conclude that algal systems producing biodiesel, animal feed, and succinic acid at the same time could be beneficial to the environment when compared to algal systems that produces only biodiesel. Thus, algal systems could be a potential alternative or an additional renewable source to mitigate environmental impact (Gnansounou and Raman, 2016).

4.3 Livestock feed market and related policies

The global market of cereals grain and oilseed of oleaginous crops has grown substantially in recent years and their increased use for biofuel production has not led to reduced availability for feed or food use. Actually, the feedstock for ethanol and biodiesel moderates in accordance with slowing national mandates. Economic viability of the biofuel industry depends on ability of industry to derive high value-added products from the biofuel chain when it produces as well co-products. Bioenergy could enable farmers in poor regions to introduce agricultural technologies and improve infrastructure to increase farm productivity (Lynd et al., 2015). Bioenergy and food production can co-exist in the same country by enhancing each other by advanced technologies and increasing yield (Dale et al., 2014).

Feedstock generally accounts for around 70% of production costs including processing and transportation. Declining feedstock prices, industry was helped by reducing overall production costs. Increased hectares for corn farming in the US are driven by the significant increase in corn prices since 2005. However, several key factors caused increases in corn price including corn ethanol production, weather events, grain demand increases, and diet changes in emerging economies and markets (Trostle et al., 2011). Some economic models showed that biofuel use can result in higher crop prices, though the range of estimates in the literature is very wide. Estimated trends for the effect of biofuel on corn prices in 2015 ranged from 5% to 53% in increasing (Zhang et al., 2013). The possible impact of the developed countries on global food prices in grains and oilseeds due to biofuel policies became a significant concern in 2007 when global grain prices reached historic heights (De Gorter and Drabik, 2012a,b; De Gorter et al., 2013). A factor emphasizing how biofuel policy is only responsible for part of that fraction of price increases in food grain commodities is due to biofuels price (Durham et al., 2012). Another study has estimated that impact of the EU biofuels demand from 2000 to 2010 increased world grain prices of about 1–2% and oilseed prices by around 4%. It also has been estimated that without any cap on crop-based biofuels, the EU policy could raise grain prices by 1% and those of oilseed by 10% in 2020 (Hamelinck, 2013). Land sparing is more effective than land sharing to increase bioenergy potential. Maize ethanol, often portrayed as the villain of the piece in the food versus fuel debate, may in fact have been key in stimulating yield improvement that has resulted in increased exports of grain from the USA while providing a buffer in drought years. In the US, for biofuels to expand their role in the transportation sector they must be produced from cellulosic feedstocks. In the EU, new legislation limits the contribution of biofuels derived from sugars, starch, and oleaginous crops.

Biofuel co-products often replace traditional feeds in animal rations for their higher price. The increased use of agricultural commodities for biofuel has led to higher costs for animal feeds, however the increased substitution of traditional feedstuffs with biofuel co-products in feed rations has mitigated the major costs faced by livestock and poultry producers. The grain and oilseed market is expected to grow more significantly in the long term period. Growth in the use of agricultural commodities for biofuel is expected to continue in the next 10 years, but with growth rates slowing in the key producing countries as government-imposed limits on grain use for biofuels are reached and new feedstocks are commercialized (FAO, 2015; USDA, 2015). Soybean and rapeseed meals have always been the best components for animal feed because they are an excellent source of protein. The impact of increased ethanol production is largely felt through competing crops as the increased production of DDGS and oilseed meals can jointly reduce the demand of livestock industry for maize and oil meal, thereby offsetting the increase in the demand of ethanol due to the US and EU biofuel mandates. By increasing biofuel production, the output of co-products also increases because prices of such co-products are correlated with prices of feedstock and represent an important component of total industry revenues. For example, DDGS and corn prices are highly correlated among them, and as a result co-products prices fall relative to other feed ingredients. This encourages livestock producers to use more biofuel co-products as shock absorber and price adjuster (Taheripour et al., 2010). The price of DDGS relative to maize has fallen by 50% between 1983 and 2006. This has provided a strong incentive for livestock producers to use more DDGS in their production cycles and has also enhanced the US exports of them. Further, the relative contribution of distillers' grains to gross returns has varied over time as the price of DDGS varied. The net contribution of DDGS to net returns for ethanol production is a function of the ratio of the price of DDGS versus the price of maize. When the ratio is high, distillers' grains provide a larger contribution to net return and vice versa. But, with all other prices unchanged, a decline in the price of DDGS would result in a corresponding decline in net return (Irwin and Good, 2015).

4.4 Environmental impact of GHG emission and directives for a sustainability challenge

It has been established from previous studies that the environmental impact of bio-based products on GHG emissions strongly depends on feedstock, land-use, and technological efficiency. GHG emissions of these products using a full life cycle analysis depends on by the inclusion of emissions caused by land-use change (LUC) that affects the GHG balance by conversion of native ecosystems—often by slash and burn—and release of C in soil (Huang et al., 2013). This key factor in determining the realistic life cycle emissions was recently established and characterized into two main types (Searchinger et al., 2008). The first type of LUC, the direct land-use change (DLUC), occurs when feedstock for biorefinery purposes (e.g., soybean for biodiesel) displaces a prior land-use (e.g., forest), thereby generating possible changes in the C-stock of lands. This effect has been well studied, but the default values can often differ substantially from the actual values depending on the arbitrary choice of accounting period. However, it has been widely accepted that deforestation due to increased demand for biomass feedstock is an example of DLUC that should not occur if we are to make biomass a sustainable energy source for the future. Although the second type of LUC, the Indirect Land-Use Change (ILUC), has been longer overlooked due to the great inaccuracies in identifying and quantifying it, nonetheless this too must be estimated to maintain a healthy GHG balance. ILUC occurs if pressure on agriculture—due to the displacement of a previous activity from the production of biomass feedstock—induces LUC in other locations. The displacement of current land-use to produce biofuels and other bio-based products can trigger DLUC elsewhere. There are four types of ILUC: spatial, temporal, use, and displaced activity, all of which describe the different mechanisms by which biomass production puts pressure on land activity.

The cultivation of energy crops on arable land may lead to the displacement of food crops and to ILUC. Some of the energy crops for biofuels deliver feed as by-products, and this will have an impact on the global feed market. These by-products could be a substitute for other crops cultivated for feed elsewhere. In those cases the net land use for biofuel crops would be reduced substantially (for rapeseed and wheat from 50% to 100%, respectively, based on the substitution of soybean meal for both). The impact of this land-use effect on the overall GHG emissions can be substantial, but sensitive to assumptions on yields and regional LUC. The case of land use in Europe (for wheat and rapeseed) is usually compared to land use in Latin America (for soybean) where regional differences in ILUC emissions strongly determine the overall emissions. When maize and soybean meal are substituted in animal rations, less land for feed cultivation is needed and impact of the co-products on cropland use should be included. Several studies have attempted to include attribution of co-products which can be used as animal feed in analysis of land use requirements of biofuels. When co-products were used for process energy, the energy balance improved and feed were produced elsewhere and even ILUC effects cannot be excluded (Popp et al., 2016).

The impact of biofuel policies on GHG emissions is still too limited, however support policies that have significant impacts on global commodity prices are needed. Alternative policies may offer greater benefits including the reduction of energy demand and GHG emission. When biofuel co-products are used, "green credits" are attributed to the biofuel production chains as such credits include GHG emission savings that avoid land or energy use. The challenge is to decide how to distribute quantity of potential credits between biofuel and related co-products. Credits for biofuel co-products are an important task in the estimation of the GHG emission reductions if compared to fossil fuels. For the evaluation of GHG emission savings, the most common methods include: (1) calculation based on energy content of the co-products (allocation by energy content), (2) calculation based on market value of the co-products (allocation by market price), and (3) calculation based on substitution of one co-product by another (allocation by substitution) (Table 2). For reasons of feasibility, the Renewable Energy Directive (RED) recommended to apply the method of allocation by energy content to determine GHG emission savings when compared to fossil fuels (European Commission, 2009a).

Table 2 Methods for allocating the environmental impacts (such as GHG emissions and LUC) of one process or production chain to the different products (such as biofuel and livestock feed).

Methodology	Advantage	Disadvantage
Allocation based on the energy content of the products.	– Relatively simple. – Physical parameters are (quite) constant. – For biofuels, energy is the main quality aspect. – Producer controls the result.	– It does not simulate the actual processes. – Products with applications other than bioenergy are not valued in a proper way.
Allocation based on the economic value of the products (market price).	– Relatively simple. – All products are valued on an equal basis.	– Economic values fluctuate strongly. – Producer does not control the result. – It does not simulate the actual processes.
Allocation based on substitution of one product by another.	– The best simulation of the actual impact. – All products are valued in a comparable way, based on their physical properties and applications. – Producer controls the result.	– A detailed analysis can be very complicated. – Simplification threatens to erode the first two advantages.

At the present time, there is still much uncertainty about ILUC and current methodology relies on arbitrary choices of production periods of biofuels with significant implications for the final results (Mueller, 2008). Depending on methodology LUC have either positive or negative effects on the GHG balance of biorefinery outputs. The conversion of forests, wetlands, and grasslands to cropland usually results in a net emission of C from the biomass and soil into the atmosphere, thereby determining negative effects. However, where sparsely vegetated or disturbed lands are converted to cropland, this can result in a net gain in both overall biomass production and C-sequestration in soil. However, it is not possible to previously know what the substituted product really is for a fuel supplier, therefore GHG emissions of substituted products are often unpredictable. Moreover, animal feed based on co-products can replace many different products which differ for the GHG emission impact. Change in land use, principally those associated with deforestation and expansion of food production, contribute about 15% of global GHG emissions. Currently, less than 3% of global agricultural land is used for cultivating biofuel crops and LUC associated with bioenergy represents only 1% of the total emissions, most of which are produced by change in land use for food, fodder production, or other reasons (European Commission, 2009b). Most biofuel production systems have significant reduction in GHG emission to fossil fuels if no ILUC effect was counted. However, land use in agriculture has actually grown very slowly in the past decades since the practice of multiple cropping systems has spread particularly in those areas where land is relatively scarce (Asia). Rising concerns and governmental regulations over climate change, and biodiversity and resource management will push agriculture toward greater sustainability and lower environmental costs and risks, while competition between foods and biofuels could also become more intense (FAO, 2009). Change in the patterns of land use may increase GHG emissions by releasing C stocks into the atmosphere (Searchinger et al., 2008). Biofuel feedstock grown on deforested lands, often illegally such as soybean in Brazil (Amazon) and oil palm in South-East Asia (Indonesia), generates particularly high GHG emissions (Fargione et al., 2008). Even, the use of cellulosic feedstocks can spur higher crop prices that encourage the expansion of agriculture into undeveloped land, thereby leading to major GHG emissions and biodiversity loss (Melillo et al., 2009). Although biofuel production is only one of the numerous contributors to LUC, Fritsche and Wiegmann (2011) estimated that the overwhelming majority of LUC and ILUC by 2020 could be caused by biofuel feedstock.

Thanks to its indirect nature, ILUC is not a local phenomenon only that can be observed in a given place and time, since it cannot be monitored for individual feedstock. Biofuel production can be displaced anywhere in the world and it can be distributed through global trading where significant time lags were occurred. Depending on location of production, the total amount of area converted, as well as GHG emissions per hectare converted, can vary greatly and the global distribution of production depends on the assumptions made about the role of geography in international trade. Yield response for commodities are even more important than yield response for a specific feedstock/region. Furthermore, the consumers' demand elasticity for food products plays a crucial role for determining the necessary expansion in global land-use following expansion of ethanol production (Golub and Hertel, 2012). The demand for land to produce biofuel feedstock can cause shifts in land use patterns that can affect agricultural lands. Searchinger et al. (2008), by investigating LUC associated with biofuel crop production and subsequent GHG emissions, have outlined that timing and understanding of biofuel-associated LUC improved greatly because better land use data are available and economic models for LUC have been improved significantly. Results on LUC-GHG emissions (Dunn et al., 2013) indicated that other authors (Searchinger et al., 2008) have previously overestimated LUC-GHG emissions significantly. Later, the same authors (Searchinger and Heimlich, 2015) have underlined that additional biomass for bioenergy is the key for bioenergy-derived GHG emissions reduction. The concept of additionality for GHG reduction has really needs of the scenario by which bioenergy does not exist or exists only at current levels. These authors concluded that any biomass growth would result in double counting by assuming that bioenergy is provided by lands where biomass would be grown anyway with the same yields without differences in land productivity of biomass between the two scenarios. Difference for land biomass productivity could be caused by different cropping systems in relation to the suitability of using marginal or degraded lands, for instance by changing from a single cropping system to double/triple cropping systems for biofuel production (Babcock and Iqbal, 2014). A careful characterization of bioenergy scenarios to address the biomass additionality issue has been conducted to compare biofuels marginally produced with high-GHG-intensity fossil fuels (Van den Bos and Hamelinck, 2014). According to Searchinger and Heimlich (2015), phasing out bioenergy that uses crops or that otherwise makes dedicated use of land is a sound step toward a sustainable food future. The US farm hectares for different crops illustrate that, while total corn areas increased in parallel with the build-up of the corn ethanol industry between 2004 and 2013, total crop lands remained fairly constant in the same period (Taheripour and Tyner, 2013; Wang et al., 2014). Another mechanism that should be taken into account in further estimates is likely the conversion of grasslands, wetlands, and other lands since the key point is that agricultural hectares in the USA has not significantly increased when compared to increased biofuel use. The "US renewable fuel standard states" noted that biofuel feedstocks must come from land that was not forested before 2007. In the case of woody feedstock, the forests from which they derive must have been managed plantations before 2007. This provision strongly limits the expansion of cultivated lands into forested lands for biofuel production. Reduced deforestation and increased agricultural production can occur simultaneously in tropical forest (Macedo et al., 2012).

Second and third generation biofuels have significant potential to reduce GHG emissions if compared to first generation biofuels because feedstock can be produced using marginal land (Hertel et al., 2010). Moreover, in the case of agricultural residues and plant-waste, no additional agricultural land is required and indirect market-mediated GHG emissions can be minimal if the biomass waste have no other productive uses. Biofuel production and consumption will not reduce GHG and other conventional pollutant emissions, lessen petroleum imports, or alleviate pressure on exhaustible resources. Biofuel production and biofuel use should coincide with reduction in the production and use of fossil fuels. Benefits would be mitigated if biofuel emissions and resource demands augment rather than displace with those of fossil fuels. Biofuel processing releases GHG during the various production steps, for example, fertilizer application in soil releases nitrous oxide, a potent GHG, and most biorefineries operate using fossil fuels that release CO_2. Some authors suggested that GHG emissions resulting from the biofuel production and use, including those from ILUC, may be higher than those generated by fossil fuels by depending on the time of the analysis (Mosnier et al., 2013). Regarding non-GHG environmental impacts, these authors suggested that production of biofuel from corn and soybean could increase water pollution from nutrients, pesticides, and sediments. Air quality could also decline in some regions if the impact of tailpipe emissions surplus the additional emissions generated by biorefinery, thereby increasing net conventional air pollution.

Biofuel co-products can help to mitigation of the environmental consequences of biofuel expansion. For example, DDGS, CGF, CGM, and oil cakes have substituted wheat, maize, and soybean meal in many livestock feed rations. It means that not all the feedstock used for bioethanol production should be subtracted from the supplies. Based on the land-use efficiencies, land-use for biofuel production would need to increase from about 30–35 million ha to around 100 million ha in 2050. This corresponds to an increase from 2.5% of total arable land in 2016 to around 6% in 2050. This dramatic expansion would include some croplands as well as pastures and currently unused lands for production of lignocellulosic biomass (IEA, 2013). This substitution reduces the land use impact for biofuel production as well as the increased demand for chemical fertilizers and pesticides. According to the conventional assumption, ethanol producers return a one-third of the maize processed back to the feeding sector which is the difference between the gross and net volume of maize used for ethanol. While ethanol expansion raised demand for maize, DDGS partly offsets the impacts on feed market. Consequently, the net effect of one ton of maize being used for ethanol production is less than one ton in domestic feed market. Furthermore, cereals grain for ethanol is expected to moderate in accordance with slowing national mandates in the US (USDA, 2015) and the EU. By aggregating the data actually available, one ton of DDGS can replace 1.22 tons on average of feed consisting of corn and soybean meal in the US. In fact, the amount of maize and soybean meal replaced by DDGS represents 38% weight of the maize used in the associated ethanol production process for a given yield per year. If such co-products are taken into account, the net use of feedstock will decline. More complicated, but no less important, is the impact of DDGS on LUC and GHG emissions associated with ethanol production from maize. Most biofuel regulations significantly undervalue the contribution of DDGS by comparing the net GHG impacts of maize ethanol with no substitution of soybean meal and assuming that one ton of DDGS replaces only one ton of maize. The importance of DDGS is being overlooked by the regulatory bodies of agencies that require a GHG assessment of ethanol. In the future scenarios, an accurate DDGS accounting is of increasing importance (Popp et al., 2016). Ignoring co-product generation in early biofuel impact assessments has led to an overestimation of land requirements by 40% or even more. The contribution of feed co-products is relatively high in the US, China, and the EU due to large share of cereals with high feed yields, but it is low in Brazil where ethanol production is dominated by sugarcane that generates no feed co-products (Croezen and Brouwer, 2008). Implications for land use will also depend on the role of yield improvement. Reports have estimated that the use of co-products generated from rapeseed, soybean, wheat, and maize can reduce net land use from 11% to 25% (Langeveld et al., 2014).

5. Conclusions and future market trends

The potential global of lignocellulosic bioenergy supply ranges from less than 100 to more than 1.500 EJ/year by 2050 according to recent estimates. Besides biomass availability, the future trends on biomass for energy production is determined by its interaction with other energy sources and relative costs. A number of studies have given insights on the potential contribution of bioenergy into future energy mix. Besides the uncertainties of cost-effective use of bioenergy application, large-scale energy crops production may create conflicts with other aspects of sustainability likely food security, water availability, climate change mitigation, and biodiversity preservation. The need to increase renewable and sustainable energy systems both in developed or developing countries is widely accepted, tough future yield improvement and land use expansion remains still unclear. Producing more sustainable foods require crops that make better use of limited and hazardous resources including land, water, chemical pesticides, and mineral fertilizers. With the increased demands of energy has become more evident how the continued GHG emissions in atmosphere and the C sinks loss in soil have influenced the world climate change. The increasing demand for fertile and arable lands that compete with food production causes conflict between land use for foods and energy crops. These concerns are amplified by climate change for land productivity that has caused erosion, water stress, biodiversity loss, and progressive desertification. Though the global potential for biomass energy production is very large in absolute terms, nevertheless it is still too far to replace the current energy source. In fact, the increasing biomass energy production beyond a certain level would have significant negative effects on land use and conventional agricultural markets. On the other hand, by replacing partly fossil fuels with biofuels has the potential to generate larger benefits by reducing the demand and price for petroleum, thereby determining economic advantages for domestic consumers.

Next-generation biofuel capacity is expected to expand only slowly in the next years, though the majors commercial-scale plants in the US and EU were commissioned about 15 years ago. While the US should remain the largest biofuel producer, technical and economic challenges related to blending with more than 10% ethanol in the gasoline pool raise uncertainty over the outlook. In Brazil, the more optimistic sugarcane harvest conditions and new governmental policies have supported specific measures and incentives that should drive the expected growth, although the ethanol sector shows still difficulties. High feedstock prices and poor profit margins have addressed to challenge biofuel producers in Europe. The current energy mix in the EU-27 is still not favorable to environment, while a broader mix of crops should reduce the environmental impacts. For example, grasses or perennial crops would enhance the ecosystem services provided by farmland, such as flood prevention and water filtration. The increasing use of agricultural waste for biofuel is expected to be continued by 2022, but the growth rates will slowed in the key producing countries due to government-imposed limits on grain use for biofuel, therefore biofuels capacity is expected to expand only slowly. The second reason for a moderation in the increasing use of agricultural residues for biofuels is the future growth of this sector that very probably will come from new feedstock that currently have no or limited application in the animal feed market, such as perennial grasses, agricultural/forestry residues, and microalgae. Food crop-based feedstock are expected to dominate the global biofuel market over the coming decade. The biofuel production growth has been accompanied by the increased outputs of animal feed co-products since they are growing in volume and importance. While the increased use of agricultural commodities for biofuels is generally expected to contribute to slightly higher input costs for certain livestock and poultry feeds, the impacts are expected to be modest and mitigated partly by the increased substitution of biofuel co-products with traditional feedstuffs. The use of traditional energy crops for biofuel that results in a small increase in livestock feed costs will be offset to some extent by using co-products as feed that will increase crop yield over time. The feed co-product output is however expected to grow more slowly in the coming years since emerging technologies and new market targets will improve the nutritional quality of feed co-products.

The global climate change indicates an overall warming trend for the earth as a whole in association with rising levels of GHG emissions. While the natural sources affect the concentration of GHG over time, more studies have indicated that human sources of GHG can contribute to global warming. There is no scientific consensus on whether bioenergy as a whole contributes to global climate change, though first generation biofuels emit more CO_2 than second generation biofuels. For instance, corn cropping requires more input of fertilizer and pesticide when compared to giant reed or miscanthus in resulting greater soil depletion and leading to land use-induced CO_2 emission. Moreover, marginal lands and set-aside policies for biodiversity conservation tends to be converted into cropping systems along the time, leading to DLUC-induced CO_2 emissions when agricultural commodity prices become too high. In comparison to first generation biofuels, those of second generation offers a great potential for GHG mitigation through use of lignin-cellulosic feedstock originated from cropping systems that have less land-use related GHG emissions. While third generation biofuels can potentially reduce GHG emissions, they are not yet widely available at the commercial scale because future demand of them is still uncertain and production costs too high. The biofuel supply chains give rise to co-products which are largely suitable for animal feed, but they are supposed to be credited with the area of cropland required to produce the amount of feed they substitute, since if co-products are taken into account the net use of feedstocks decline. Most existing biofuel regulations significantly undervalued the contribution of co-products when assessing the net land use and GHG impacts of biofuel production as it is fundamental to take into account co-products in the GHG assessment models. The bioenergy potential can have both positive or negative environmental impacts that affect the overall net impact. Many of the impacts due to bioenergy potential are associated with DLUC in connection to biomass production. Bioenergy is often considered C-neutral in the balance models as CO_2 released in combustion is assumed to be compensated by CO_2 absorbed during plant growth. However, ILUC can exclude any GHG savings from biofuel production based on energy crops. This is due to the displacement of crop production onto previously unused land which can lead to deforestation (conversion of forests into extensive crops). Two indirect effects of bioenergy production will receive much attention in the future assessment models: (1) ILUC, leading to GHG emissions and biodiversity loss; and (2) the indirect impact on food prices, determining the availability of food. There are no fixed indirect emission factors for a specific bioenergy product because the indirect effects of biofuels does not have determined by the characteristic of co-products, being resulted from the interaction with the dynamic global economic and physical systems. Furthermore, indirect GHG emissions determine scientific uncertainty and wide variability in the time. Emissions from ILUC caused by bioenergy products ranged from 30% to more than 100% if compared to fossil fuel emissions. Intensification of cropping systems is a better way to minimize ILUC when higher yields are the resultant of increasing fertilizer use only and indirect emissions might occur. So that, efficiency in fertilizer use should be drastically improved with the crop potential to further reduce GHG emissions. Indirect implications on biodiversity of soil are strongly related to land conversion since will indeed take more than

100 years to restore the soil biodiversity loss in the short term due to direct land conversion for energy crops. In case of ILUC, the long time needed for soil restoration can be shortened by stimulating global agricultural intensification. Indeed, biofuel co-products are supposed to be credited with croplands to produce the amount of feed they substitute. By considering biofuel co-products as valid substitutes for grains and oilseeds in livestock feed, the net land needed for cultivation of feedstock will decline from 2% to 1.5% of the net global cropland area.

References

Abo, B.O., Odey, E.A., Bakayoko, M., Kalakodio, L., 2019. Microalgae to biofuels production: a review on cultivation, application and renewable energy. Rev. Environ. Health 34 (1), 91–99.

Allwright, M.R., Taylor, G., 2016. Molecular breeding for improved second generation bioenergy crops. Trends Plant Sci. 21, 43–54.

Alshaal, T., Domokos-Szabolcsy, E., Marton, L., Czako, M., Katai, J., Balogh, P., Elhawat, N., El-Ramady, H., Fari, M., 2013. Phytoremediation of bauxite-derived red mud by giant reed. Environ. Chem. Lett. 11, 295–302.

Alshaal, T., Domokos-Szabolcsy, E., Marton, L., Czako, M., Katai, J., Balogh, P., Elhawat, N., El-Ramady, H., Gerocs, A., Fari, M., 2014a. Restoring soil ecosystems and biomass production of *Arundo donax* L. under microbial communities-depleted soil. Bioenergy Res. 7, 268–278.

Alshaal, T., Elhawat, N., Domokos-Szabolcsy, E., Katai, J., Marton, L., Czako, M., El-Ramady, H., Fari, M.G., 2014b. Giant reed (*Arundo donax* L.): a green technology for clean environment. In: Ansari, A.A., Gill, S.S., Gill, R., Lanza, G.R., Newman, L. (Eds.), Phytoremediation. vol. 1. Springer International Publishing, Cham, pp. 3–20.

Arora, S., Wu, M., Wang, M., 2010. Estimated displaced products and ratios of distillers' coproducts from corn ethanol plants and the implications of lifecycle analysis. Biofuels 1 (6), 911–922.

Babcock, B.A., Iqbal, Z., 2014. Using Recent Land Use Changes to Validate Land Use Change Models. Center for Agricultural and Rural Development, Ames, IA, pp. 1–34.

Bai, A., Stundl, L., Barsony, P., Feher, M., Jobbagy, P., Herpergel, Z., Vaszko, G., 2012. Algae production on pig sludge. Agron. Sustain. Dev. 32, 611–618.

Becker, E.W., 2007. Microalgae as a source of protein. Biotechnol. Adv. 25, 207–210.

Cai, X.M., Zhang, X.A., Wang, D.B., 2011. Land availability for biofuel production. Environ. Sci. Technol. 45, 334–339.

Corno, L., Pilu, R., Adani, F., 2014. *Arundo donax* L.: a non-food crop for bioenergy and bio-compound production. Biotechnol. Adv. 32, 1535–1549.

Croezen, H., Brouwer, F., 2008. Estimating Indirect Land Use Impacts From by-Products Utilization. CE Delft, Delft, pp. 1–31.

Dale, B.E., Anderson, J.E., Brown, R.C., Csonka, S., Dale, V.H., Herwick, G., Jackson, R.D., Jordan, N., Kaffka, S., Kline, K.L., et al., 2014. Take a closer look: biofuels can support environmental, economic and social goals. Environ. Sci. Technol. 48, 7200–7203.

Darzins, A., Pienkos, P., Edye, L., 2010. Current Status and Potential for Algal Biofuels Production. National Renewable Energy Laboratory, Golden, CO, pp. 1–131.

De Corato, U., De Bari, I., Viola, E., Pugliese, M., 2018. Assessing the main opportunities of integrated biorefining from agro-bioenergy co/by-products and agroindustrial residues into high-value added products associated to some emerging markets: a review. Renew. Sust. Energ. Rev. 88, 326–346.

De Gorter, H., Drabik, D., 2012a. Policy update: the effect of biofuel policies on food grain commodity prices. Biofuels 3, 21–24.
De Gorter, H., Drabik, D., 2012b. Policy update: biofuel policies and grain crop price volatility. Biofuels 3 (111), 113.
De Gorter, H., Drabik, D., Just, D.R., 2013. Biofuel policies and food grain commodity prices 2006-2012: all boom and no bust? AgBio Forum 16, 1–13.
Diltz, R., Pullammanappallil, P., 2013. Biofuels from algae. In: Fang, Z. (Ed.), Liquid, Gaseous and Solid Biofuels: Conversion Techniques. InTech Open Access Publisher, Rijeka, pp. 431–449.
Dunn, J.B., Mueller, S., Kwon, H.Y., Wang, M.Q., 2013. Land-use change and greenhouse gas emissions from corn and cellulosic ethanol. Biotechnol. Biofuels 6, 51.
Durham, C., Davies, G., Bhattacharyya, T., 2012. Can Biofuels Policy Work for Food Security? An Analytical Paper for Discussion. UK Department for Environment, Food and Rural Affairs (DEFRA), Devon, pp. 1–29.
Elhawat, N., Alshaal, T., Domokos-Szabolcsy, E., El-Ramady, H., Marton, L., Czako, M., Katai, J., Balogh, P., Sztrik, A., Molnar, M., et al., 2014. Phytoaccumulation potentials of two biotechnologically propagated ecotypes of *Arundo donax* in copper-contaminated synthetic wastewater. Environ. Sci. Pollut. Res. 21, 7773–7780.
Elhawat, N., Alshaal, T., Domokos-Szabolcsy, E., El-Ramady, H., Antal, G., Marton, L., Czako, M., Balogh, P., Fari, M., 2015. Copper uptake efficiency and its distribution within bioenergy grass giant reed. Bull. Environ. Contam. Toxicol. 95, 452–458.
Elobeid, A.E., Carriquiry, M.A., Fabiosa, J.F., 2012. Land-use change and greenhouse gas emissions in the fapri-card model system: addressing bias and uncertainty. Clim. Chang. Econ. 3.
El-Ramady, H.R., Abdalla, N., Alshaal, T., Fari, M., Prokisch, J., Pilon-Smits, E.A.H., Domokos-Szabolcsy, E., 2015. Selenium phytoremediation by giant reed. In: Lichtfouse, E., Schwarzbauer, J., Robert, D. (Eds.), Hydrogen Production and Remediation of Carbon and Pollutants. Springer International Publishing, Cham, pp. 133–198.
European Commission, 2009a. Directive 2009/28/EC of the European Parliament and of the Council of 23 April 2009 on the Promotion of the Use of Energy From Renewable Sources, and Amending and Subsequently Repealing the Directives 2001/77/EC and 2003/30/EC. Directive 2009/28/EC. European Commission, Brussels, pp. 16–62.
European Commission, 2009b. Directive 2009/30/EC of the European Parliament and of the Council of 23 April 2009 Amending the Directive 98/70/EC as Regards the Specification of Petrol, Diesel and Gas-Oil and Introducing a Mechanism to Monitor and Reduce Greenhouse Gas Emissions, and Amending the Directive 1999/32/EC as Regards the Specification of Fuel Used by Inland Waterway Vessels, and Repealing the Directive 93/12/EC. In Directive 2009/30/EC. European Commission, Brussels, pp. 88–113.
FAO, 2009. Agricultural Outlook 2009–2018. FAO, Paris, pp. 1–95.
FAO, 2012. In: Makkar, H.P.S. (Ed.), Biofuel Co-products as Livestock Feed—Opportunities and Challenges. FAO, Rome, ISBN: 978-92-5-107299-8.
FAO, 2015. Agricultural Outlook 2015. Available from:, https://doi.org/10.1787/data-00736-en.
Fargione, J., Hill, J., Tilman, D., Polasky, S., Hawthorne, P., 2008. Land clearing and the biofuel carbon debt. Science 319, 1235–1238.
Franscisco, L., Carlos, G.J., Antonie, P., Javier, F.M., Minerva, A.M.Z., Gil, G., 2010. Chemical and energetic characterization of species with a high-biomass production: fractionation of their components. Environ. Prog. Sustain. Energy 29, 499–509.
Fritsche, U.R., Wiegmann, K., 2011. Indirect land use change and biofuels. In: Directorate General for Internal Policies, Policy Department A: Economic and Scientific Policy Environment. Public Health and Food Safety, Brussels.

Garg, H.P., Datta, G., 1998. Global status on renewable energy: solar energy heating and cooling methods in buildings. In: Proceedings of the International Workshop University of Science and Technology, Iran, 19–20 May 1998.

Ge, X.M., Xu, F.Q., Vasco-Correa, J., Li, Y.B., 2016. Giant reed: a competitive energy crop in comparison with miscanthus. Renew. Sust. Energ. Rev. 54, 350–362.

Giessow, J., Casanova, J., Leclerc, R., MacArthur, R., Fleming, G., Giessow, J., 2011. *Arundo donax* (Giant Reed): Distribution and Impact Report. State Water Resources Control Board California Invasive Plant Council (Cal-IPC), California, CA, pp. 1–240.

Gnansounou, E., Raman, J.K., 2016. Life cycle assessment of algae biodiesel and its co-products. Appl. Energy 161, 300–308.

Golub, A.A., Hertel, T.W., 2012. Modeling land-use change impacts of biofuels in the GTAP-bio framework. Clim. Chang. Econ. 3, 1250015.

Gouveia, L., Batista, A.P., Sousa, I., Raymundo, A., Bandarra, N.M., 2008. Microalgae in novel food products. In: Papadoupoulos, K. (Ed.), Food Chemistry Research Developments. Nova Science Publishers, New York, NY, pp. 75–112.

Hamelinck, C., 2013. Biofuels and Food Security. Risks and Opportunities. ECOFYS, Utrecht, pp. 1–23.

Henman, D., Australia, R., 2012. Evaluation of Algal Meal as an Energy and Protein Source in Pig Diets. Report Prepared for the Co-Operative Research Centre for High Integrity Australian Pork. Australian Government fs Cooperative Research Centre, Corowa, NSW, pp. 1–12.

Hertel, T.W., Golub, A.A., Jones, A.D., O'Hare, M., Plevin, R.J., Kammen, D.M., 2010. Effects of US maize ethanol on global land use and greenhouse gas emissions: estimating market-mediated responses. Bioscience 60, 223231.

Hoh, D., Watson, S., Kan, E., 2016. Algal biofilm reactors for integrated wastewater treatment and biofuel production: a review. Chem. Eng. J. 287, 466–473.

Huang, H.X., Khanna, M., Onal, H., Chen, X.G., 2013. Stacking low carbon policies on the renewable fuels standard: economic and greenhouse gas implications. Energy Policy 56, 5–15.

Hughes, A.D., Kelly, M.S., Black, K.D., Stanley, M.S., 2012. Biogas from macroalgae: is it time to revisit the idea? Biotechnol. Biofuels 5, 86.

Huntley, M.E., Johnson, Z.I., Brown, S.L., Sills, D.L., Gerber, L., Archibald, I., Machesky, S.C., Granados, J., Beal, C., Greene, C.H., 2015. Demonstrated large-scale production of marine microalgae for fuels and feed. Algal Res. 10, 249–265.

IEA, 2010. Energy Technology Perspectives 2010. Scenarios & Strategies to 2050. International Energy Agency, Paris. 706 pages.

IEA, 2011. Technology Roadmap Biofuels for Transport. International Energy Agency, Paris. 52 pages.

IEA, 2013. The Renewable Energy 2013. Medium-Term Market Report. A Growing Role for Renewables in the Energy Mix. International Energy Agency, Paris. 13 pages.

IEA, 2015a. Renewable Energy. Medium-Term Market Report 2015. Market Analysis and Forecasts to 2020. International Energy Agency, Paris. 14 pages.

IEA, 2015b. World Energy Outlook 2015. International Energy Agency, Paris. 200 pages.

Irwin, S., Good, D., 2015. Ethanol production profits: the risk from lower prices of distillers grains. Farmdoc Dly. 5, 46.

Jones, M.B., Finnan, J., Hodkinson, T.R., 2015. Morphological and physiological traits for higher biomass production in perennial rhizomatous grasses grown on marginal land. Glob. Chang. Biol. Bioenergy 7, 375–385.

Khanna, M., Zilberman, D., 2012. Modeling the land-use and greenhouse-gas implications of biofuels. Clim. Chang. Econ. 3, 1250016.

Kligerman, D.C., Bouwer, E.J., 2015. Prospects for biodiesel production from algae-based wastewater treatment in Brazil: a review. Renew. Sust. Energ. Rev. 52, 1834–1846.

Komolwanich, T., Tatijarern, P., Prasertwasu, S., Khumsupan, D., Chaisuwan, T., Luengnaruemitchai, A., Wongkasemjit, S., 2014. Comparative potentiality of kans grass (*Saccharum spontaneum*) and giant reed (*Arundo donax*) as lignocellulosic feedstocks for the release of monomeric sugars by microwave/chemical pretreatment. Cellulose 21, 1327–1340.

Kretschmer, B., Allen, B., Kieve, D., Smith, C., 2013. The Sustainability of Advanced Biofuels in the EU: Assessing the Sustainability of Wastes, Residues and Other Feedstocks Set out in the European Commission. Proposal on indirect Land Use Change (ILUC). Institute for European Environmental Policy (IEEP), London. 37 pages.

Langeveld, J.W.A., Dixon, J., van Keulen, H., Quist-Wessel, P.M.F., 2014. Analyzing the effect of biofuel expansion on land use in major producing countries: evidence of increased multiple cropping. Biofuels Bioprod. Biorefin. 8, 49–58.

Licht, F.O., 2013. World Ethanol and Biofuels Report. Agra Informa, London.

Lynd, L.R., Larson, E., Greene, N., Laser, M., Sheehan, J., Dale, B.E., McLaughlin, S., Wang, M., 2009. The role of biomass in America's energy future: framing the analysis. Biofuels Bioprod. Biorefin. 3, 113–123.

Lynd, L.R., Sow, M., Chimphango, A.F.A., Cortez, L.A.B., Cruz, C.H.B., Elmissiry, M., Laser, M., Mayaki, I.A., Moraes, M.A.F.D., Nogueira, L.A.H., et al., 2015. Bioenergy and African transformation. Biotechnol. Biofuels 8, 18.

Maceda-Veiga, A., Basas, H., Lanzaco, G., Sala, M., Sostoa, A., Serra, A., 2016. Impacts of the invader giant reed (*Arundo donax*) on riparian habitats and ground arthropod communities. Biol. Invasions 18, 731–749.

Macedo, M.N., DeFries, R.S., Morton, D.C., Stickler, C.M., Galford, G.L., Shimabukuro, Y.E., 2012. Decoupling of deforestation and soy production in the southern amazon during the late 2000s. PNAS 109, 1341–1346.

Makkar, H.P.S., 2012. Biofuel Co-products as Livestock Feed. Opportunities and Challenges. FAO, Rome, ISBN: 978-92-5-107299-8. 533 pages.

Melillo, J.M., Reilly, J.M., Kicklighter, D.W., Gurgel, A.C., Cronin, T.W., Paltsev, S., Felzer, B.S., Wang, X.D., Sokolov, A.P., Schlosser, C.A., 2009. Indirect emissions from biofuels: how important? Science 326, 1397–1399.

Milledge, J.J., 2011. Commercial application of microalgae other than as biofuels: a brief review. Rev. Environ. Sci. Biotechnol. 10, 31–41.

Mosnier, A., Havlik, P., Valin, H., Baker, J., Murray, B., Feng, S., Obersteiner, M., McCarl, B.A., Rose, S.K., Schneider, U.A., 2013. Alternative us biofuel mandates and global GHG emissions: the role of land use change, crop management and yield growth. Energy Policy 57, 602–614.

Mueller, S., 2008. National dry mill corn ethanol survey. Biotechnol. Lett. 32, 1261–1264.

Nackley, L.L., Kim, S.H., 2015. A salt on the bioenergy and biological invasions debate: salinity tolerance of the invasive biomass feedstock *Arundo donax*. GCB Bioenergy 7, 752–762.

Naik, S.N., Goud, V.V., Rout, P.K., Dalai, A.K., 2010. Production of first and second generation biofuels: a comprehensive review. Renew. Sust. Energ. Rev. 14 (2), 578–597.

Ndimba, B.K., Ndimba, R.J., Johnson, T.S., Waditee-Sirisattha, R., Baba, M., Sirisattha, S., Shiraiwa, Y., Agrawal, G.K., Rakwal, R., 2013. Biofuels as a sustainable energy source: an update of the applications of proteomics in bioenergy crops and algae. J. Proteome 93, 234–244.

Nelson, K.A., Motavalli, P.P., Smoot, R.L., 2009. Utility of dried distillers grain as a fertilizer source for corn. J. Agric. Sci. 1, 3–12.

Pandey, V.C., Bajpai, O., Singh, N., 2016. Energy crops in sustainable phytoremediation. Renew. Sust. Energ. Rev. 54, 58–73.

Paulova, L., Melzoch, K., Rychtera, M., Patakova, P., 2013. Production of 2nd-generation of liquid biofuels. In: Fang, Z. (Ed.), Liquid, Gaseous and Solid Biofuels: Conversion Techniques. InTech Open Access Publisher, Rijeka, pp. 47–78.

Perlack, R.D., Eaton, L.M., Turhollow Jr., A.F., Langholtz, M.H., Brandt, C.C., Downing, M.E., Graham, R.L., Wright, L.L., Kavkewitz, J.M., Shamey, A.M., 2011. US Billion-Ton Update: Biomass Supply for a Bioenergy and Bioproducts Industry. U.S. Department of Energy, Oak Ridge, TN. 227 pages.

Popp, J., Lakner, Z., Harangi-Rakos, M., Fari, M., 2014. The effect of bioenergy expansion: food, energy, and environment. Renew. Sust. Energ. Rev. 32, 559–578.

Popp, J., Harangi-Rákos, M., Gabnai, Z., Balogh, P., Antal, G., Bai, A., 2016. Biofuels and their co-products as livestock feed: global economic and environmental implications. Molecules 21, 285.

Prieler, S., Fisher, G., 2009. Agricultural by-products associated with biofuel production chains. Report of ELOBIO WP5—Subtask D5.1: assessment of biofuel policy impacts in food and feed markets. Edited by Intelligent Energy Europe (IEE).

Rabemanolontsoa, H., Saka, S., 2013. Comparative study on chemical composition of various biomass species. RSC Adv. 3, 3946–3956.

Salerno, M., Nurdogan, Y., Lundquist, T.J., 2009. Biogas production from algae biomass harvested at wastewater treatment ponds. In: Bioenergy Engineering Conference, ASABE 2009, Seattle, Washington, October 11-14, 2009. poster 5.

Searchinger, T., Heimlich, R., 2015. Avoiding Bioenergy Competition for Food Crops and Land. World Resources Institute, Washington, DC. 44 pages.

Searchinger, T., Heimlich, R., Houghton, R.A., Dong, F.X., Elobeid, A., Fabiosa, J., Tokgoz, S., Hayes, D., Yu, T.-H., 2008. Use of U.S. croplands for biofuels increases greenhouse gases through emissions from land-use change. Science 319 (5867), 1238–1240.

Shatalov, A.A., Pereira, H., 2012. Xylose production from giant reed (*Arundo donax* L.): modeling and optimization of dilute acid hydrolysis. Carbohydr. Polym. 87, 210–217.

Sialve, B., Bernet, N., Bernard, O., 2009. Anaerobic digestion of microalgae as a necessary step to make microalgal biodiesel sustainable. Biotechnol. Adv. 27, 409–416.

Sikes, K., McGill, R., van Walwijk, M., 2011. Algae as a Feedstock for Biofuels: An Assessment of the State of Technology and Opportunities. Final Report. IEA-Advanced Motor Fuels Agreement, Annex 34, Subtask 2 USA, p. 1.23.

Silva, C.F.L.E., Schirmer, M.A., Maeda, R.N., Barcelos, C.A., Pereira, N., 2015. Potential of giant reed (*Arundo donax* L.) for second generation ethanol production. Electron. J. Biotechnol. 18, 10–15.

Spolaore, P., Joannis-Cassan, C., Duran, E., Isambert, A., 2006. Commercial applications of microalgae. J. Biosci. Bioeng. 101, 87–96.

Stolarski, M.J., Krzyzaniak, M., Luczyski, M., Zaluski, D., Szczukowski, S., Tworkowski, J., Golaszewski, J., 2015. Lignocellulosic biomass from short rotation woody crops as a feedstock for second-generation bioethanol production. Ind. Crop. Prod. 75, 66–75.

Taheripour, F., Tyner, W.E., 2013. Biofuels and land use change: applying recent evidence to model estimates. Appl. Sci. 3, 14–38.

Taheripour, F., Hertel, T.W., Tyner, W.E., Beckman, J.F., Birur, D.K., 2010. Biofuels and their by-products: global economic and environmental implications. Biomass Bioenergy 34, 278–289.

Testa, R., Fodera, M., di Trapani, A.M., Tudisca, S., Sgroi, F., 2016. Giant reed as energy crop for southern Italy: an economic feasibility study. Renew. Sust. Energ. Rev. 58, 558–564.

Triana, F., Nassi o Di Nasso, N., Ragaglini, G., Roncucci, N., Bonari, E., 2015. Evapotranspiration, crop coefficient and water use efficiency of giant reed (*Arundo donax* L.) and

miscanthus (*Miscanthus sinensis* × *giganteus* Greef *et* Deu.) in a Mediterranean environment. GCB Bioenergy 7, 811–819.
Trostle, R., Marti, D., Rosen, S., Westcott, P., 2011. Why Have Food Commodity Prices Risen Again. Economic Research Service, Washington, DC, pp. 1–29.
Ullah, K., Sharma, V.K., Dhingra, S., Braccio, G., Ahmad, M., Sofia, S., 2015. Assessing the lignocellulosic biomass resources potential in developing countries: a critical review. Renew. Sust. Energ. Rev. 51, 682–698.
USDA, 2015. Agricultural Projections to 2024. Interagency Agricultural Projections Committee, Washington, DC. 91 pages.
Van den Bos, A., Hamelinck, C., 2014. Greenhouse Gas Impact of Marginal Fossil Fuel Use. ECOFYS, Utrecht, pp. 1–29.
Viola, E., Arcieri, G., Zimbardi, F., Valerio, V., Cerone, N., De Corato, U., 2016. Evaluation of a pilot-scaled paddle dryer for the production of ethanol from lignocellulose including inhibitor removal and high-solids enzymatic hydrolysis. Biotechnol. Rep. 9, 38–45.
Wang, M., Dunn, J.B., 2015. In: Searchinger, Heimlich (Eds.), Comments on Avoiding Bioenergy Competition for Food Crops and Land. Argonne National Laboratory, Argonne, IL, pp. 1–9.
Wang, M.J., Dunn, B., Mueller, S., Qin, Z., Tyner, W., Goodwin, B., 2014. Comments on Ethanol's Broken Promise by the Environmental Working Group (May 2014). Argonne National Laboratory, Argonne, IL, pp. 1–13.
Ward, A.J., Lewis, D.M., Green, B., 2014. Anaerobic digestion of algae biomass: a review. Algal Res. 5, 204–214.
WBA, 2014. Global Bioenergy Statistics 2014. WBA, Stockholm.
Werling, B.P., Dickson, T.L., Isaacs, R., Gaines, H., Gratton, C., Gross, K.L., Liere, H., Malmstrom, C.M., Meehan, T.D., Ruan, L.L., et al., 2014. Perennial grasslands enhance biodiversity and multiple ecosystem services in bioenergy landscapes. PNAS 111, 1652–1657.
Wisner, R., 2015. November Monthly Renewable Energy Report. Distillers Grains Balance Sheet, Agricultural Marketing Resource Center, Ames, IA.
Witzke, P., Banse, M., Gomann, H., Heckelei, T., Breuer, T., Mann, S., Kempen, M., Adenauer, M., Zintl, A., 2008. Modelling of Energy-Crops in Agricultural Sector Models. A Review of Existing Methodologies. Joint Research Centre (JRC). Institute for Prospective Technological Studies Luxembourg, Ispra, Varese. 59 pages.
Yap, T.N., Wu, J.F., Pond, W.G., Krook, L., 1982. Feasibility of feeding *Spirulina maxima*, *Arthrospira platensis* or *Chlorella* sp. to pigs weaned to a dry diet at 4 to 8 days of age. Nutr. Rep. Int. 25, 543–552.
Zhang, W., Yu, E.A., Rozelle, S., Yang, J., Msangi, S., 2013. The impact of biofuel growth on agriculture: why is the range of estimates so wide? Food Policy 38, 227–239.
Zhou, W.G., Chen, P., Min, M., Ma, X.C., Wang, J.H., Griffith, R., Hussain, F., Peng, P., Xie, Q.L., Li, Y., et al., 2014. Environment-enhancing algal biofuel production using wastewaters. Renew. Sust. Energ. Rev. 36, 256–269.

CHAPTER 14

Education and training in agriculture and the Bioeconomy: Learning from each other

Yaprak Kurtsal, Giacomo Maria Rinaldi, Marco Maria Grande, and Davide Viaggi
Department of Agricultural and Food Sciences, University of Bologna, Bologna, Italy

1. Introduction

In both agriculture and Bioeconomy literature, education is regarded as an issue of paramount importance. Education is often used as an explanatory variable of behavior by consumers (e.g., for food, bio-based products) or entrepreneurs (farmers, bio-based start-uppers etc.). Innovation, that is a cross cutting issue in the Bioeconomy, is also closely connected to education processes.

Nevertheless, in the current practice, education is usually deemed to be insufficient or ineffective. This is due to several factors including: (a) the fact that education is treated as a standalone issue, disconnected from industry needs and research practice; (b) the fact that education is still largely carried out in a sector perspective, rather than in a holistic view.

In addition, for the Bioeconomy, a major issue is that the Bioeconomy education is often addressed in isolation from the most traditional sectors of the Bioeconomy, in particular agriculture, that have, on the contrary, a strong tradition with educational and training issues. On the other hand, agriculture education is often carried out without considering the broad picture of the Bioeconomy.

The COVID-19 outbreak highlighted the need that Bioeconomy and food systems become more resilient, and emphasized issues such as social coordination and consensus, fast adaptation of skills and the role of science in society. All of these topics are highly connected to different levels of education, which hints at the fact that education will become even more important in the future.

Agricultural Bioeconomy
https://doi.org/10.1016/B978-0-323-90569-5.00002-0

The aim of this paper is to discuss current and perspective issues of agriculture education in the framework of the Bioeconomy and to identify promising conceptual pathways to overcome the current lack of integration. The assumption of the chapter is that this can be achieved by reconsidering education in the wider perspective of innovation and knowledge exchange.

The structure of the chapter is organized around an analysis of the literature in the direction of meeting this objective. In Section 2, we review selected issues in Bioeconomy education. In Section 3, we review the literature in agricultural education. In Section 4 we review the literature on Agricultural Knowledge and Innovation Systems (AKIS) and discuss how this can be used as an overarching framework to address the topic of agricultural education in the Bioeconomy.

2. Bioeconomy education needs and approaches

2.1 Educational needs

Education and training have prominent positive impacts on the quality of people's lives, on social and economic progress, and on regional development (Gregáňová et al., 2017). Education is also a prerequisite for the development of a sustainable Bioeconomy, together with research, knowledge and social participation (Golowko et al., 2019). In the agricultural sector, a higher level of education of farmers leads to greater success and income for farms, the willingness to be sustainable in their own practices, and it is usually correlated with higher crop yields (Lewandowski, 2018). Many institutions around the world have sustained biotechnology education with specific initiatives to support the development of the Bioeconomy, both in developed regions, such as the European Union and the United States, and in emerging and developing countries, such as India (Ray et al., 2016).

The transition toward a sustainable Bioeconomy needs new education and skills (European Commission, 2018; Lewandowski, 2018; Urmetzer et al., 2020), including soft skills (Pubule et al., 2020; Urmetzer et al., 2020) such as skills for communication, education, policy and decision-making (Urmetzer et al., 2020). Due to this transformation, numerous gaps become evident in Bioeconomy education that have to be identified and bridged (Global Bioeconomy Summit 2020, 2021). Public awareness and education are essential to adequately train the new workforce (Pascoli et al., 2021) as well as public decision-makers, investors, and other types of stakeholders (Adamseged and Grundmann, 2020). To meet these needs, education programs need to be adjusted and developed (Lewandowski, 2018), which requires time, experience, and financial and human resources (Golowko et al., 2019).

2.2 Education levels

The contribution that education can make to the development of the Bioeconomy concerns all levels of the education and training systems (Viaggi, 2018). It is argued that it is important to start Bioeconomy education already at the school level (Pascoli et al., 2021), providing general knowledge on Bioeconomy and technologies (Viaggi, 2018). The consequent awareness raised in young people should allow them to make informed decisions both for sustainable consumption and for pursuing a career in the Bioeconomy (Pascoli et al., 2021).

University education, on the other hand, is considered essential to allow for a change and transition toward a sustainable Bioeconomy and innovation (Pubule et al., 2020; Urmetzer et al., 2020). Programs that lead toward a career in Bioeconomy mainly concern areas of life sciences (science, biotechnology, agriculture) and other fields (business, finance, policy, marketing, etc.) (Pascoli et al., 2021); but they are often focused on a specific field (Viaggi, 2018). This could limit the graduates' ability to be specialists with a capability of systemic and interdisciplinary thinking (Pascoli et al., 2021). Diversely, it seems that the doctoral level and leadership roles are characterized by a more holistic approach (Viaggi, 2018).

Also, vocational training is promoted to develop the professional skills required in the Bioeconomy (Viaggi, 2018). However, different motivational factors have been found between the communities of universities of applied sciences and vocational schools that should be considered and addressed: in vocational training, many demotivating factors may exist regarding problems with students' vocational identity and teachers' career status, while in universities of applied sciences social motivation seems to be more expressed (Daukilas and Keibienė, 2017).

To update the skills of workers in the Bioeconomy sectors, life-long learning is promoted (Viaggi, 2018). Life-long learning is considered crucial both for workers and industries to enable the transition toward a circular and sustainable Bioeconomy, also allowing for greater diversity in career paths (Global Bioeconomy Summit 2020, 2021). Global Bioeconomy Summit 2020 (2021) suggests that "Life-long learning is only possible where educational material and scientific findings are available to the public." According to this principle, life-long learning needs to be professionally recognized and made accessible, through for example the digitalization of education, training on the job, and university training programs for career development and in collaboration with external stakeholders (Global Bioeconomy Summit 2020, 2021).

Hakovirta and Lucia (2019) and Pagliaro (2019) claim that the change and transformation in the Bioeconomy are too rapid to make conventional education systems keep pace during young people's education. Therefore, formal education (such as school and higher education) needs to be accompanied by informal education (Pascoli et al., 2021). Informal learning is based on hands-on and interactive activities, such as conversation and experimentation; for STEM disciplines, these could be joined through, for example, museums, libraries, nature centers, exhibitions, games, and communities of practice, stimulating young people toward innovation and sustainability (Hakovirta and Lucia, 2019). It is also argued that informal education could be significant to involve more under-represented and rural students (Hakovirta and Lucia, 2019).

2.3 Digitalization

Digital transition and automation are disrupting the Bioeconomy, fostering a transformation in production and consumption, and helping to find solutions to societal challenges (Ryymin, 2021). Considering the agricultural sector, new digital applications can lead farmers to a change in decision-making, business models, knowledge exchange, and learning (Ryymin, 2021).

While digitalization is leading to new opportunities and modalities of learning and education (Ray et al., 2016; Global Bioeconomy Summit 2020, 2021; Ryymin, 2021), the consequent renewal of industries requires new skills in the workforce to sustain this transformation (Autor et al., 2020; Ryymin, 2021).

In this regard, the digitalization of Bioeconomy education can provide many technical skills to a larger community of people, even beyond developed countries (Global Bioeconomy Summit 2020, 2021). One example in this context is Massive Open Online Courses (MOOCs), which could be attended online as part of life-long learning programs. Besides, interactive learning and experimentation activities at home are increasingly used with the support of digital technologies, such as virtual labs, virtual tours and interactive video tutorials (Ray et al., 2016; Hakovirta and Lucia, 2019; Pagliaro, 2019; Global Bioeconomy Summit 2020, 2021), and there are examples of platforms and initiatives that are providing successful outcomes (Ramachandran et al., 2019).

While the significance of digitalization in the Bioeconomy continues to increase at a fast pace, educational institutions, such as schools, risk falling behind these massive societal changes, due to being too conservative in some cases (Sá et al., 2021). The Covid-19 pandemic has clearly evidenced the existence of the digital divide, causing greater inequalities due to the lack of adequate infrastructures and skills; for sustainable development, it is necessary to bridge this gap by providing digital literacy (Sá et al., 2021).

It is further claimed that digitalization allows the democratization of education, in the Bioeconomy context as well (Global Bioeconomy Summit 2020, 2021), and digital fluency and equity are both required (Sá et al., 2021). Toward this goal, open access publishing (Global Bioeconomy Summit 2020, 2021) and open digital educational resources (Tang, 2021) are fundamental to making knowledge available and reducing cost of education for students. One example is provided by some open universities in developing countries, where they aim to include rural students through virtual labs (Ray et al., 2016).

2.4 Contents, competencies and skills

Sustainability is needed in the Bioeconomy, and both can be based on similar competencies (Lewandowski, 2018; Pubule et al., 2020). Sustainability and Bioeconomy require transdisciplinary approaches, knowledge and skills (Lewandowski, 2018; Golowko et al., 2019; Hakovirta and Lucia, 2019; Masiero et al., 2020; Pubule et al., 2020; Pascoli et al., 2021), combining science fields with social sciences (Reddy et al., 2011). In this paradigm, Bioeconomy education needs to be interdisciplinary (Drejerska, 2017; Lewandowski, 2018; Golowko et al., 2019; Masiero et al., 2020; Pubule et al., 2020; Pascoli et al., 2021; Sá et al., 2021). Hence, combining STEM and social disciplines, such as ethics, law and economics (Drejerska, 2017; Masiero et al., 2020; Global Bioeconomy Summit 2020, 2021) would allow students to address problems from a broader perspective (Lewandowski, 2018). Lewandowski (2018), argues that interdisciplinary education at the university level could be pursued by acquiring disciplinary expertise during the bachelor's degree and extending this with interdisciplinary expertise in postgraduate courses (supporting the development of a "T-shaped" profile, instead of an "I-shaped" profile). A further element that Urmetzer et al. (2020) add in this context is the need to provide transformative knowledge for the transition to a sustainable Bioeconomy. With this pedagogical approach, educators attempt to make students be able to imagine the disruption of current systems in order to guide society toward sustainability.

2.5 Students

Zulkifli (2018) states that education "refers to an activity that is aimed to make students able to achieve success in their lives, become knowledgeable and skilled people, and have good attitudes based on which they can face the surrounding environment." Consequently, education empowers students as citizens (Tilak, 2006; Sá et al., 2021) and raises their quality of life, broadening their interests and potential, fostering personal and professional development (Tilak, 2006), and leading to an increase in their income (Tilak, 2006; Lewandowski, 2018). From a different perspective, "human capital theorists place emphasis on education's role in transforming human beings into human capital, into instruments of production and economic growth and thus forwarding the economic well-being of peoples and societies" (Tilak, 2006, p. 237). Beyond this dichotomy, sustainability education also trains for social activism and political engagement (Lozano-Díaz and Fernández-Prados, 2020), while higher education at large allows for greater social integration (Tilak, 2006).

Focusing on Bioeconomy, Masiero et al. (2020) show that in some cases students may have already heard about the term "Bioeconomy" (mainly through university courses), and they seem curious and interested in it (Stern et al., 2018). Moreover, the changes in education driven by digitalization should lead to more active involvement of students and assign them a more important role in education (Di Sia, 2020; Sá et al., 2021). However, it has to be taken into account that one of the implications of the digital transition in education pushed by the Covid-19 pandemic is the effect on students' health of being connected for too long in video calls (Sá et al., 2021).

Eventually, education systems should make accessible the available resources (Global Bioeconomy Summit 2020, 2021; Sá et al., 2021), while at the same time education design should consider the students' needs (European Students' Union, 2020) and involve them in co-designing the programs.

The inclusion of under-represented students in Bioeconomy education is crucial and advantageous for the affected people's lives, and also for sustainability, as greater diversity allows for better coping with complex problems (Schusler et al., 2021).

To include under-represented and rural students, pursuing equity is required, also removing the causes of digital divide (Sá et al., 2021) and using informal education (Hakovirta and Lucia, 2019). Addressing gender balance in bioeconomy education and workplaces, it is necessary to encourage women toward higher education (Roos et al., 2021). To do this, female role models can be used and represented, while at the same time, specific knowledge sharing and networking activities for women can be implemented to inspire young women to take part in Bioeconomy studies and careers (Roos et al., 2021). Furthermore, a potential for involving students of underrepresented ethnicities in Bioeconomy has been found (McAlexander et al., 2021). This can be achieved through greater equity and inclusion, as well as fostering the diversity of student representatives (Schusler et al., 2021). Toward the goal of being inclusive, higher education should also set management policies to support students with disabilities (Amka, 2017).

3. Agricultural education and the epistemology of inter- and trans-disciplinarity

The European Union statistical office, Eurostat, defines three levels of agricultural training as follows (EC, 2018):

- Practical agricultural experience: experience acquired through work on a farm.
- Basic agricultural training: any courses completed at a general agricultural college or an institution specializing in certain subjects (horticulture, viticulture, sylviculture, pisciculture, veterinary science, agricultural technology); a completed agricultural apprenticeship is considered basic training.
- Full agricultural training: any course continuing for the equivalent of at least 2 years full-time training after the end of compulsory education, completed at an agricultural college or at university, in agriculture, horticulture, viticulture, sylviculture, pisciculture, veterinary science, agricultural technology.

According to the most recent EU data available (EC, 2018), only 8.9% of the present generation of European farmers (EU27) have received full agricultural training, while 68.3% have only practical experience (Eurostat, 2016) and the remaining 22.7% followed some kind of basic agricultural training. Instead, looking among the young farmers (less than 35 years), the percentage of people who attended a full agricultural training was as high as 21.6%. While, this rate observed among young farmers is promising, especially in terms of ensuring better management of European farms, a significant advancement is still required in order to create a young generation of farmers equipped with the skills and competencies needed in the sector.

Furthermore, the grand challenges of our time (climate change, energy transition, digital transition, sustainability, etc.) require an urgent transformation of the current educational systems. The fact that farmers older than 55 years of age consist more than 57% of total farmers in the EU, while farmers younger than 35 years are just over 5%, calls for a need to train not only the young and future farmers, but also to adapt the pedagogical methods to train the current (and older) farmers (Medvedev and Molodyakov, 2019).

However, joining courses provided by an educational institute (e.g., University), where the courses are held mostly during the day, may not be affordable for most farmers. This entails two preliminary considerations: the importance of recognizing the needs of the target audience (i.e., farmers); and the importance of flexibility (e.g., recorded lessons in remote learning) in the education of the future.

Moreover, the transformation of the educational systems in this direction—due to the global nature of these goals -, involves not only the agricultural actors like farmers, researchers, policy-makers and other stakeholders in agriculture, but also civil society, associations, NGOs and consumers, which have a great role to play in issues related to agricultural production, animal welfare, food management and consumption, and all the topics directly related with agriculture. An example of this role can be the increased attention provided in the last years to the purchase of food products that are good for the environment (e.g., organic food), or to labor and social conditions of farmers/workers (e.g., Fairtrade) or for human health (e.g., quinoa).

3.1 New gaps in agricultural education

To date, as the world is re-thinking and re-shaping itself in a post-pandemic scenario, the interest in bio-based materials is increasing, leading the agricultural sector to a new position as a source of row—and renewable—materials for new industries (Fritsche et al., 2021). Although this does not represent a novelty, it indicates a convergence of the agriculture toward other new industries and sectors, like energy sector, apparel industry, or chemical and pharmaceutical industries (Boehlje et al., 2011).

The "industry convergence" is a phenomenon studied in several sectors (Bröring et al., 2006; Curran et al., 2010; Katz, 1996) and can be defined as the blurring of boundaries between industries by converging value propositions, technologies and markets (Bröring, 2010; Choi and Valikangas, 2001). Hence, the aforementioned convergence of the agricultural sector toward other industries brings new business opportunities, while on the other hand, leads to a higher degree of uncertainty (Boehlje et al., 2011). This degree of uncertainty matched with the blurred boundaries between industries imply new gaps, including regulation and standard gaps, legal gaps, or, from a human resources point of view, competence and skills gaps.

The need of bridging these education and training gaps will become increasingly important in the future of agriculture. Indeed, in the described scenario, it is presumable that the professionals of tomorrow (i.e., farmers) will act and work in a dynamic context where the connections with other actors (e.g., researchers, engineers, advisors, biotechnologists, fashion managers, computer scientists, etc.) will be daily. Hence, it is important to prevent these competence gaps from evolving into communication gaps, which would lead, inevitably, to failures and waste of resources (Galmiche-Tejeda, 2004).

This aspect emphasizes the centrality of curriculums' content in developing and shaping the agriculture of the future and raises questions about the arising agricultural education and training needs, namely:

- What can be the epistemic of the future?
- How to find a balance between theory and practice?
- How will the different levels of education (i.e., school, university, vocational education and training, and life-long learning) be instrumental in providing the right competencies and skills?
- What are the disciplines that will be fundamental for the "fully trained" farmers?

Some of these questions are not new in the agricultural education history. Indeed, yet in 1988, Peterson et al. wrote: "The problem is that applied research contributes directly to the discipline, but seldom contributes to improved professional training. [...] This issue is particularly troublesome in agricultural economics where research has become highly quantitative and mathematical while the changing undergraduate audience increasingly demands practical, business-oriented instruction." (Peterson et al., 1988, p. 28).

However, more recently, Migliorini and Lieblein (2016) noted that bringing university students closer to stakeholders in society as part of their learning process is highly important, because of its applied approach that is necessary toward a transition to sustainable agriculture. These two examples indicates how the debate on these issues is long-standing and lasting in agriculture.

3.2 Finding coral solutions in the complexity

It was argued that the convergence of different disciplines and professionals was required in the agricultural sector by the premise of sustainability. Indeed, there is a wide literature that underlines how the environmental and sustainability issues increased the request of interdisciplinary and transdisciplinary methods (Fletcher et al., 2020; Galmiche-Tejeda, 2004; Harris et al., 2009; Parr et al., 2007). This is due to the idea that traditional disciplinary approaches are not able to address critical issues and provide the holistic perspective needed to face sustainability problems (Harris et al., 2009).

Indeed, sustainability was defined as a "wicked problem" (Dentoni and Peterson, 2011), id est a problem that, for its complex nature, seems difficult or impossible to solve (Rittel and Webber, 1973). Hence, the need of seeing the "whole picture" brought along new research questions, as complexity requires different expertise, approaches, and perspectives.

However, inter- and transdisciplinarity have not a unique and a common understanding across the literature. In this article we define them as follow:

- Interdisciplinary research integrates social and natural science methods, with the consequent integration of ideas, theories and ways of viewing the world (Lawrence et al., 2020; Sillitoe, 2004);
- Transdisciplinary research consists in uniting academic and non-academic perspectives, going beyond academic disciplines and including groups like farmers, other business, practitioners, policymakers, advisors, and public (Harris et al., 2009).

The possibilities enabled by inter- and transdisciplinary research are, in some way, similar to the possibilities enabled by the industry convergence: they open the door to new possibilities of research and development of new ideas, technologies, approaches, and answers. Nevertheless, just like industry convergence, inter- and trans-disciplinarity lead to new gaps, such as competence and skill gaps, or language and communication gaps. Furthermore, it was underlined that the underlying premises and motivations that each discipline and professional brings to research are different, and that the development of a common understanding about these issues (gaps and differences) requires time and trust (Harris et al., 2009).

From an educational point of view, the growing interest in inter- and transdisciplinary approaches leads to the creation of new professions and/or the updating of the old ones. Indeed, the agriculture of tomorrow will require specialized people that can manage different languages, perspectives and competencies (mainly from natural and social sciences). This aspect is clear in the agricultural education institutes, which initiated a discussion way back on their re-organization according to the changing world and the requests—from inside and outside—of new teaching disciplines, new pedagogy and epistemology. Indeed, some agricultural education institutions have not maintained terms as "agriculture" or "agricultural" in their names. Presently, agricultural curricula may offer different disciplines such as geo-informatics, biotechnology, farm management, business management, economics, sociology, consumers, food and nutrition, health and communication and environmental studies (Barbinov, 2018).

The variety of these disciplines points to a substantial convergence, directly or indirectly, toward Bioeconomy. Nevertheless, to date, agricultural education is often carried out without considering the broad picture of Bioeconomy, because of the dominant paradigm of a sector perspective. This approach limits the possibilities of integrating agricultural education into the framework of the Bioeconomy, hampering the development of both. Thus, identifying conceptual pathways to reconsider education in the wider perspective of innovation and knowledge exchange, can be a first step to overcome the current lack of integration, allowing new opportunities in terms of new jobs, new business opportunities, and, subsequently, sustainable development.

4. Transforming the AKIS framework toward collaborative learning and innovation: Can we extend it further to cover Bioeconomy

4.1 AKIS in a nutshell

The need to speed-up innovation has been repeatedly emphasized in recent years and is now a core element of the European Commission's communication on the future of food and farming. For this reason, the necessity of a good policy framework that could drive the education of the future professionals in the agrifood and forestry sector has become urgent.

The term Agricultural Knowledge and Innovation Systems (AKIS) is used to describe the whole knowledge exchange system: the ways how people and organizations join together and interact to promote mutual learning, to generate, share, and use agriculture-related knowledge and information within a country or a region. Farmers, advisors, researchers, education and training providers (secondary, tertiary, or life-long learning levels), input suppliers, retailers, farmer organizations, NGOs, business and enterprises, media services and ministries are all part of national or regional AKIS, since they all either need, produce or exchange knowledge and innovation for agriculture and interrelated fields. Dockès et al. (2011) group the actors within the AKIS as follows: (i) information and knowledge system actors (research institutions, advisors, schools, farmers' unions), (ii) socio-economic system actors (farmers, food processors, networks), (iii) public decision-making system actors (authorities at regional and local level), and (iv) end-user actors (consumers, NGOs). Hence, using the AKIS as a contextual framework, it is possible to identify the roles of these actors and how change and transformation may occur (Spendrup and Fernqvist, 2019). Moreover, the AKIS framework serves to understand and describe the differences between countries, regions, and sectors in the field of innovation and knowledge production (EU SCAR, 2012). Meanwhile, the AKIS is not an isolated entity and, like most systems, interacts with the external environment (Spendrup and Fernqvist, 2019); thus, external effects may derive through the linkages to system drivers (e.g., climate change, Green Deal, SDGs), linkages to general science and technology, linkages to other economic sectors and linkages to the wider policy framework. Fig. 1 presents the AKIS framework.

Moreover, the AKIS concept was operationalized into policy mainly by the Standing Committee of Agricultural Research (SCAR) Strategic Working Group (SWG) on AKIS (EU SCAR, 2012) in consultation with the European Commission's (EC) Directorate-General for Agriculture and Rural Development (DG AGRI) (Fieldsend et al., 2020). The EIP-AGRI collaborates with several actors in public and private sectors at EU, national and regional levels, different knowledge players and Operational Groups for pooling and exchanging existing policies and instruments to achieve societal benefits and a fast modernization of the agricultural system. Fieldsend et al. (2020) suggest that central to the EIP-AGRI approach is the interactive innovation model, defined as: "the collaboration between various actors to make best use of complementary types of knowledge (scientific, practical, organisational etc.) in view of co-creation and diffusion of solutions/opportunities ready to implement in practice" (EIP-AGRI SP, 2017, p. 3).

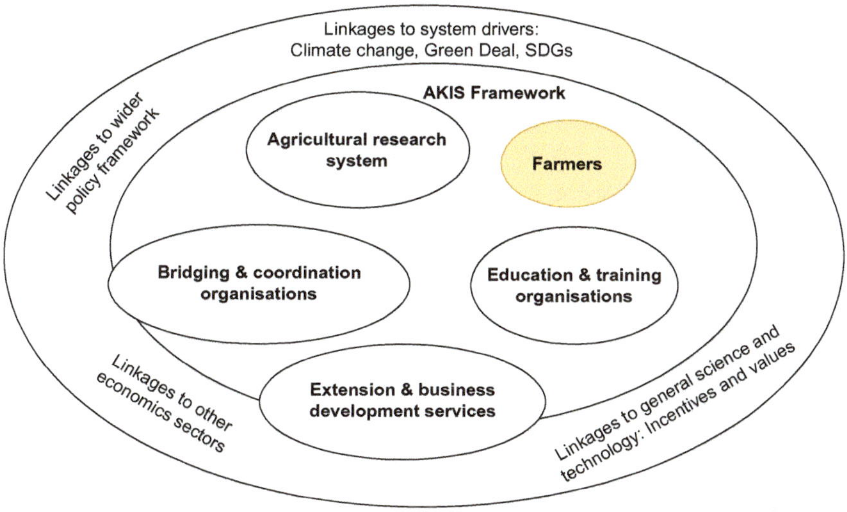

Fig. 1 The AKIS framework. *(Sources: Adapted by EU SCAR, 2015. Agricultural Knowledge and Innovation Systems Towards the Future—A Foresight Paper, Brussels; Dockès, A.C., Tisenkopfs, T., Bock, B., 2011. Collaborative working group agricultural knowledge and innovation systems. In: WP1: Reflection Paper on AKIS. Sub-Deliverable of the AKIS CWG—WP1—April 2011. European Commission, Brussels; Rivera, W.M., Qamar, M.K., Mwandemere, H.K., 2005. Enhancing Coordination Among AKIS/RD Actors: An Analytical and Comparative Review of Country Studies on Agricultural Knowledge and Information Systems for Rural Development (AKIS/RD). http://www.fao.org/3/y9087e/y9087e00.htm; Arnold, E. & Bell., M., 2001. Some Ideas About Research for Development—Unpublished Background Paper Provided to the Commission on Development Related Research in Denmark, Science and Technology Policy Research (SPRU), Technopolis,; Spielman, D. & Birmer, R., 2008. How Innovative is Your Agriculture? Using Innovation Indicators and Benchmarks to Strengthen National Agricultural Innovation Systems.)*

4.2 Transformation of the AKIS to date: Toward integration and collaboration

The lack of education and skills in the agri-food sector is regarded as a main bottleneck that is hampering innovation. This gap has been exacerbated by the use of out-dated methods of learning and education. In the same line, initially a top-down conceptualization of learning in the agri-food sector was prevalent, in the context of which scientific research was positioned at the top, and knowledge and innovation were believed to flow in a downstream way to reach the final applications in agricultural production (from the lab to the field), through the use of education and extension (Esposti, 2012).

In this vein, the "linear knowledge transfer" model—where researchers, trainers and technical experts develop solutions to agricultural problems and then pass them down to farmers—has been challenged considerably toward a stricter coordination and integration between the components of the "knowledge triangle" (research, education and extension) (Esposti, 2012). Oliveira et al. (2019) argue that producing R&D is not the only main goal, because innovation with no knowledge transferred to farmers does not enhance agriculture innovation; hence, innovation is to be fostered by various rural development measures, such as "knowledge transfer," "cooperation," and "investments in physical assets." In other words, this linear view of innovation in agricultural contexts are being replaced with systems approaches, where agricultural producers are also seen as important actors rather than merely consumers of the technologies that are generated by agricultural research and transferred by education and extension services for subsequent adoption (Home and Rump, 2015). Materia (2012), in this direction, argues that such changes implemented in the last decade indicate a general movement from the traditional linear top-down approach to an innovation systems approach (Hall et al., 2006), which is more reactive and interactive, and where agents contribute together to find innovative solutions.

In parallel to the transformation of the knowledge transfer models, the AKIS framework also experienced a transformation. The transition from the old concept of Agricultural Knowledge and Information System (Rivera et al., 2005) to Agricultural Knowledge and Innovation System represents the expansion to a wider concept that can incapsulate a higher number of actors involved in the system (actors related with innovation: input suppliers, food processors, retailers, accountants, banks, media) (Knierim et al., 2015) and reflects the complexity of the actual agri-food and forestry sector. Hence, the AKIS framework, thanks to its transformation to date, has been instrumental to demonstrate that knowledge and innovation in fact do not flow in a linear way—from researchers to farmers—but rather is consisted of an intricate network in which the flows of knowledge and innovation are often unpredictable (Arzeni et al., 2021; Hermans et al., 2015; Klerkx et al., 2012). According to the AKIS, it is critical to share knowledge and innovation for the development of agriculture and rural areas, and to promote mutual learning through the involvement of farmers, advisors, trainers, researchers, media and other agricultural experts operating at EU, national, regional and local levels. However, at the same time, incentives need to be in place for the systems' actors to generate, develop and exchange new technologies, knowledge and experiences (OECD, 2012).

In the literature, the AKIS framework was largely used by researchers as a lens to analyze the national knowledge and innovation structures (Knierim et al., 2015; Zahran et al., 2020), but also as a starting point for a deeper analysis of the sector. The research that use the AKIS framework evaluates the knowledge and innovation dynamics mainly by using three approaches: Actors-oriented research, Relationships-oriented research, and External factors-oriented research.

The Actor(s)-oriented research is mainly interested in one (or more) actor of the AKIS ecosystem, exploring the skills required from that actor in modern agriculture, and the typologies of approaches toward innovation and knowledge, information, power and goals (Long, 2001; McDonald and Macken-Walsh, 2016). The actors-oriented literature is comprised of two principal groups: in the first, the attention is given to a core group of actors (farmers, advisory services, private sector, private and public research institutes, policymakers, NGOs); and in the second, the main focus is rather on consumers, financial and economic entities, technology developers and the media. Spendrup and Fernqvist (2019) note that a considerable number of studies that focus on different actors in the AKIS, analyze the methods of knowledge transfer, education and training, as well as technology involved in supporting firms with knowledge, mainly through extension and advisory services. However, other methods studied include teaching in vocational schools, universities or farm schools, participatory activities by different stakeholders, farm demonstrations and champion farmers, good practice models and communities of practice and exchange of experience (Spendrup and Fernqvist, 2019).

The relationship-oriented research, on the other hand, aims to understand the quality and the different types of interactions, relationships, partnerships, contracts, networks, alliances, platforms (Zahran et al., 2020), but also shared values, tacit knowledge, common objectives (Klerkx et al., 2012; Lioutas et al., 2019) between AKIS actors and across different networks (organization, performances, etc.), underlining that the knowledge and innovation systems are foremost all social systems. Generally, the main focus is to point out the strengths and weaknesses that can affect an effective collaboration. Furthermore, studies that analyze how knowledge, in particular, is shared and exchanged focus on processes of learning, with the involvement of actors in the AKIS. These studies include reflexive approaches in learning (Alroe et al., 2017), life-long learning approaches (von Munchhausen and Haring, 2012) and co-learning (Quiedeville et al., 2017). Despite the growing interest in the nature of interactions among actors (Hilkens et al., 2018), this type of exploration of the AKIS framework is still rare.

Last but not least, despite the personal efforts of each actor in an AKIS to enhance their skills and competencies and to collaborate through an improved networked approach, there are also external factors that have an impact on the AKIS. In this direction external factor-oriented research mainly focuses on the analysis of a new technique or a new technology, and their possible impacts on a certain AKIS, and the relationships that are inherent among the relationships of the actors. In particular, digitalization and its consequences on the different systems (Ingram and Maye, 2020; Klerkx and Begemann, 2020; Rijswijk et al., 2019) has been a key topic, which also gained importance with the current realities of our day (including the COVID-19 pandemic).

4.3 Proposals for a further transformation of the AKIS: From a sectoral approach to a wider perspective of Bioeconomy

In accordance with the new challenges and paradigms faced in our day, a further need to reform the learning approaches utilized, and hence to re-conceptualize the AKIS proves to be prevalent. In this direction, as mentioned in the previous section, the necessity to switch from a strictly agricultural perspective to a system open to all traditional sectors—now converging into Bioeconomy—is being stressed by an increasing number of studies (Guillou, 2012; EU SCAR, 2012; Esposti, 2012; Moschitz et al., 2021).

Gava et al. (2017), arguing that flaws in the AKIS—including missing stakeholders, missing links between relevant stakeholders or ineffective knowledge transfer that hinder farmers' ability to build their knowledge-base (EU SCAR, 2015)—makes it crucial to transform the AKIS, acknowledging its crucial role in the sustainable transition toward a Bioeconomy. In this vein, Esposti (2012) argues that to enter the age of the Bioeconomy, the traditional definition of the AKIS has been updated to introduce more heterogeneous, less formal and more autonomous ways of knowledge diffusion, which suggest a more complex interaction between stakeholders of the knowledge triangle (research, education, extension) and potential adopters of innovation in agriculture (Esposti, 2012). Besides, bearing in mind the complex dynamics of this expanded ecosystem, some research also points to the necessity to extend this "knowledge triangle" into a "knowledge rectangle," in the context of which "the community" aspect is also integrated (Viaggi et al., 2021). Hence, opening-up the strictly agricultural perspective to that of a system of innovation that encompasses all sectors that composes the Bioeconomy (and a wider range of actors) may introduce the possibility of capturing the wider perspective and the complexities of the interwoven intersectoral relationships and exchanges that shape the innovation ecosystems in real life (Gava et al., 2017). For this reason, the European Commission also empowers member states to effectively support and re-organize regional AKIS via their Bioeconomy Action Plans (European Commission, 2012).

In this direction, Guillou (2012) proposes a transformation of the AKIS framework to that of a KISB: Knowledge, technology and innovations for a bio-based economy. This new wider perspective may also be helpful to better understand the emergence of new practices and approaches that have been prevalent in the last two decades. Most of the major progress in successful cases, including agrotourism, organic agriculture, direct selling or agroenergy, involved multiple sectors and a wider ecosystem of exchanges. In the same line, Vehvilainen et al. (2021) who provide policy messages in order to deploy Bioeconomy in the EU, propose to encourage diffusion of bio-based knowledge, innovation and technological advances to support rural and regional development. Hence, they underline that a prototype AKIS approach tailored for the Bioeconomy, to support engagement by primary producers and farm and rural economy innovation intermediaries and entrepreneurs, would help member states to integrate Bioeconomy into their strategic plans. While, the KISB framework is far from being complete, special emphasis needs to be placed to address the challenges of policy coordination and harmonization in the EU policy.

In addition to the idea of the KISB, there are other studies in the literature, who underline the importance of transforming the AKIS in order to capture the wider perspective of sectoral interactions. Moschitz et al. (2021) argue that with a view of the necessary transformation of the agri-food system, the AKIS should also be extended to a Food and Agricultural Knowledge and Innovation System. Fieldsend (2020) introduces the idea of ARKIS: Agricultural and Rural Knowledge and Information Systems, which can work as a multi-stakeholder platform of civil society networks and organizations that are pushing for a CAP reform to encompass an integrated rural agenda. Lawrence et al. (2020), on the other hand, introduce the concept of Forestry Knowledge and Innovation System, FOKIS, to better fit with the forestry context, and Klerkx and Begemann (2020) develop the concept of mission-oriented Agricultural Innovation System (MAIS). Last but not least, Viaggi et al. (2021), by focusing on the Education and Training component of the AKIS framework in the agri-food and forestry sectors, propose to integrate the policy aspect into the framework in order to explain the transition process of current education and training policies toward

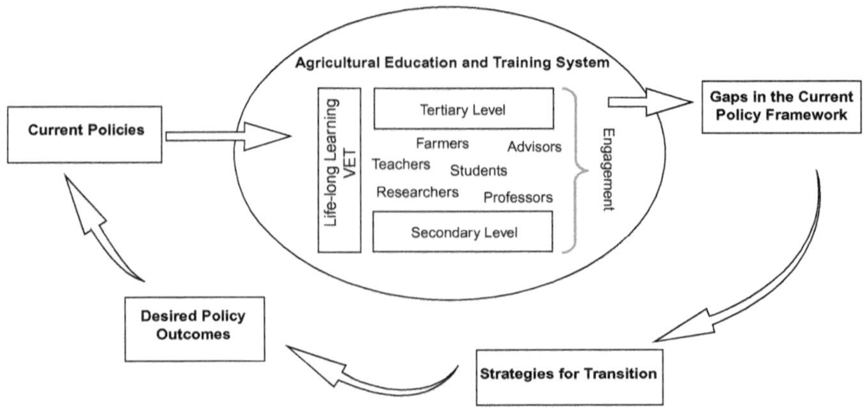

Fig. 2 The proposed framework for educational policy transition process in the agrifood and forestry sector in relation to AKIS. *(Sources: Adapted by Standing Committee of Agricultural Research (SCAR) Policy Brief on New Approaches on Agricultural Education, 2017.)*

desired policy outcomes. The framework proposes that strategies, that are to be developed, would address the identified gaps in the current policy framework toward creating desired policy outcomes, which then feed into the current educational policy system, in an iterative and circular manner. The proposed framework can be seen below (Fig. 2).

5. Discussion: Lessons learnt and the way forward

To date, agricultural education has been often carried out without considering the broad picture of Bioeconomy, mainly because of the dominant paradigm of embracing a sector-specific perspective. However, adopting inter- and transdisciplinary approaches, knowledge and skills which can allow students to address problems from a broader perspective is becoming more urgent in the era of complex sustainability challenges (Lewandowski, 2018). Hence, identifying pathways to reconsider integration of educational systems in a holistic way, considering the wider perspective of the Bioeconomy, can be an important step to tackle these challenges, while also creating opportunities in terms of new jobs and business areas.

In order to enhance the skills of current and future professionals of Bioeconomy, bridging institutions and different levels of education (i.e., Pre-University, University, Life-long Learning) is necessary. While formal education (especially University education) is considered essential to drive this transition, the changes in the Bioeconomy are too rapid to make conventional education systems keep pace to equip students with the necessary skills to match the needs of the everchanging systems (Hakovirta and Lucia, 2019; Pagliaro, 2019). In this regard, promoting especially life-long learning is of critical importance, including Vocational Education, which is regarded as a key element of life-long learning systems (European Commission, 2018).

Life-long learning is considered as a crucial tool for the transition in the Bioeconomy—given the flexibility that it can provide both for workers and industries—also allowing for greater diversity in career paths in a circular and sustainable Bioeconomy (Global Bioeconomy Summit 2020, 2021). For this reason, life-long learning is seen as the right tool to respond to the rapid changes of our society and the business environment.

Life-long learning is also seen as fundamental to include the citizens in the Bioeconomy, especially young, that do not intend to attend university courses or graduate students that cannot find a job. In other words, it can be a solution to minimize the number of NEETs (Not in Education, Employment, or Training people) in Europe.

On the other hand, accompanying formal education with informal education, is also critical. Informal learning is based on hands-on and practical experiences, such as training on-site, or peer-to-peer learning, and could be significant to involve more under-represented and rural students in education (Hakovirta and Lucia, 2019).

Meanwhile, from the point of view of skills, the inter- and transdisciplinary nature of Bioeconomy will require the creation of new professions and/or the updating of the old ones. Indeed, to sustain this transformation the future professionals of Bioeconomy will need new skills, including soft skills such as skills for communication, co-working, policy and decision-making, and will be required to manage different languages, methods, perspectives and competencies (Urmetzer et al., 2020; European Commission, 2018; Lewandowski, 2018). In this regard, digitalization requires special emphasis, since it is a cross-cutting issue that is transforming Bioeconomy, leading to new opportunities and modalities of learning, education and ways of doing business.

While digitalization creates many opportunities, it also requires the development of digital skills, and necessitates the provision of right tools and capabilities to manage modern technologies and techniques.

The Covid-19 pandemic has also shown the magnitude of the importance of digitalization, while also proved the existence of a large digital divide across people as well as countries. In order to close this gap and to equip students and professionals with necessary skills, providing digital literacy and fluency in Bioeconomy will be even more critical.

On the other hand, the digitalization of Bioeconomy education can deliver technical skills to a larger community of people, allowing the democratization of education in the Bioeconomy context.

To give an example, digitalization of education will be able to bring more distance learning opportunities, the need of which became also much more prominent following the pandemic. Distance learning is an available option that can provide education to people who would not otherwise have the chance, while addressing also the gender gap in Bioeconomy education.

In order to transform education systems and equip students with the transdisciplinary approaches, knowledge and skills required in the Bioeconomy, it will also be necessary to broaden the innovation perspective, opening-up the strictly sectoral approach (of agricultural) to that of a system of innovation that encompasses all sectors that composes the Bioeconomy.

In this direction, adopting multi-stakeholder approaches and establishing inter-sectoral coordination mechanisms in the education, training and research systems will be of significant importance.

UNESCO underlines the need to involve governmental and non-governmental organizations and the private sector, and universities to develop bilateral or multilateral partnerships between sectors and stakeholders in order to share resources and increase the availability of learning opportunities, and to enhance innovation.

Toward this goal, the AKIS framework acts as a useful tool to identify all key actors of the education, training and the research ecosystems and to demonstrate that knowledge and innovation in fact do not flow in a linear way—from researchers to farmers—but rather is consisted of an intricate network in which the flows of knowledge and innovation are often unpredictable (Arzeni et al., 2021; Hermans et al., 2015; Klerkx et al., 2012). While AKIS framework is one of the most instrumental tools to support the innovation process in the agri-food sector, there are also lessons that needs to be learnt from the AKIS literature up to date. In particular:

- AKIS has difficulties in reflecting farmers' innovation needs or address users' needs, gaps and knowledge differences between researchers and practitioners and sub-optimal institutional and policy arrangements (Spendrup and Fernqvist, 2019).

- Monitoring of the knowledge and innovation systems is generally fragmented, and for the moment a major inconsistency exists between the high level of attention to innovation in the policy domain and the lack of data and research for evidence-based policy (Materia, 2012). Furthermore, the studies that attempt to discuss existing policies in the area of agricultural education and skills development are mostly conducted on a local or national level, through specific and fragmented case studies and this does not allow to have a European vision.
- Education is often regarded as a standalone issue, while disconnected from industry needs and research practices. Instead, it needs to be placed in the wider system of the AKIS, considering the connections to all of its components (i.e., research system, education and training organizations, extension and business development services, bridging and coordination organizations, farmers, as well as external factors), and to all key actors of the system.
- The multi-stakeholder and multi-sectoral diversity encountered in a real-world context is too wide to be unambiguously captured with the current analytical level of the AKIS concept (Knierim et al., 2015). While, currently the AKIS framework is directed to target the agricultural sector alone, there are proposals to widen its perspective to introduce the possibility of capturing the complexities of the interwoven intersectoral relationships and exchanges that shape the innovation ecosystems in real life. One of these proposals was put forth by Guillou (2012), who suggests a transformation of the AKIS framework to that of a KISB: Knowledge, technology and innovations for a bio-based economy. In the meantime, the European Commission empowers member states to effectively support and re-organize regional AKIS via their Bioeconomy Action Plans (European Commission, 2012) in order to capture this wider perspective.
- Despite the efforts to develop the AKIS framework to meet the needs of our day, there is still much work to do ahead, and it is important to note that every significant change in the AKIS brings with technical, social, economic, ethical and practical questions with significant implications for how these dynamics are structured, practiced and governed (Ingram and Maye, 2020). In this vein, integrating social and natural science methods, different fields of research, and different sectors (interdisciplinary), and uniting different actors (transdisciplinary) can be steps to enhance the AKIS framework and to improve its practical consequences.

6. Concluding remarks

Education is often mentioned as an issue of critical importance in both agriculture and Bioeconomy literature. Innovation, that is a cross cutting issue in the Bioeconomy, is also closely linked to education processes. Especially in the current context, characterized by the sustainability challenges, the urgency to combat climate change and the re-shaping of our society in the post-pandemic scenarios, enabling innovative solutions and uptake of skills are even more important.

In this paper we identified a number of open issues in the role of education in agriculture and the Bioeconomy, highlighting two key areas of research. The first concerns the need to bring the topic beyond the mere consideration of delivering appropriate skills for industry needs and rather looking at the broader perspective, including epistemological visions of the Bioeconomy and visions of sustainable transitions in society. The second concerns the need to address more clearly training and education in the framework of innovation and knowledge sharing systems. An important step forward in this direction can come from the evolution and the operationalization of the AKIS concept into the wider Bioeconomy innovation systems.

References

Adamseged, M.E., Grundmann, P., 2020. Understanding business environments and success factors for emerging bioeconomy enterprises through a comprehensive analytical framework. Sustainability 12 (21), 9018. https://doi.org/10.3390/su12219018.

Alroe, H.F., Sautier, M., Legun, K., Whitehead, J., Noe, E., Moller, H., Manhire, J., 2017. Performance versus values in sustainability transformation of food systems. Sustainability 9 (3). https://doi.org/10.3390/su9030332.

Amka, M., 2017. Inclusive Education in College (Strengthening Multicultural Values Among Students with Disabilities and Non-Disability). Atlantis Press, https://doi.org/10.2991/icsse-17.2018.7.

Arzeni, A., Ascione, E., Borsotto, P., Carta, V., Castellotti, T., Vagnozzi, A., 2021. Analysis of farms characteristics related to innovation needs: a proposal for supporting the public decision-making process. Land Use Policy 100. https://doi.org/10.1016/j.landusepol.2020.104892.

Autor, D., Mindell, D., Reynolds, E., 2020. The Work of the Future: Building Better Jobs in an Age of Intelligent Machines. MIT Task Force on the Work of the Future. https://workofthefuture.mit.edu/wp-content/uploads/2021/01/2020-Final-Report4.pdf.

Barbinov, V., 2018. Vocational training of future agricultural specialists: European experience. Comp. Prof. Pedag. 8, 160–165. https://doi.org/10.2478/rpp-2018-0034.

Boehlje, M., Roucan-Kane, M., Stefanie, B., 2011. Future agribusiness challenges: strategic uncertainty, innovation and structural change. Int. Food Agribus. Manag. Rev. 14, 53–82.

Bröring, S., 2010. Developing innovation strategies for convergence—is "open innovation" imperative? Int. J. Technol. Manag. 49, 272–294. https://doi.org/10.1504/IJTM.2010.029421.

Bröring, S., Martin Cloutier, L., Leker, J., 2006. The front end of innovation in an era of industry convergence: evidence from nutraceuticals and functional foods. R&D Manag. 36 (5), 487–498. https://doi.org/10.1111/j.1467-9310.2006.00449.x.

Choi, D., Valikangas, L., 2001. Patterns of strategy innovation. Eur. Manag. J. 19 (4), 424–429. https://doi.org/10.1016/S0263-2373(01)00045-7.

Curran, C.-S., Bröring, S., Leker, J., 2010. Anticipating converging industries using publicly available data. Technol. Forecast. Soc. Chang. 77 (3), 385–395. https://doi.org/10.1016/j.techfore.2009.10.002.

Daukilas, S., Keibienė, R., 2017. Framework of Vocational School and University of Applied Sciences teacher activities: Motivational Factors, Professional Roles and Application of Educational Technologies in Correlation with Traits. Aleksandras Stulginskis University, https://doi.org/10.15544/RD.2017.238.

Dentoni, D., & Peterson, H., 2011. Multi-stakeholder sustainability alliances in agri-food chains: a framework for multi-disciplinary research. Int. Food Agribus.Dentoni, D., Peterson, H., 2011. Multi-stakeholder sustainability alliances in agri-food chains: a framework for multi-disciplinary research. Int. Food Agribusiness Manag. Rev. 14 (5), 83–108.

Di Sia, P., 2020. Looking at the new digital school: didactics 2.0 and school 2.0. E-Methodology 6 (6), 9–21. https://doi.org/10.15503/emet2019.9.21.

Dockès, A.C., Tisenkopfs, T., Bock, B., 2011. Collaborative working group agricultural knowledge and innovation systems. In: WP1: Reflection Paper on AKIS. Sub-Deliverable of the AKIS CWG—WP1—April 2011. European Commission, Brussels.

Drejerska, N., 2017. Employment in vs. Education for the Bioeconomy. Aleksandras Stulginskis University, https://doi.org/10.15544/RD.2017.245.

EC, 2018. CAP Context Indicators 2014–2020.

EIP-AGRI SP, 2017. Horizon 2020 Multi-Actor Projects. EIP-AGRI Service Point, Brussels. https://ec.europa.eu/eip/agriculture/sites/agri-eip/files/eip-agri_brochure_multi-actor_projects_2017_en_web.pdf. (Accessed 24 March 2020).

Esposti, R., 2012. Knowledge, Technology and Innovations for a Bio-Based Economy: Lessons From the Past, Challenges for the Future., p. 34.

EU SCAR, 2012. Agricultural Knowledge and Innovation Systems in Transition—A Reflection Paper, Brussels.

EU SCAR, 2015. Agricultural Knowledge and Innovation Systems Towards the Future—A Foresight Paper, Brussels.

European Commission, 2012. Enhancing Innovation and the Delivery of Research in the EU Agriculture. In: Background Paper for the Participants to the Conference "Enhancing Innovation and the Delivery of Research in the EU Agriculture", European Commission, 7 March, Brussels.

European Commission, 2018. A Sustainable Bioeconomy for Europe: Strengthening the Connection Between Economy, Society and the Environment—Updated Bioeconomy Strategy. Available from: https://op.europa.eu/en/publication-detail/-/publication/edace3e3-e189-11e8-b690-01aa75ed71a1/.

European Students' Union, 2020. Bologna with Student Eyes 2020. European Students' Union, Brussels.

Eurostat, 2016. Eurostat. https://ec.europa.eu/eurostat.

Fieldsend 2021. "'Sharing the space' in the agricultural knowledge and innovation system: multi-actor innovation partnerships with farmers and foresters in europe." J. Agric. Educ. Ext. 27 (4): 423–442. doi:https://doi.org/10.1080/1389224X.2021.1873156.Fieldsend, A, 2020. Agricultural Knowledge and Innovation Systems in European Union policy discourse: Quo vadis? Studies in Agricultural Economics 122 (3), 115–123. ISSN 1418-2106.

Fieldsend, A.F., Cronin, E., Varga, E., Biró, S., Rogge, E., 2020. Organisational innovation systems for multi-actor co-innovation in European agriculture, forestry and related sectors: diversity and common attributes. NJAS: Wagening. J. Life Sci. 92 (1), 1–11. https://doi.org/10.1016/j.njas.2020.100335.

Fletcher, J., Gamliel, A., Gullino, M.L., McKirdy, S.J., Smith, G.R., Stack, J.P., 2020. A fresh look at graduate education in Plant Pathology in a changing world: global needs and perspectives. J. Plant Pathol. 102 (3), 609–618. https://doi.org/10.1007/s42161-020-00509-2.

Fritsche, U., Brunori, G., Chiaramonti, D., Galanakis, C.M., Matthews, R., Panoutsou, C., 2021. Future transitions for the Bioeconomy towards Sustainable Development and a Climate-Neutral Economy—Bioeconomy Opportunities for a Green Recovery and Enhanced System Resilience.

Galmiche-Tejeda, A., 2004. Who is interdisciplinary? Two views, two goals, professionals and farmers. Interdiscip. Sci. Rev. 29 (1), 77–95. https://doi.org/10.1179/030801804225012464.

Gava, O., Favilli, E., Bartolini, F., Brunori, G., 2017. Knowledge networks and their role in shaping the relations within the Agricultural Knowledge and Innovation System in the agroenergy sector. The case of biogas in Tuscany (Italy). J. Rural. Stud. 56, 100–113. https://doi.org/10.1016/j.jrurstud.2017.09.009.

Global Bioeconomy Summit 2020, 2021. How to Shape Education for a Sustainable Circular Bioeconomy? Available from: https://gbs2020.net/wp-content/uploads/2021/03/Report_Workshop_education_GBS2020.pdf.

Golowko, N., Marquardt, K., Budz, S., Foerster-Metz, U.S., 2019. German students' perception of bioeconomy—an exploratory study. Amfiteatru Econ. 21 (50), 138–151. https://doi.org/10.24818/EA/2019/50/138.

Gregáňová, R.H., Országhová, D., Horváthová, J., 2017. The Role of Tertiary Education in Regional Development in Conditions of Slovak Republic. Aleksandras Stulginskis University, https://doi.org/10.15544/RD.2017.126.

Guillou, M., 2012. EU agricultural research and innovation: needs and priority are- as. In: Presentation Made at Conference "Enhancing Innovation and the Delivery of Research in the EU Agriculture", 7 March, Brussels.

Hakovirta, M., Lucia, L., 2019. Informal STEM education will accelerate the bioeconomy. Nat. Biotechnol. 37 (1), 103–104. https://doi.org/10.1038/nbt.4331.

Hall, A., Janssen, W., Pehu, E., Rajalahti, R., 2006. Enhancing Agricultural Innovation: How to Go Beyond the Strengthening of Research Systems. World Bank, Washington.

Harris, F., Lyon, F., Clarke, S., 2009. Doing interdisciplinarity: motivation and collaboration in research for sustainable agriculture in the UK. Area 41 (4), 374–384. https://doi.org/10.1111/j.1475-4762.2008.00859.x.

Hermans, F., Klerkx, L., Roep, D., 2015. Structural conditions for collaboration and learning in innovation networks: using an innovation system performance lens to analyse agricultural knowledge systems. J. Agric. Educ. Ext. 21 (1), 35–54. https://doi.org/10.1080/1389224X.2014.991113.

Hilkens, A., Reid, J.I., Klerkx, L., Gray, D.I., 2018. Money talk: how relations between farmers and advisors around financial management are shaped. J. Rural. Stud. 63, 83–95. https://doi.org/10.1016/j.jrurstud.2018.09.002.

Home, R., Rump, N., 2015. Evaluation of a multi-case participatory action research project: the case of SOLINSA. J. Agric. Educ. Ext. 21 (1), 73–89.

Ingram, J., Maye, D., 2020. What are the implications of digitalisation for agricultural knowledge? Front. Sustain. Food Syst. 4. https://doi.org/10.3389/fsufs.2020.00066.

Katz, M.L., 1996. Remarks on the economic implications of convergence. Ind. Corp. Chang. 5 (4), 1079–1095. https://doi.org/10.1093/icc/5.4.1079.

Klerkx, L., Begemann, S., 2020. Supporting food systems transformation: the what, why, who, where and how of mission-oriented agricultural innovation systems. Agric. Syst. 184. https://doi.org/10.1016/j.agsy.2020.102901.

Klerkx, L., van Mierlo, B., Leeuwis, C., 2012. Evolution of systems approaches to agricultural innovation: concepts, analysis and interventions. In: Farming Systems Research Into the 21st Century: The New Dynamic., https://doi.org/10.1007/978-94-007-4503-2_20.

Knierim, A., Boenning, K., Caggiano, M., Cristóvão, A., Dirimanova, V., Koehnen, T., Labarthe, P., Prager, K., 2015. The AKIS concept and its relevance in selected EU member states. Outlook Agric. 44 (1), 29–36. https://doi.org/10.5367/oa.2015.0194.

Lawrence, A., Deuffic, P., Hujala, T., Nichiforel, L., Feliciano, D., Jodlowski, K., Lind, T., Marchal, D., Talkkari, A., Teder, M., Vilkriste, L., Wilhelmsson, E., 2020. Extension, advice and knowledge systems for private forestry: understanding diversity and change across Europe. Land Use Policy 94. https://doi.org/10.1016/j.landusepol.2020.104522.

Lewandowski, I. (Ed.), 2018. Bioeconomy—Shaping the Transition to a Sustainable, Bio-based Economy. Springer, https://doi.org/10.1007/978-3-319-68152-8.

Lioutas, E.D., Charatsari, C., Černič Istenič, M., la Rocca, G., de Rosa, M., 2019. The challenges of setting up the evaluation of extension systems by using a systems approach: the case of Greece, Italy and Slovenia. J. Agric. Educ. Ext. 25 (2), 139–160. https://doi.org/10.1080/1389224X.2019.1583818.

Long, N., 2001. Development Sociology: Actor Perspectives, first ed. Routledge.

Lozano-Díaz, A., Fernández-Prados, J.S., 2020. Educating digital citizens: an opportunity to critical and activist perspective of sustainable development goals. Sustainability 12 (18), 7260. https://doi.org/10.3390/su12187260.

Masiero, M., et al., 2020. Bioeconomy perception by future stakeholders: hearing from European forestry students. Ambio 49, 1925–1942. https://doi.org/10.1007/s13280-020-01376-y.

Materia, V.C., 2012. The agricultural knowledge and innovation system in Italy: dynamics, incentives, monitoring and evaluation experiences. Stud. Agric. Econ. 114 (2), 71–78. https://doi.org/10.7896/j.1214.

McAlexander, S.L., et al., 2021. Measuring undergraduate students' beliefs about and career interest in bioproducts and bioenergy. Bioresources 16 (3), 5679–5693. https://doi.org/10.15376/biores.16.3.5679-5693.

McDonald, R., Macken-Walsh, A., 2016. An actor-oriented approach to understanding dairy farming in a liberalised regime: a case study of Ireland's New Entrants' Scheme. Land Use Policy 58, 537–544. https://doi.org/10.1016/j.landusepol.2016.08.025.

Medvedev, B., Molodyakov, S., 2019. Internet of Things for Farmers: Educational Issues., https://doi.org/10.22616/ERDev2019.18.N058.

Migliorini, P., Lieblein, G., 2016. Facilitating transformation and competence development in sustainable agriculture university education: an experiential and action oriented approach. Sustainability 8 (12), 1243. https://doi.org/10.3390/su8121243.

Moschitz, H., Muller, A., Kretzschmar, U., Haller, L., de Porras, M., Pfeifer, C., Oehen, B., Willer, H., Stolz, H., 2021. How can the EU Farm to Fork strategy deliver on its organic promises? Some critical reflections. EuroChoices 20, 30–36. https://doi.org/10.1111/1746-692X.12294.

OECD, 2012. Improving Agricultural Knowledge and Innovation Systems: OECD Conference Proceedings, OECD Publishing. https://read.oecd-ilibrary.org/agriculture-and-food/improving-agricultural-knowledge-and-innovation-systems_9789264167445-en#page2.

Oliveira, M.D., da Silva, F.G., Ferreira, S., Teixeira, M., Damásio, H., Ferreira, A.D., Gonçalves, J.M., 2019. Innovations in sustainable agriculture: case study of Lis Valley Irrigation District, Portugal. Sustainability 11 (2). https://doi.org/10.3390/su11020331.

Pagliaro, M., 2019. Preparing for the future: solar energy and bioeconomy in the United Arab Emirates. Energy Sci. Eng. 7 (5), 1451–1457. https://doi.org/10.1002/ese3.440.
Parr, D.M., Trexler, C.J., Khanna, N.R., Battisti, B.T., 2007. Designing sustainable agriculture education: academics' suggestions for an undergraduate curriculum at a land grant university. Agric. Hum. Values 24 (4), 523–533. https://doi.org/10.1007/s10460-007-9084-y.
Pascoli, D.U., et al., 2021. The US bioeconomy at the intersection of technology, policy, and education. Biofuels Bioprod. Biorefin. https://doi.org/10.1002/bbb.2302.
Peterson, E.W.F., Ruppel, F.J., Padberg, D.I., 1988. Assessing agricultural education: agricultural economics at a crossroads. Agric. Hum. Values 5 (4), 26–33. https://doi.org/10.1007/BF02217645.
Pubule, J., et al., 2020. Education for advancing the implementation of the bioeconomy goals: an analysis of master study programmes in bioeconomy. Environ. Clim. Technol. 24 (2), 149–159. https://doi.org/10.2478/rtuect-2020-0062.
Quiedeville, S., Barjolle, D., Mouret, J.C., Stolze, M., 2017. Ex-post evaluation of the impacts of the science-based research and innovation program: a new method applied in the case of farmers' transition to organic production in the Camargue. J. Innov. Econ. Manag. 1 (22), 145–170. https://doi.org/10.3917/jie.022.0145.
Ramachandran, R., Sparck, E.M., Levis-Fitzgerald, M., 2019. Investigating the effectiveness of using application-based science education videos in a general chemistry lecture course. J. Chem. Educ. 96, 479–485. https://doi.org/10.1021/acs.jchemed.8b00777.
Ray, S., et al., 2016. Delivering on the promise of bioeconomy in the developing world: link it with social innovation and education. In: Srivastava, S. (Ed.), Biomarker Discovery in the Developing World: Dissecting the Pipeline for Meeting the Challenges. Springer, New Delhi, pp. 73–81, https://doi.org/10.1007/978-81-322-2837-0_6.
Reddy, P.J., et al., 2011. Personalized medicine in the age of pharmacoproteomics: a close up on India and need for social science engagement for responsible innovation in post-proteomic biology. Curr. Pharmacogenomics Person Med. 9 (1), 67–75.
Rijswijk, K., Klerkx, L., Turner, J.A., 2019. Digitalisation in the New Zealand agricultural knowledge and innovation system: initial understandings and emerging organisational responses to digital agriculture. NJAS - Wagening. J. Life Sci. 90–91. https://doi.org/10.1016/j.njas.2019.100313.
Rittel, H.W.J., Webber, M.M., 1973. Dilemmas in a general theory of planning. Policy. Sci. 4 (2), 155–169. https://doi.org/10.1007/BF01405730.
Rivera, W.M., Qamar, M.K., Mwandemere, H.K., 2005. Enhancing Coordination Among AKIS/RD Actors: An Analytical and Comparative Review of Country Studies on Agricultural Knowledge and Information Systems for Rural Development (AKIS/RD). http://www.fao.org/3/y9087e/y9087e00.htm.
Roos, A., et al., 2021. The digitalisation of the Nordic bioeconomy and its effect on gender equality. Scand. J. For. Res. https://doi.org/10.1080/02827581.2021.1996629.
Ryymin, E., 2021. Perspectives from higher education: applied sciences university teachers on the digitalization of the bioeconomy. Technol. Innov. Manag. Rev. 11 (2), 24–32. https://doi.org/10.22215/timreview/1420.
Sá, M.J., Santos, A.I., Serpa, S., Miguel Ferreira, C., 2021. Digitainability—digital competences post-COVID-19 for a sustainable society. Sustainability 13, 9564. https://doi.org/10.3390/su13179564.
Schusler, T.M., et al., 2021. Students of colour views on racial equity in environmental sustainability. Nat. Sustain. 4, 975–982. https://doi.org/10.1038/s41893-021-00759-7.
Sillitoe, P., 2004. Interdisciplinary experiences: working with indigenous knowledge in development. Interdiscip. Sci. Rev. 29, 6–23. https://doi.org/10.1179/ 030801804225012428.
Spendrup, S., Fernqvist, F., 2019. Innovation in agri-food systems—a systematic mapping of the literature. Int. J. Food Syst. Dyn. 10, 402–427.

Stern, T., et al., 2018. Understanding perceptions of the bioeconomy in Austria—an explorative case study. Sustainability 10 (11), 4142. https://doi.org/10.3390/su10114142.
Tang, H., 2021. Implementing open educational resources in digital education. Educ. Tech. Res. Dev. 69, 389–392. https://doi.org/10.1007/s11423-020-09879-x.
Tilak, J.B., 2006. Higher education between the state and the market. In: Neave, G. (Ed.), Knowledge, Power and Dissent: Critical Perspectives on Higher Education and Research in Knowledge Society. United Nations Educational, Scientific and Cultural Organization, Paris, https://doi.org/10.4324/9781315146386-3.
Urmetzer, S., Lask, J., Vargas-Carpintero, R., Pyka, A., 2020. Learning to change: transformative knowledge for building a sustainable bioeconomy. Ecol. Econ. 167 (106435). https://doi.org/10.1016/j.ecolecon.2019.106435.
Vehvilainen, J.D., Inglis, B.K., Barrett July, P., 2021. Deploying the Bioeconomy in the EU: A Framework Approach for Bioeconomy Strategy Development. European Commission Directorate-General for Research and Innovation. Publications Office of the European Union. ISBN: 9276401431, 9789276401438.
Viaggi, D., 2018. The Bioeconomy: Delivering Sustainable Green Growth. CABI, ISBN: 9781786392756.
Viaggi, D., Kurtsal, Y., Rinaldi, G., Savini, F., 2021. Report on Identification of Strategies for Improvement. Deliverable 4.2 of the Next Food project funded under the European Union's Horizon 2020 Research and Innovation Programme GA No: 771738.
von Munchhausen, S., Haring, A.M., 2012. Lifelong learning for farmers: enhancing competitiveness, knowledge transfer and innovation in the eastern German state of Brandenburg. Stud. Agric. Econ. 114 (2), 86–92.
Zahran, Y., Kassem, H.S., Naba, S.M., Alotaibi, B.A., 2020. Shifting from fragmentation to integration: a proposed framework for strengthening agricultural knowledge and innovation system in Egypt. Sustainability 12 (12).
Zulkifli, M., 2018. Teacher's Role in the Implementation of Character Education on Students. Atlantis Press, https://doi.org/10.2991/iccite-18.2018.71.

Further reading

Arnold, E., Bell, M., 2001. Some Ideas About Research for Development. unpublished background paper provided to the Commission on Development Related Research in Denmark, Science and Technology Policy Research (SPRU), Technopolis.
Spielman, D., Birner, R., 2008. How Innovative Is Your Agriculture? Using Innovation Indicators and Benchmarks to Strengthen National Agricultural Innovation Systems.

CHAPTER 15

Urban agriculture and agro-environment: A case study of homestead food gardeners in the City of Tshwane, Gauteng Province, South Africa

Phokele Maponya and D. Oelofse
Agricultural Research Council-Vegetable, Industrial and Medicinal Plants, Pretoria, Gauteng, South Africa

1. Introduction

Urban agriculture is a substantial economic activity that affects the lives of tens of millions of people worldwide (UNDP, 1996). In addition urban agriculture is an industry that produces, processes, and markets food and fuel, largely in response to the daily demand of customers inside a town, city, or metropolis, on land and water dispersed (UNDP, 1996). Urban farmers work on smaller plots of land than rural farmers, frequently in unoccupied, underutilized, or unsuitable for urban development open spaces. Around 800 million individuals are projected to be actively involved in urban agriculture around the world (UNDP, 1996). The percentage of urban households involved in agriculture ranges from 10% in some large North American towns to as high as 80% in some smaller Siberian and Asian cities (UNDP, 1996). Furthermore, women are more likely to undertake most of the urban agriculture in countries and cultures where women do most of the rural agricultural. For example, women account for 64% of African urban farmers, 80% of home gardens in Lima (Peru), and 67% of hydroponics cultivators in Bogota (Colombia) (UNDP, 1996).

Urban agriculture plays a significant role in supplying healthy, inexpensive, and accessible food to impoverished urban households in South Africa through domestic food gardens. According to a study by Madlala (2014), food gardening is usually thought to be a rural activity. On the contrary, evidence shows that even people living in urban and peri-urban settings use agriculture. Due to severe unemployment and food insecurity, many people

Agricultural Bioeconomy
https://doi.org/10.1016/B978-0-323-90569-5.00001-9

in rural areas and townships have turned to food gardening as a way to supplement their income (Maponya, 2019). The absence of natural resources in townships, such as land, has hampered people's efforts to begin producing crops on a long-term basis (Maponya et al., 2021). Many residents in townships farm tiny plots of land in their neighborhood, which are frequently held by the municipality, to produce food for their families (Madlala, 2014).

Poor communities and households can benefit from homestead food gardens by having direct access to nutritious foods that are not widely available or within their financial reach. Food gardening in poor communities, both community and homestead, can help to alleviate food insecurity and poverty. According to a research done in the Hlanganani and Bergville districts by Mpanza (2008), there is proof that community gardening benefits communities in a variety of ways, including physical and mental health, as well as socioeconomic benefits. In addition, a report by the International Fund for Agricultural Development underlined the importance of households in relation to ecosystems, finding that well-functioning ecosystems had an impact on household productivity (IFAD, 2012).

According to a study by Schaible et al. (2015), households in the United States who followed ecosystem management and environmental conservation practices gained a profit. To be deemed environmentally sustainable, households is required to preserve soil fertility, sustain water, maintain biodiversity, enhance air, reduce climate change impacts, and manage energy resources and waste.

South Africa is considered a "food-secure" nation, producing enough calories to adequately feed every one of its 57.73 million people (StatsSA, 2014). At country level, South Africa exceeds most global benchmarks for amounts of food produced and exported. However, national figures hide the reality at the household level (De Cock et al., 2013). The Gauteng Province population is also the highest in South Africa: (1) 14, 7 million out of 57, 73 million South African population live in the Gauteng Province, (2) 300,000 people migrate annually to Gauteng Province due to urbanization. (3) This adds more strain on the food security in the province. Studies have shown that almost 20% of households in the Gauteng Province go to bed hungry due to food insecurity and unsustainable income. In fact, the latest GDARD funded study indicated that 25% of households were food secure in relation to 75% of households who were food insecure pre COVID-19 pandemic (Maponya, 2019). The study further indicated the levels of food insecurity: mild (23%), moderate (32%) and severe (20%) (Maponya, 2019). It was further emphasized before the COVID-19 pandemic that between November 2015 and April 2017, the cost of a food basket comprising of the most basic of items has risen from R1 648.10 to R2 053.98 (PACSA, 2017). Items which were included in the basket focused on the staples that people buy on a monthly basis in order to service their food needs. These included starchy foods such as mealie meal and rice, vegetables like onions and cabbage, fruit such as oranges and bananas, dry beans, meat and poultry, milk, oils, sugars and salt.

During COVID-19 pandemic, South Africans are paying nearly R300 more for the average food basket in May 2021 compared to September 2020 (PMBEJD, 2021). This is according to the latest Household Affordability Index report compiled by the Pietermaritzburg Economic Justice & Dignity Group (PMBEJD, 2021), which tracks food price data from 44 supermarkets and 30 butcheries in Johannesburg, Durban, Cape Town, Pietermaritzburg and Springbok in the Northern Cape. Furthermore, while the overall cost of the average household food basket has decreased slightly, the cost of core foods like maize meal, cake flour, sugar beans, samp, cooking oil, potatoes and onions continues to rise, "albeit more slowly" (PMBEJD, 2021). Furthermore, there is an increase in food prices, with sunflower, tomatoes and dry beans increasing by 30.3%, 29.4% and 27%, respectively (StatsSA, 2021). The problem is also realized in the studies by Maponya (2019), PACSA (2017), PMBEJD (2021) and StatsSA (2021), which indicated the high costs of core staple foods (low stability) and the deteriorating food affordability, availability, accessibility and unsustainable conditions (environmental, economic and social). This situation of deteriorating affordability has a negative consequence for overall household health and well-being, which in turn will affect the developmental and sustainability outcomes, hence the need for agencies to address the problem. In addition, it is against the above conditions that the South African president announced the Presidential Employment Stimulus Initiatives (PESI) which is a COVID-19 subsistence farmer's relief fund initiative. Through PESI, the Department of Agriculture, Land Reform and Rural Development (DALRRD) identified subsistence producers/homestead food gardeners and asked them to indicate which of the DALRRD prioritized commodity areas need support.

Furthermore, the Presidency requested the DALRRD to support employment in the country through a short-term (6 months) support intervention in the agricultural sector. The DALRRD was asked to initiate a producer relief fund for subsistence producers/homestead food gardeners, which would help them retain self-employment and support the food value chains.

The study's major aim is to determine how urban agriculture contributes to food security in a changing environment. The study's objectives are to (1) evaluate socioeconomic variables among homestead food gardeners, (2) determine land and crop suitability among homestead food gardeners, and (3) determine natural resource management techniques among homestead food gardeners.

2. Methodology

2.1 Study area

As indicated in Fig. 1, the City of Tshwane Metropolitan Municipality (also known as the City of Tshwane) forms the local government of the northern Gauteng Province, South Africa. The metropolitan area is centered in the city of Pretoria with surrounding towns and localities included in the local government area. The municipality has seven regions and the study was conducted in all regions. The regions are as follows: (1) Region 1—Soshanguve, Mabopane, Garankuwa and Winterveldt; (2) Region 2—Eesterust, Stinkwater, Walmansdall and Hammanskraal; (3) Region 3—Atteridgeville; (4) Region 4—Olivenhoutbosch and Laudium; (5) Region 5—Cullinan, Refilwe and Onverwacht; (6) Region 6—Mamelodi; and Region

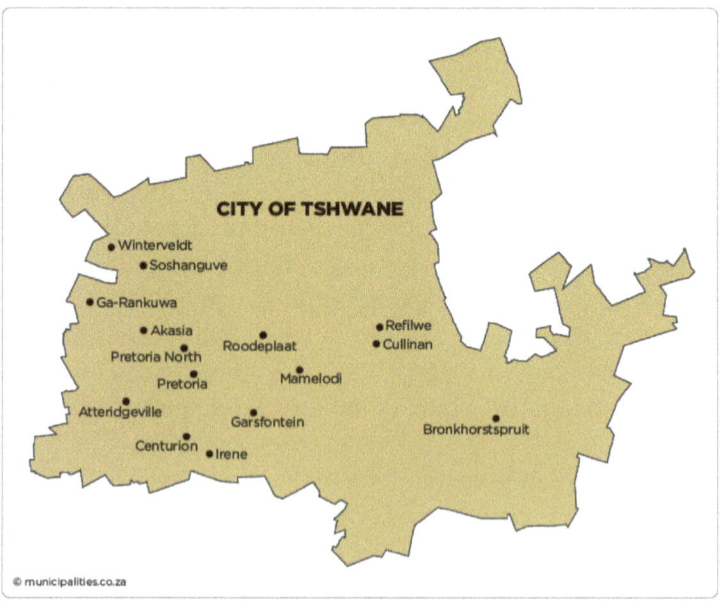

Fig. 1 City of Tshwane Metropolitan Municipality map.

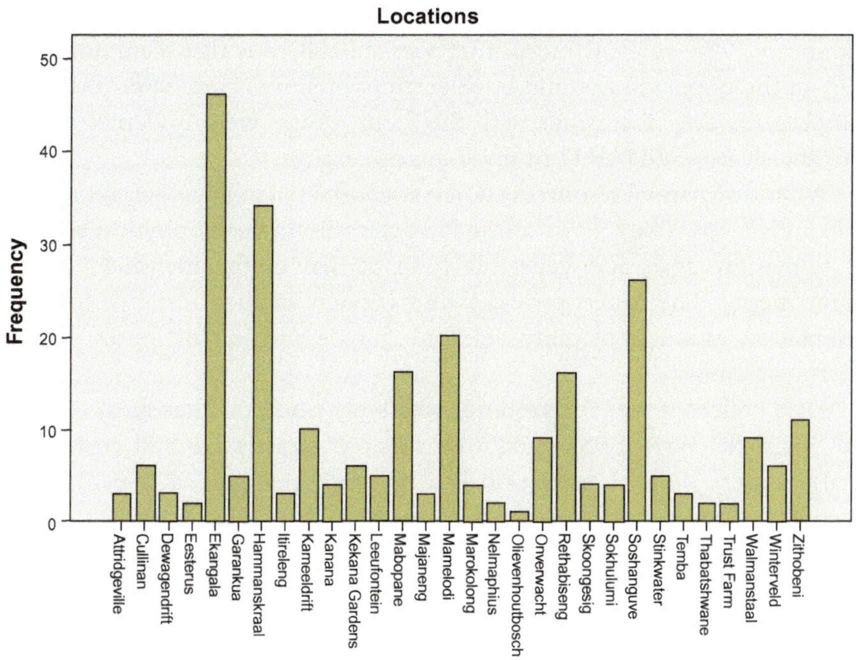

Fig. 2 City of Tshwane Metropolitan townships and informal settlements visited.

7—Zithobeni, Rethabiseng, Ekangala and Sokhulumi. As indicated in Fig. 2, the researcher and team visited 30 locations, including townships and informal settlements.

2.2 Study design

Quantitative and qualitative research designs were employed as a structured questionnaire written in English was developed to elicit information for the study. The study also included group discussions, stakeholder discussions, as well as field observations. As part of the standard protocol for conducting the study, meetings were held with the Gauteng Department of Agriculture and Rural Development (GDARD) officials. The aim of the meetings was to introduce and explain the aims of the research and, subsequently, GDARD approved the study, and all participants agreed voluntarily to participate.

2.3 Sampling procedure and analytical technique

A purposive sampling technique was used to select 270 homestead food gardeners from a database of GDARD. The research sample size was agreed on with the stakeholder, and that the size of the sample should statistically be significant. The rule of thumb that was applied, was that a minimum of 10% of the population would be selected, which was considered as a good sample size. The discussions with the focus groups were undertaken with 24 officials from GDARD in the Tshwane region.

A desktop natural resources audit was conducted to develop a set of crop suitability maps. The activity assisted in determining the biophysical viability of homestead fruit and vegetable food garden in the identified district municipality. The results provided an overview of the district in terms of orientation, water availability, selected climate and soil attributes, as well as crop suitability.

Data collected was analyzed quantitatively using the Statistical Package for Social Sciences (SPSS) windows version. Descriptive and correlation analyses were done. Correlation is a bivariate analysis that measures the strengths of association between two variables and the direction of the relationship. In terms of the strength of the relationship, the value of the correlation coefficient varies between $+1$ and -1. When the value of the correlation coefficient is around ± 1, then it is said to be a perfect degree of association between the two variables. As the correlation coefficient value moves toward 0, the relationship between the two variables will be weaker. The direction of the relationship is simply the + (indicating a positive relationship between the variables) or—(indicating a negative relationship between the variables) sign of the correlation. Usually, in statistics, four types of correlations are measured: the Pearson correlation, the Kendall rank correlation, the Spearman correlation, and the Point-Biserial correlation. In this example, Spearman correlations were used and the following variables were found to be significant: Availability of a garden, household members, age, household income and gender.

3. Results and discussion

3.1 Land and crop suitability

Crop suitability maps for the Gauteng Province, including the City of Tshwane Metropolitan (upper area on the maps), are shown in Figs. 3–5 and are self-explanatory. These will aid the Department of Agriculture and Rural Development in assisting homestead food gardeners with input distribution and total crop output under various climatic circumstances.

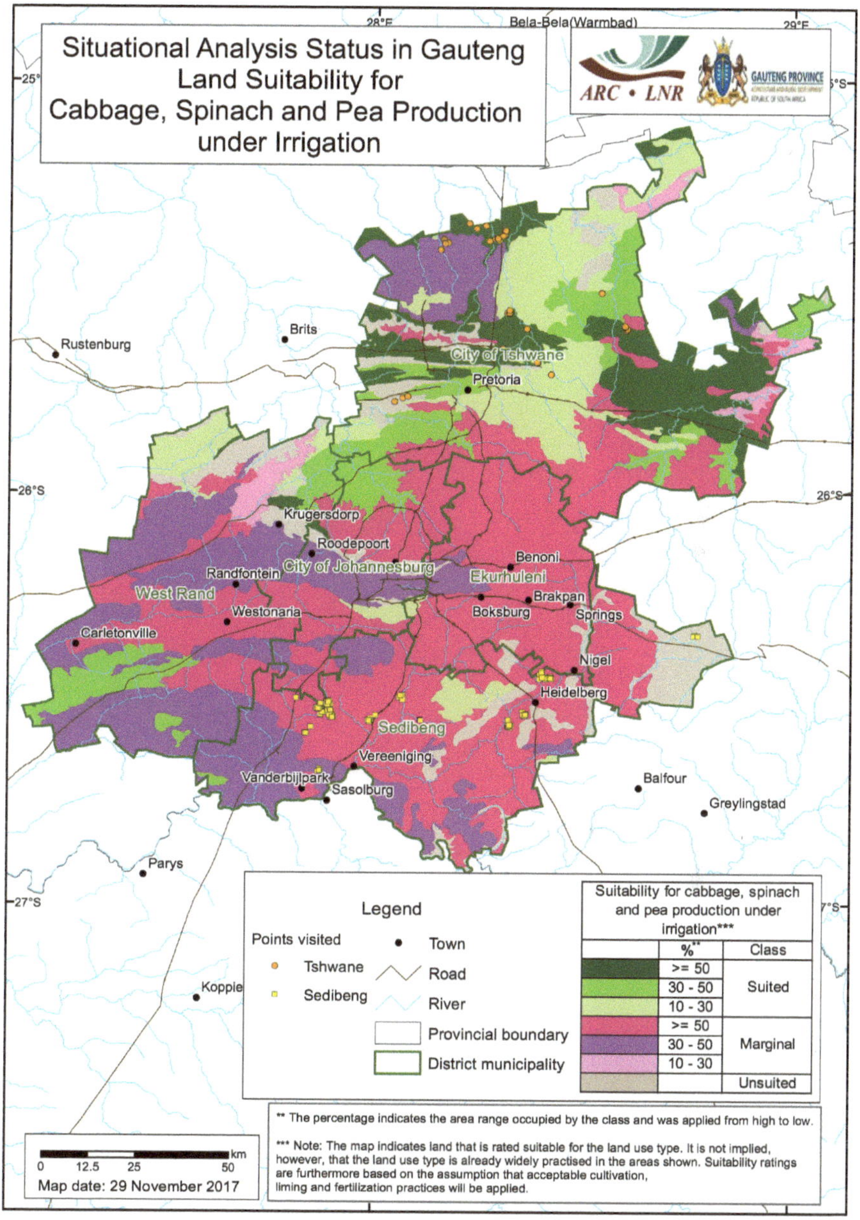

Fig. 3 Land suitability for cabbage, spinach and pea production under irrigation (ARC-SCW, 2017).

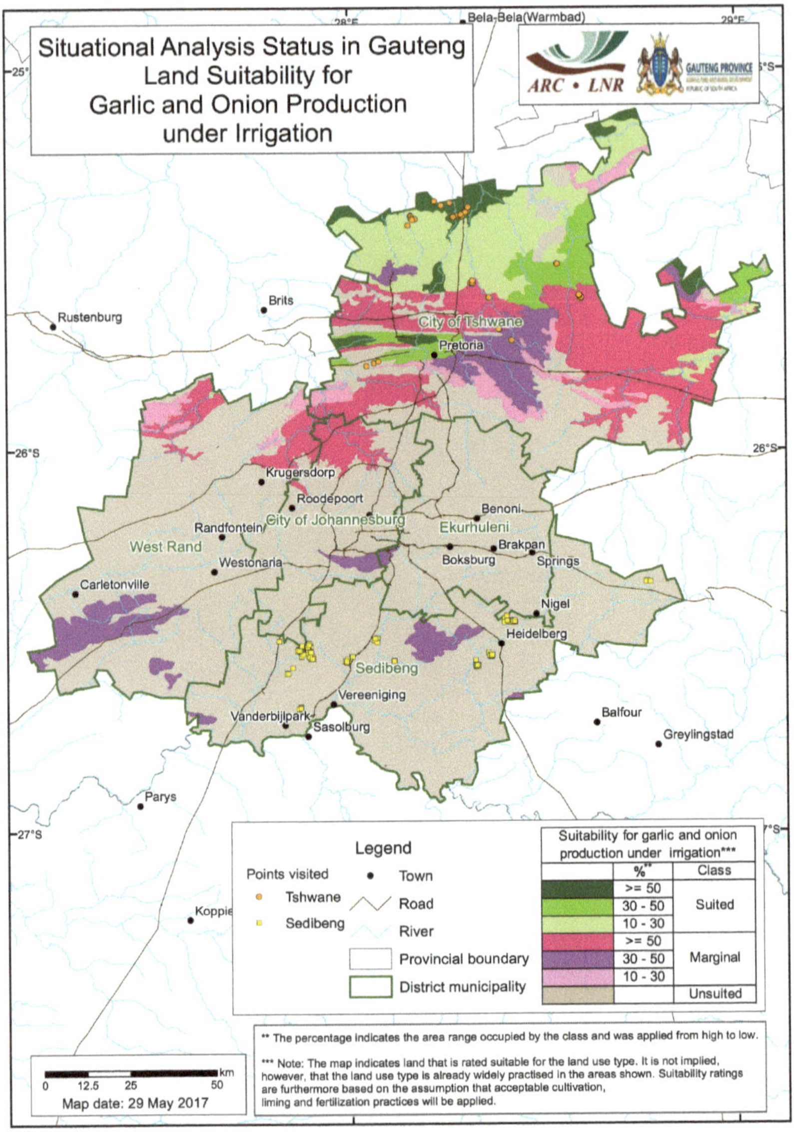

Fig. 4 Land suitability for garlic and onion production under irrigation (ARC-SCW, 2017).

On the maps, all of the coordinates for the homestead food gardeners were plotted. The methodology used to determine crop adaptability is essentially an adaptation and simplification of the procedures used by the Food and Agriculture Organization (FAO). Some of the techniques offered in several editions of the United States Department of Agriculture's (USDA) National Soils Handbook. On the maps, all of the coordinates for the

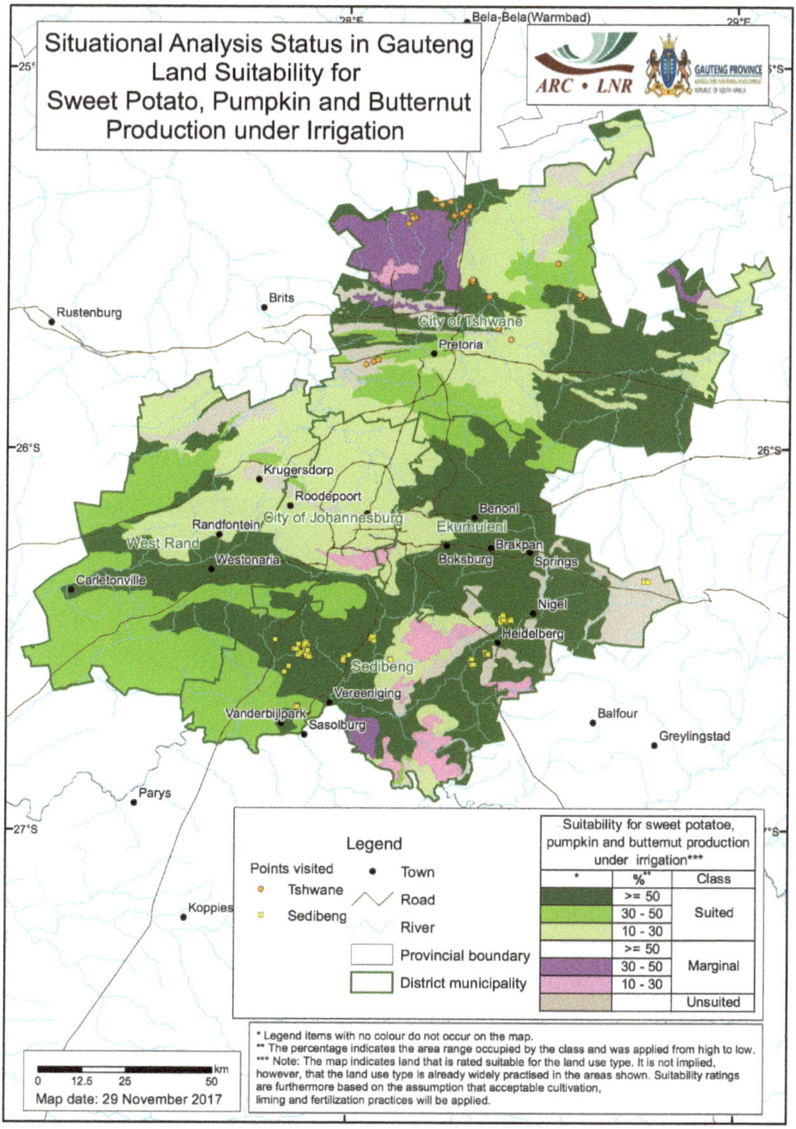

Fig. 5 Land suitability for sweet potato, pumpkin and butternut under irrigation (ARC-SCW, 2017).

homestead food gardeners were plotted. The methodology used to determine crop adaptability is essentially an adaptation and simplification of the procedures used by the Food and Agriculture Organization (FAO). Some of the techniques offered in several editions of the United States Department of Agriculture's (USDA) National Soils Handbook.

Only land-related factors of sustainability, such as technical feasibility and environmental sustainability, are included in the results. Separately, the aspects of social acceptability and economic feasibility must also be addressed (Van Schalkwyk et al., 2004, 2009).

The FAO pioneered the biophysical land use appropriateness method, which includes the following key concepts:

- Each land use type (LUT) of cluster that are related LUTs is evaluated separately.
- It is possible to receive information on a land use type's environmental and other requirements (LUR).
- It is possible to collect data and information on crucial land attributes and characteristics.

3.2 Homestead food gardeners socio-economic characteristics

In the City of Tshwane Metropolitan Municipality, the majority of homestead food gardeners interviewed were female-headed (212). The findings are also consistent with a study conducted in Limpopo Province by Maponya and Moja (2012), which found that female-headed households make up a significant portion of the economically active population, and that female-headed households are disproportionately vulnerable, marginal, and food insecure. The varying ages of households are shown in Table 1's results. As seen in Table 1, the majority of households (134) are over 60 years old, while only a few are between the ages of 18 and 35 years old (18). This indicates that the community's population is aging, which could pose a threat to household food security.

The majority of community members (125) earned secondary education, with only a few attaining tertiary education (Table 1) (13). Education's importance is undeniable, and numerous studies have demonstrated the advantages of investing in education. According to Hanley et al. (2007) improving education and employment are critical needs for encouraging local engagement in agricultural production and natural resource management activities. Furthermore, as shown in Table 1, the majority of community members (115) earn between R1001 and R2500 per month, while a significant proportion of community members have no income (95).

The majority of community members spent more than R601 of their monthly income on food. This situation confirms the findings of (PACSA, 2017), which found that the cost of a food basket including the most basic of commodities increased from R1 648.10 to R2 053.98 between

Table 1 Distribution of homestead food gardeners according to their socio-economic characteristics.

	Homestead food gardeners	% Homestead food gardeners socio-economic characteristics
Gender		
Female	212	79
Male	58	21
Total	270	100
Age categories		
18–35	18	6
36–45	49	18
46–60	69	26
>60	134	50
Total	270	100
Level of education		
No education	27	10
Primary education	105	39
Secondary education	125	46
Tertiary education	13	5
Total	270	100
Employment status		
Unemployed and community grower/ farmer	270	100
Total	270	100
Household members size		
1–5	177	66
6–10	82	30
>10	11	4
Total	270	100
Household monthly income		
<R500	8	3
R501–R1000	41	15
R1001–R2500	115	43
>R2501	11	4
No income	95	35
Total	270	100
Household monthly expenditure on food		
<R200	8	3
R201–R400	27	10
R401–R600	78	29
>R601	157	58
No expenditure	0	0
Total	270	100

November 2015 and April 2017. The basket's contents centered on the necessities that people buy on a regular basis in order to feed their families. Starchy foods like mealie meal and rice, vegetables like onions and cabbage, fruit like oranges and bananas, dry beans, meat and chicken, milk, oils, sweets, and salt were all on the list. In May 2021, South Africans will pay roughly R300 more for the average food basket than they did in September 2020 (PMBEJD, 2021). The majority of the households had one to five members. According to Amaza et al. (2009) family size is a major contributing component of household agricultural production and thus food security, which has been widely established globally, through the supply of labor.

3.3 Homestead food garden initiative

To combat food insecurity, malnutrition, poverty, and hunger, the South African government has launched homestead food garden (HFG) programs. Homestead food gardeners receive some training as well as starter kits in the practice of HFG. Training is also offered, and after successful completion of the training portion of the program, the beneficiary receives the benefit. Beneficiaries of the program attend meetings with local officials (ward councilors, ward committees, and so on) and program implementers to debate the program's long-term viability. Public meetings are also arranged to inform community members about the program's benefits and to encourage them to participate. Regarding the initiative, some homestead food gardeners, as shown in Fig. 6, thought it was fair (137), poor (51) and good (45), very good (31) and very poor (31). The main reason for this is that some people have lost interest, sold their tools, had high expectations, relocated, and there was little supervision because there are so many homestead food gardeners to assist.

As shown in Fig. 7, the majority of homestead food gardeners thought the support they received was very good (216), good (31) or fair (15). A tiny proportion of homestead food gardeners thought the support they received was inadequate (4) or non-existent (4). A spade, fork, rake, hand hoe, two 30 dcm^3 bags of compost, a 10 L watering container, and six types of seeds (spinach 10 g, beetroot 10 g, onion 7 g, carrot 8 g, beans 15 g, and tomato 2.5 g) were provided to homestead food gardeners.

3.4 Homestead food gardeners natural resource management

As shown in Fig. 8 (98) homestead food gardeners had access to irrigation water, while 172 homestead food growers did not. The majority of the 172 homestead food gardeners said it's difficult to get irrigation water

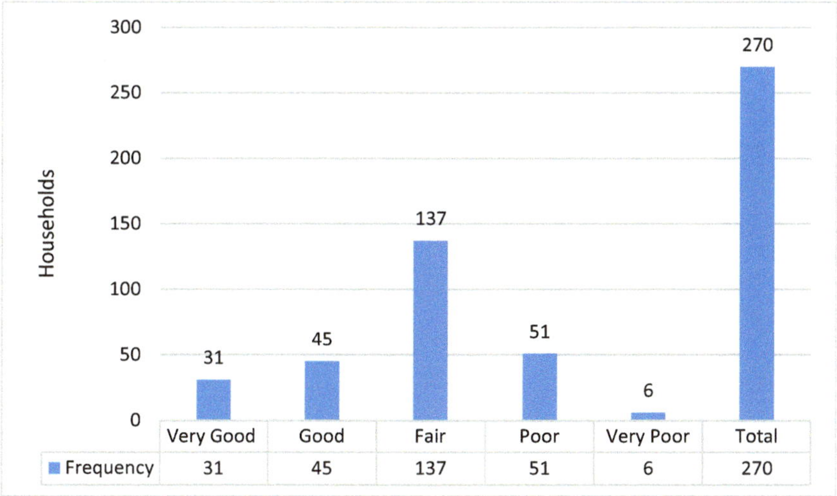

Fig. 6 Homestead food gardeners overall feeling of the homestead food garden initiative.

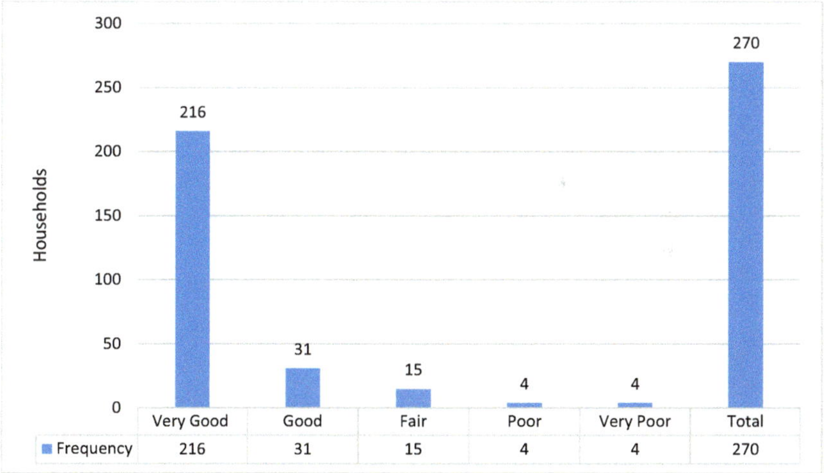

Fig. 7 Homestead food gardeners impression of support received.

because they live in informal settlements or regions where service delivery is still an issue. This issue must be addressed since water for irrigation is critical to the success and long-term viability of food farming.

Fig. 9 shows that the majority (71) of homestead food gardeners use municipal piped water to irrigate their crops, with only a handful

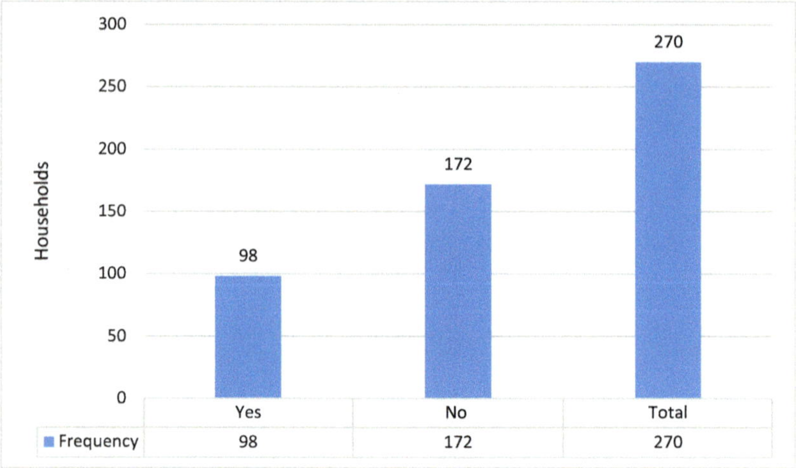

Fig. 8 Homestead food gardeners access to irrigation water.

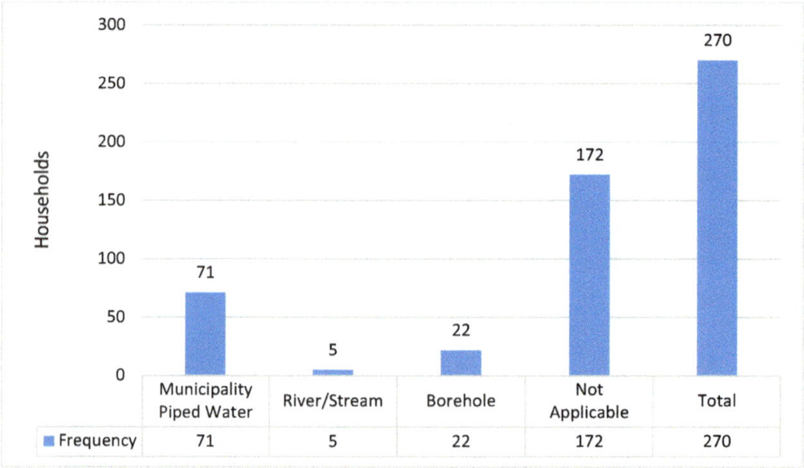

Fig. 9 Homestead food gardeners irrigation water source.

mentioning rivers or boreholes as water sources. During the focus group sessions, officials also emphasized that domestic food gardeners should use rainwater collecting and gray water. Fig. 10 shows that nearly 77 homestead food gardeners said they had access to water all year, while 193 said they don't have access to water all year.

Fig. 11 shows that around 232 household food gardeners were unaware of environmental hazards, while 38 were. All homestead food gardeners should make environmental awareness and food gardening a priority since

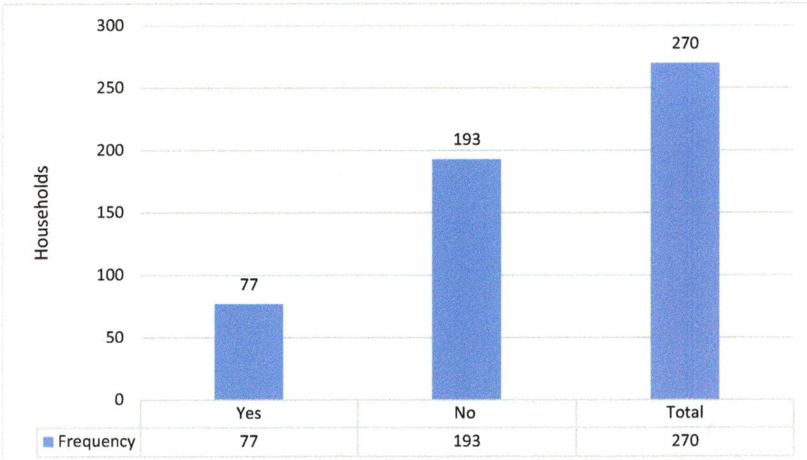

Fig. 10 Homestead food gardeners water availability year round.

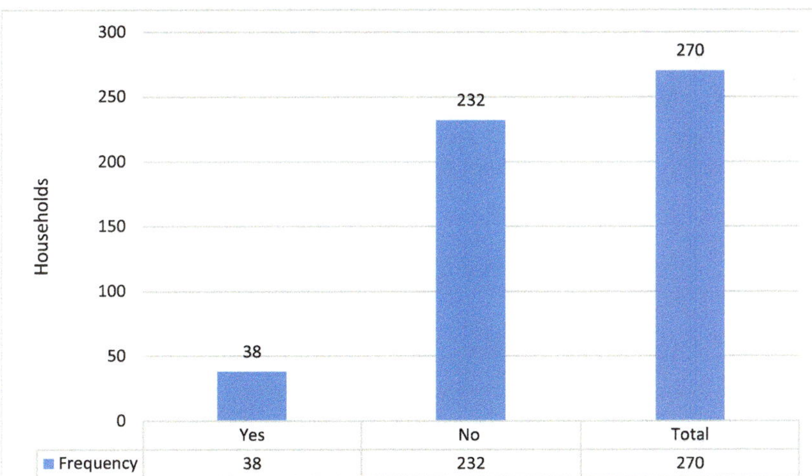

Fig. 11 Homestead food gardeners environmental issues awareness.

it will encourage the production of fresh, healthy food with low carbon and water footprints, while conserving land, decreasing emissions, and reducing waste and providing nutritious, inexpensive, and easily available food. Furthermore, recycling water and waste nutrients, limiting shade, dust, and erosion, and maintaining or growing local biodiversity will all improve the environment.

3.4.1 *Selected homestead food gardeners pictures in the field*

See Pictures 1–6.

Picture 1 To ensure reduced soil erosion and saving water resources from less evapotranspiration, some homestead food gardeners produce under a shade net structure.

Picture 2 Crops are grown on raised seedbeds to prevent irrigation water loss through run off and reducing soil loss from water erosion during heavy rains.

Picture 3 To reduce municipal water utilization for food gardening purposes, the homestead food gardener are harvesting rainwater from the roofs of churches, houses and school buildings, therefore contributing toward sustainable water management.

Picture 4 Homestead food gardeners practice mulching to suppress weeds from growing, to preserve the moisture in the soil and to reduce evaporation and the grass will decompose with time to form compost.

3.5 Correlations among variables

As indicated in Table 2, there is a positive correlation between gender and the availability of gardens, which can indicate that any gender can participate in homestead food gardening. These results are in line with Ndobo and Sekhampu (2013), who indicated that households headed by females are more likely to face mild-to-moderate, as well as severe forms of food

Picture 5 The homestead food gardeners also make their own compost from plant debris, which is mixed into the soil to keep it healthy and fertile. This organic farming is not a health hazard to them and poses no threat to the environment as well.

Picture 6 Homestead food gardeners prepare an organic mixture (garlic, sunlight liquid, onion, aloe and khakhibos) instead of purchasing commercial chemical pesticides. The mixture is not hazardous and does not degrade the soil by killing important soil microorganisms.

Table 2 Spearman correlation coefficients among variables.

	GAR[a]	GEN[b]	HOU[c]	AGE[d]	HOUI[e]
GAR	1.00	0.011[f]	0.040[g]	0.027[g]	0.013[f]
GEN	0.011	1.00	0.009	0.180	−0.19
HOU	0.040	0.009	1.00	−0.76	−0.025
AGE	0.027	0.180	−0.76	1.00	−0.57
HOUI	0.013	−0.19	−0.025	−0.57	1.00

[a]Garden still available.
[b]Gender.
[c]Household members.
[d]Age.
[e]Household income.
[f]1% significance level.
[g]5% significance level.

insecurity than those headed by males. Similar results were found by Mohammadi-Nasrabadi et al. (2014) who reported that severe food insecurity was more frequent in female-headed households. The study further emphasized that age category and garden availability are positively correlated. As shown in Table 2, both variables are significant at 5% and showed that any age category can participate in homestead food gardening provided there is a garden available. Furthermore, there is a positive correlation between household income and gardens that are still available. It is thus expected that households with gardens can make extra income from surplus production. It is also expected that an increase in a household income can reduce food insecurity among households. According to Maponya and Moja (2012), the combination of income generation and production improves food security among households. There is also a positive correlation between household size/members and gardens that are still available. Maponya and Moja (2012) indicated that households who are economically active with more members generally are more food secure, compared to those who are less active economically. Maponya (2019) further emphasized that as household size increases, the number of people to be fed from a given amount of food increases, hence the need to have a food garden.

4. Conclusions and recommendations

Urban agriculture has a major role to play in providing healthy, affordable and accessible food to poor urban households in South Africa through homestead food gardens. In conclusion, the study promotes the inclusion and promotion of home food gardens as an eco-friendly sustainable agricultural practice to improve food security and enhance economic growth. The structure, functions, and contributions ofhome food gardens vary in geographic regions and home food gardens fulfill social, cultural and economic needs, while providing a number of ecosystem services. The homestead food gardens are also making a difference in the community as some use the starter packs given to grow vegetables. Correlation results also indicated a positive association among the following variables: Availability of a garden, household members, age, household income and gender. The study recommends an increase in urban agriculture that includes the following: Monitoring and training of households in terms of agricultural production and natural resources, channeling of policies to benefit the community, intersectoral and interdepartmental collaboration, and youth involvement, to mention a few.

Declaration statement

Funding (information that explains whether and by whom the research was supported)

The research was supported by Agricultural Research Council and Gauteng Department of Agriculture and Rural Development.

Conflicts of interest/competing interests (include appropriate disclosures)

No conflict of interests that are directly or indirectly related to the work submitted for publication.

Availability of data and material (data transparency)

Not applicable.

Code availability (software application or custom code)

Not applicable.

Authors' contributions (optional: Please review the submission guidelines from the journal whether statements are mandatory)

The author contributed to the study as a project leader and main author.

Additional declarations for articles in life science journals that report the results of studies involving humans and/or animals

Not applicable.

Ethics approval (include appropriate approvals or waivers)

Ethical approval was received from the Gauteng Department of Agriculture and Rural Development.

Consent to participate (include appropriate statements)

Consent to participate was received from the Gauteng Department of Agriculture and Rural Development.

Consent for publication (include appropriate statements)

The author gives consent for publication.

Acknowledgments

The authors wish to acknowledge the following: Agricultural Research Council and Gauteng Department of Agriculture and Rural Development. A special thanks to Ms. Marjan Van Der Walt (ARC ISCW) for her excellent & professional work in drawing the climate maps.

References

Amaza, P., Abdoulaye, T., Kwaghe, P., Tegbaru, A., 2009. Changes in Household Food Security and Poverty Status in PROSAB Area of Southern Borno State, Nigeria. International Institute of Tropical Agriculture (IITA), pp. 11–13.

ARC (Agricultural Research Council-SCW), (ARC-SCW), 2017. Pretoria, South Africa.

De Cock, N., D'Haese, M., Vink, N., Van Rooyen, C.J., Staelens, L., Schonfeldt, H.C., D'Haese, L., 2013. Food security in rural areas of Limpopo Province, South Africa. Food Secur. 5, 269–282. https://doi.org/10.1007/s12571-013-0247-y.

Hanley, N., Wright, R.E., Alvarez-Farizo, B., 2007. Estimating the economic value of improvements in river ecology using choice experiments: an application to the water framework directive. In: Environmental Value Transfer: Issues and Methods. Springer, Dordrecht, pp. 111–130.

IFAD, 2012. Sustainable Smallholder Agriculture: Feeding the World, Protecting the Planet. IFAD, Rome, Italy.

Madlala, P.J., 2014. The Role of Food Gardens in Providing Sustainable Livelihoods in the Msunduzi Municipality (Doctoral dissertation). University of Kwazulu Natal.

Maponya, P., 2019. Water resource and food security: a case of households in Gauteng Province, South Africa. In: Oral Presentation & Full Paper Publication, BEST PAPER AWARDED Under Conference Sub Theme 1 Enabling Policy Environment for Water, Food and Energy, 3rd World Irrigation Forum, 01–09 September. International Commission on Irrigation and Drainage (ICID), Bali, Indonesia.

Maponya, P., Moja, S., 2012. Asset portfolios and food accessibility in Sekhukhune District, Limpopo province. J. Agric. Sci. 4 (12), 144–153.

Maponya, P., Venter, S.L., Du Plooy, C.P., Oelofse, D., Van Den Heever, E., Manyaga, C., Nyirenda, O., Masuku, P., Mahlangu, M., 2021. The mechanism for improving the sustainability of homestead gardens in Gauteng Province, South Africa. In: Ramanan, V.V., Shah, S., Prasad, R. (Eds.), Sustainable Bio Economy—Pathways to Sustainable Development Goals. Springer, Singapore, ISBN: 978-981-15-7320-0, pp. 303–323, https://doi.org/10.1007/978-981-15-7321-7_14.

Mohammadi-Nasrabadi, F., Omidva, N., Khoshfetrat, M., Kolahdooz, F., 2014. Household Food Insecurity in the Islamic Republic of Iran: A Systematic Review and Meta-analysis.

Mpanza, P.T., 2008. An Evaluation of the Perceived Benefits and Constraints of Community Gardens Established by the KwaZulu-Natal Department of Agriculture and Environmental Affairs. University of KwaZulu-Natal.

Ndobo, F., Sekhampu, T.J., 2013. Determinants of Vulnerability to Food Insecurity in a South African Township: A Gender Analysis.
Pietermaritzburg Agency for Community Social Action (PACSA), 2017. Households Rising Food Prices. PACSA, Pietermaritzburg.
Pietermaritzburg Economic Justice & Dignity Group (PMBEJD), 2021. Struggling South Africans Are Paying Nearly R300 More for Food Than Nine Months Ago. PMBEJD, Pietermaritzburg.
Schaible, G.B., Mishra, K., Lambert, A.K., Panterov, D.M., 2015. Factors Influencing Environmental Stewardship in US Agriculture, Conservation Program Participants vs. Non-participants.
Statistics South Africa, 2014. Mid-Year Population Estimates, Pretoria.
Statistics South Africa, 2021. Changes in Food Prices, Pretoria.
UNDP, 1996. Urban Agriculture: Food, Jobs and Sustainable Cities. UNDP, New York, USA.
Van Schalkwyk, H., Carstens, J.P., Cloete, P.C., Botha, L., Vadeveer, S., Van der Walt, M., Schoeman, J.L., 2004. A Comprehensive Agricultural Plan for Local Municipalities Within the Motheo District Municipality in the Free State Province. Digitally Available From ARC-ISCW, Pretoria.
Van Schalkwyk, H., Carstens, J.P., Cloete, P.C., Van der Walt, M., Schoeman, J., 2009. North West Agricultural Master Plan and Decision Support System. Digitally Available From ARC-ISCW, Pretoria.

Index

Note: Page numbers followed by *f* indicate figures and *t* indicate tables.

A

Abiotic stresses, tolerance to, 157
Abundance paradox, 18–19
Active pharmaceutical ingredients (API), 51–52
Advanced Spaceborne Thermal Emission and Reflection Radiometer (ASTER), 109
Aerospace investment in research and innovation, 104
AGO301 Vaccine, 57–58
Agricultural biotechnology
  in Bangladesh, 145–148
    cotton, 153–154
    genome editing, 154–159
    regulatory framework for GMOs adoption, 159–160
    rice, 152–153
    vegetables, 150–152
  Indonesia, 169
Agricultural/forestry residues, 258
Agricultural industrialization, 3
Agricultural products and households, impact prices, 170–171
Agricultural Research Council (ARC), 86
Agricultural supply chain, 18
  management, Indonesia, 194–195
Agriculture of today, 205–206
"Agro-bioeconomic citizenship", 13
Agro-bioeconomy, 5–8
  tourism, 15
Agrobiotechnology, 5
Agroforestry systems, 122
  average monthly rainfall, 127
  average monthly temperature, 127
  Eastern Cape in South Africa, 121–139, 124*f*
  soil drainage, 127
  soil types, 127, 128*t*
  structural and ecological diversity, 128–130
Agrosilviculture community growers, 121–122
Air quality improvement, after the Covid-19 lockdown in China, 184
Air quality index (AQI), 189
Animal production, antibiotics, 211–212
Animal welfare, 211
  technology systems, 225
Antibiotic resistance, 211–212
Antimicrobial peptides, 77–78, 79*f*
  and antiviral effects, 78–80
Antiviral peptides, 77–80, 79*f*
Arable land, future agricultural use, 207–208
ARCoV Vaccine, 65–66
ARCT-021 Vaccine, 64
Artificial intelligence (AI)-driven systems, 112–114
ASTER (Advanced Spaceborne Thermal Emission and Reflection Radiometer), 109
Asymmetric information and alternative costs, 17
Automatic tire inflation systems, 219

B

Bangladesh Biosafety Clearing House (BBCH), 159–160
Bangladesh Delta Plan (BDP), 144
Bangladesh's economy
  agricultural biotechnology
    cotton, 153–154
    genome editing, 154–159
    regulatory framework for GMOs adoption, 159–160
    rice, 152–153
    vegetables, 150–152
  agricultural production, 144
Best available technology (BAT), 212
Bio-based businesses, 231
Bio-based solutions, 240–242
Bio-based value chains, 242

Biodiesel production, 246–248
Biodiversity
conservation efforts, 187
protection, 206–207
Bioeconomy
business opportunities, 238, 241
characteristics, 3–5, 6*t*
collaboration and entrepreneurship, 235–236
collaboration issues, 242
collaboration networks, 232–233
collaborative behavior, 238, 241–242
collective food sector based on planning and partnership, 13
conventional education systems, 290
digitalization of, 306
education, 288–293
inclusion of under-represented students, 292
knowledge dissemination, 241–242
limited access to information, 239
on natural resources and ecosystem services, 232
patents or intellectual property rights, 239
policies around the World, 3–5, 7*f*
primary production, 236–237
pursuing equity, 293
qualitative and case-study-based research, 242
by stakeholders, 238–239
strategy, 231
students, 292–293
systematic approach, 232–233
validity and sustainability, 20–21
VENN diagram, 3–5, 4*f*
Bio-economy of nature-based resources management, 184
Indonesia, 192–194
Bioenergy systems, 246–248
Biofuel co-products, 270–271
generation in early biofuel, 248–249
price and environmental impact, 245–246
Biofuels, 246–248
from biomass and related value added co-products, 263
cellulosic biofuel production, 260–261
co-products for livestock feed, 264–278, 264*f*
feedstocks, 249–251
policies on GHG emissions, 272–273
for sustainable market, 249–263
Biological control, forms of, 218–219
Biomass recycling
classification, 258, 259*t*
hierarchical distribution, 263, 263*f*
plantations, 245–246
Biosafety Core Committee (BCC), 159–160
Biosafety Guidelines of Bangladesh, 2006, 159–160
Biotechnology
child labor, agricultural sector, 14
depth and prevalence, 11–12
BRICS (Brazil, Russia, India, China and South Africa) vaccines, 51–52, 54*t*

C

Carbon capturing in soil, 219
Catastrophe risk mitigation, 104
Centrality of curriculums, 295
Challenge-based innovation, 205–212
Circular bio-economy model, 175
City of Tshwane Metropolitan Municipality
map, 318–319, 318*f*
townships and informal settlements, 318–319, 319*f*
Climate change, 18
prerequisites for crops, 208–209
Clinical Trial Protocols (CTP), 52
Collaborative behavior, 241–242
Common Agricultural Policy (CAP), 19
Communication, lack of infrastructure for, 210
Competencies, 291–292
and skill development, 291–292
Connected systems, 223–224
Constitution of the Corpus of analysis and descriptive analysis, 234–235, 235*t*
Contents, 291–292
Coopid, 234, 240
Coping with disruptive X factors, 212
Coral solutions in complexity, 296–297
Corn ethanol, 264–265

Corona tax, 16
Coronavac, 49–50
Coronomy, 19–20
Correlation analysis among variables, 91–92
Cortellis Drug Discovery Intelligence (CDDI), 42, 44–48*t*
Cotton *(Gossypium hirsutum)*, 153–154
Cotton Development Board (CDB), 153–154
COVAXIN (Bharat Biotech's vaccine), 49–50
Covid-19 pandemic, 2
  disruption on socio-economic and nature, 171–174
  in Indonesia (*see* Indonesia)
  knowledge-based development, 174–175
  and sustainable agro-bioeconomy, 8–10
Covid-19 vaccines, 15
  global distribution and related variants, 40–42, 43*f*, 44–48*t*
CRISPR/Cas9 gene-editing technology, 157
Crop and cultivation concepts, 217–220
Crude glycerin, 268
Crypteins, 78–80
Cryptides, 78–80
C-stock of lands, 271–272
CvnCoV Vaccine, 63–64
Cyclotides, 78–80

D

Data analysis, 126–127
Decision-making process, 240–241
Deep crisis, 29–35
Deforestation, 176
Department of Environment, Fisheries and Forestry (DEFF) for agroforestry integration, 139
Desktop natural resources audit, 320
Digital control systems, 113
Digitalization, 290–291, 306
Digital land type information, 127, 128*t*
Digital Navigation Technology, 113
Digital skills, 305
Digital transition and automation, 290–291
Direct land-use change (DLUC), 271–272
Disruptive X factors, coping with, 212
Dissemination of public health information, 115
Dissemination strategies and activities, 239
Distillers' co-products, 264–265
  displacement ratio, 264–265
DNA repair mechanisms, 156
DNA vaccines in clinical trials, 53–58, 53*f*
  AGO301 Vaccine, 57–58
  GX-19 Vaccine, 55–56
  INO-4800 Vaccine, 55
  ZyCoV-D Vaccine, 56–57
Downstream impacts, 115
Drone-based airborne bioremediation, 114
Dry milling, 264–268
Dystopia, 10, 21, 22–24*t*
  scenarios for world and Turkey, 16–20

E

Earth-orbiting satellites, 113
Eastern Cape in South Africa, 121–139, 124*f*
  annual rainfall, 130–139, 132–135*f*
  soil depth, 127, 131*f*
  soil drainage, 128, 129*f*
Ecosystem services, 12
Education
  levels, in agriculture and bioeconomy, 289–290
  needs, in agriculture and bioeconomy, 288–289
  policy transition process, 302–304, 304*f*
  and training, in agriculture and the Bioeconomy, 293–297
Eggplant, 151
Enablers and disablers, fifth innovation era, 226
Energy, 104
  crops cultivation, 272
  independence, 207
Enforcement of Restrictions on Community Activities (ERCA), 168
Entrepreneurial culture, 239
Entrepreneurial orientation, 239
Environmental pollution, 17
Environmental services, 171

Environment Policy, 1992, 159–160
Ethanol production, 246–248, 265*f*, 266–268
Eukaryotic expression systems, 77–78
European bioeconomy, 233–234, 236
  initiatives in, 240–241
European strategy, 231
Evolution, 212

F

Family of Hope Program, 168
Farm animal diseases, 209
Farmer Support Programme, 88
Farming, 104, 210
  methods, 29
Farm to Fork Strategy, 19
Field Level Biosafety Committee (FBC), 159–160
Field robotics, 220–221
Fifth innovation era, 217–225
  animal welfare technology systems, 225
  enablers and disablers, 226
  field robotics, 220–221
  fossil-free energy and products, 222–223
  new crop and cultivation concepts, 217–220
  smart connected systems, 223–224
Financial assets, 177
Financial systems, 110
Finite resource of phosphorus, 210
First generation biofuels, 249–251
  feedstock for, 252–255
First innovation era (approximately 10,000 B.C. to 1750 A.D), 214
Food and Agriculture Organization (FAO), 9, 322–323
Food production, in world, 205–206
Food safety, 104
  lack of trust, 208
Food security, 87, 161, 316–318, 324–326, 331–333
Food supply chain, 11–12
Food waste, 175, 211
Foresight studies of organic systems, 34–35
Forest and environmental afforestation, 171
Forest-based sector, 176–177
Forestry and agricultural sectors, in Indonesia, 185
Forestry enterprises, 123
Formal education with informal education, 305
Fossil-free energy and products, 207, 222–223
Foundations of "agro-bioeconomic citizenship", 13
Fourth innovation era (approximately 2000–today), 216–217, 216*f*
Fruit and shoot borer (FSB), 151
Fuel ethanol, 253–254
Functional genomic tools, 144

G

Gauteng Department of Agriculture and Rural Development (GDARD), 319
Genome-edited legislation, 145*f*, 161–162
Genome editing, 144–145
  agricultural improvement in Bangladesh, 154–159
  crops, 157
Geographical awareness, 103–104
Geographic information system (GIS), 114
Geographic information system-based health mapping, 114
Geospatial technologies, management through COVID-19, 103–115
Giant reed (Arundo donax) cultivation, 256–258, 256*f*
Global agro-bioeconomic diplomacy, justice and security, 12–13
Global bioenergy sources, 251–252, 252*f*
Global climate change, 279–281
Global emissions, 105
Global Navigation Satellite System (GNSS), 113
Global population, 206
Global Positioning System (GPS), 113
Global warming, 18
Government agencies, 103
Gqukunqa Forestry Enterprise, 125
Green budgets, agriculture, 12

Greenhouse gases (GHG) emissions, 246–249
balance of biorefinery outputs, 274–276
biofuel policies on, 272–273
environmental impact, 271–278, 273*t*
GX-19 Vaccine, 55–56

H

Harry Gwala (previously Sisonke) District Municipality (HGDM), 85–99
Healing and bioeconomic prospect, tropical ecosystems, 187–192
High titer neutralizing antibodies, 52
Homestead food gardeners
between gender and availability of gardens, 331–333, 333*t*
initiative, 326, 327*f*
natural resource management, 326–330, 328–329*f*
pictures in field, 330, 330–332*f*
socio-economic characteristics, 324–326, 325*t*
Homoeconomicus, 15
Household consumption patterns, 173
Household food security, 169
Household socioeconomic resilience, in Covid-19 era, 167–178
Human population growth and technological developments, 187–188

I

IKONOS satellite, 106–109, 108*t*
Image processing technology, spectral response, 109
Imitation, 212
Indirect land-use change (ILUC), 271–272, 274–276
Indonesia
Covid-19 pandemic in
agricultural supply chain management, 194–195
bio-economy of nature-based resources management, 192–194
environmental improvement indicators, 190–191, 191*t*
healing of tropical ecosystem, 187–192
La-Nina phenomenon in, 185–187
on marine environment, 190–192
pollutant reduction, 190–191, 191*t*
precision and smart agriculture, 195–198
on terrestrial environment, 189–190
tropical ecosystem, 167–178
Industrial impacts, 110
Industry convergence, 294–295
Innovation, 212–213
INO-4800 Vaccine, 55
Insect-resistant crops, 160–161
Insects, 208
Institutional Biosafety Committee (IBC), 159–160
Integrated bio-cycle farming system (IBFS), 176, 193–194
in post Covid-19 era, 175–177
Inter- and transdisciplinarity, agricultural education and epistemology, 293–297
International Labour Organization (ILO), 9–10
International production, 110–111
In vitro methods, 78–80
Italian cooperative Caviro, 240–241
Izinini Forestry Enterprise, 125

K

Knowledge-based development, 174–175
natural resource-based to, 176
Knowledge transfer process, 232
Kwazulu Natal (KZN) Province, South Africa, 85–99

L

Land and crop suitability, 320–324, 321–323*f*
Land redistribution for agricultural development (LRAD), 95–96
Landsat 7 and 8 bands, 106–109, 108*t*
LANDSAT satellite, 106–109, 108*t*
Land-use change (LUC), 271–272, 273*t*, 274–276
La-Nina phenomenon in Indonesia, 185–187

Large-scale cultivation of energy crops, 245–246
Large-scale social restrictions (LSSR), 178
in Indonesia, 184–185
Learning process, 295
Life-long learning, 305
Lignocellulosic biomass, 276–278
Linear economy method, 175
Linear knowledge transfer model, 300
Liquid biofuels, 246–248
Literature analysis, 233–234
Livestock feed market, and related policies, 269–271
Livestock feed potential, 266–269
Live virus neutralization assay, 49–50
LNP-mRNA-1273 Vaccine, 59–61
LNP-nCoVsaRNA Vaccine, 64–65
LUC-GHG emissions, 274–276
Lusikisiki Forestry Enterprise, 126

M

Machine Learning, Coronavirus suppression, 115
Maize, 88–89
Management crisis of global organic systems, 34–35
Manpower, lack of, 209
Manufacturing, 110–111
Mapping and Geoinformation (GIS)-driven infrastructure, 112–113
Mapping, organic systems, 30
Medicinal proteins, in tobacco and potatoes, 78
Mega-biodiversity, 183
Microalgae biomass, into livestock feed industry, 268–269
Microalgal biorefinery, 258, 260–261, 260*f*
Migration waves, 16
Mkambathi Forestry Enterprise, 123–125
Moderate Resolution Imaging Spectroradiometer (MODIS), 109
Modern biotechnological tools and techniques, development and adoption of, 144
MODIS (Moderate Resolution Imaging Spectroradiometer), 109
Molecular farming
with antiviral effects, peptides, 77–82
for production drugs and vaccines, 78
Monitoring changes
in port and airport activities, 111
in production, 110–111
mRNA BNT-162b2 Vaccine, 62–63
mRNA vaccine platform, 52
Multi- and interdisciplinary, agricultural education and epistemology, 293–297
Multifunctional (annual) crops, 252–255

N

National Biodiversity Strategy and Action Plan for Bangladesh, 2004, 159–160
National biosafety framework (NBF), 159–160
National Biotechnology Policy, 2006, 159–160
National Committee on Biosafety (NCB), 159–160
National Development Plan (NDP), 85–86
National Guidelines for Fish and Animal Biotechnology, 2006, 159–160
National Guidelines on Medical Biotechnology, 2006, 159–160
National Technical Committee on Biosafety (NTCB), 159–160
Natural resource-based to knowledge-based development, 176
Natural resource-rich countries, 18–19
Nature-based resources management, Indonesia, 192–194
NDVI (Normalized Difference Vegetation Index), 109
Neoliberalism, 14–15
Networks dynamics, 239
New breeding techniques (NBTs), 144, 146–148
consumer acceptance and future regulation, 160–161
New crop and cultivation concepts, 217–220
New food habits, 206
New Space economy, 104–105
Next-generation biofuel capacity, 278–279

Nitrogen (N), 261–263
Nitrogen dioxide, 111–112
Non-compliance with LSSR, 178
Non-timber forest products, 171
Normalized Difference Vegetation Index (NDVI), 109
NOVAVAX's vaccine with recombinant nanoparticle technology, 49–50
Nucleic acids vaccines, 52

O

Online shopping, 173
Open-pit mining overexploitation, 176
Organic cycles, 192
Organic phenomena, 29–35
Organic processes, 30, 31*t*
Organic products, 30, 31*t*
Organic systems
  components of, 30, 31*t*
  global dynamics of major changes, 30
  growth barriers, 30, 32–34*t*
  under major changes, 29–35
  mapping, 30
  relative importance and growth of, 30, 30*t*
  types, 30, 31*t*
  working hypothesis, 30, 31*t*
Organisation for Economic Co-operation and Development (OECD), 5
Osmoprotectants, 158
Osmoregulation, 158

P

Pandession, 16–17
Passive systems, 106
Patent protection for vaccines, 51–52
Peptide-based drugs, 82
Peptide heterologous expression host systems, 80–81
Peptide production, 78–80
  by molecular farming with antiviral effects, 77–82
  in plants, 81
Peptides, challenges of, 82
Perennial grasses, 255–258
Person-based relationships, 232
Pests, 208
Phosphorus (P), 261–263
  finite resource of, 210
Photosynthesis via genome editing, 157
Physical distancing policies, 178
Plant biotechnology, 161–162
  consumer acceptance, 160–161
Plant-derived medicinal and non-medicinal proteins, 81
Plant expression system, 80–81
Plant genetic resources, 144
Plant genome editing, 157
Plant systems, 77–78
Pollution, types of, 188
Polyphenol oxidase (PPOs), 157–158, 160–161
Port and airport activities, 111
Potato, 151–152
Precision, Indonesia, 195–198
Preferred Reporting Items for Systematic Reviews (PRISMA) guidelines, 233, 234*f*
Preharvest losses, 208
Primary breadwinners, 171
Primary production, 232
  entrepreneurial activity of, 232
Principal sensors, characteristics and applications, 105–110
Private-indirect agricultural bio-economics sector, 15
Production systems, 77–78
Prokaryotic expression systems, 77–78
Protease degradation, 82
Protectionism and domestic preparation, 206
Protein-rich crops, 264–265
Public financial management (PFM) framework, 12
Public health information, dissemination of, 115
Purification of recombinant antimicrobial peptides, 81–82
Purposive sampling technique, 91, 319–320

Q

Quantitative and qualitative research designs, 91, 319

R

Rainfall availability, Eastern Cape in South Africa, 121–139, 124*f*, 132–135*f*
Rapeseed oil, 254
Recombinant antimicrobial peptides, purification, 81–82

Recombinant proteins, 77–78, 80–81
Regulatory models for genome-edited crops, 161
Relationship-oriented research, 301
Remote sensing
  Earth observation economics, 104, 107*t*
  as tool to survey endemic diseases, 111–115
Renewable and sustainable agriculture, 245–246, 247*f*, 251–252, 255, 268–269, 272–273, 278
Repeat in vivo/in vitro tests, 78–80
Research approach, 126
Resource curse, 18–19
Rice, 152–153
RNA-based technology, BRICS' countries, 58, 60*t*
RNA vaccines, 52
  in clinical trials, 58–66, 59*f*, 60*t*
    ARCoV Vaccine, 65–66
    ARCT-021 Vaccine, 64
    CvnCoV Vaccine, 63–64
    LNP-mRNA-1273 Vaccine, 59–61
    LNP-nCoVsaRNA Vaccine, 64–65
    mRNA BNT-162b2 Vaccine, 62–63
Robotic biotech, 11–12
Robotics, field, 220–221

S

Sampling method, 126
SARS-CoV-2 variants and genomic vaccines, 42–50, 43*f*, 44–48*t*
Satellite-based positioning and navigation technology, 113
Satellite remote sensing
  data, 104–105
  and earth observation, 114
Second generation biofuels, 251–252
  feedstock for, 255–258
  GHG emissions, 276
Second innovation era (approximately 1750–1900), 214–215
Security issues, 210
Sentinel constellation, 105
Shipping, 111
Sinawo Forestry Enterprise, 123
Sixhotyeni Forestry Enterprise, 125–126
Skills, 291–292
Smallholder farmers
  agriculture sector, 88–89
  beneficiaries per local municipality, 93, 93*f*
  community-driven agricultural production and processing, 96–97
  crops planted, 96–97, 96*t*
  extension services, 97–98, 97*t*
  farming experience, 95–96, 95*f*
  in Harry Gwala district, South Africa, 85–99
  institutional bottlenecks, 87
  land acquisition, 95–96, 95*f*
  land sizes, 98, 98*f*
  market access, 98, 99*f*
  Pearson correlation coefficients among variables, 98–99, 99*t*
  permission to occupy, 95–96, 95*f*
  socio-economic characteristics, 94–95, 94*t*
  socio economic factors, 87
  source of water, 96, 96*t*
  supply inputs, 97, 97*t*
  sustainability of, 85–87
  in Umzinyathi district, South Africa, 85–99
  visited per local municipality, 93, 93*f*
Smart agriculture, Indonesia, 195–198
Smart connected systems, 223–224
"Snowball" search strategy, 233
Social distancing policies, 168, 178
Social sustainability, 211
Socio-economic activities, 170
Soil availability and suitability, 127–128
Soil conditions, 127–130
Soil depth, 127, 131*f*
Soil fertility, 210–211
South Africa
  "food-secure" nation, 316–317
  Kwazulu Natal (KZN) Province, 85–99
Soybean oil, 254
Space-based tools, 112–113
Space exploration technologies, 104
Spatial technologies, 112–113

Spike protein (antigen), 52
SPOT satellite, 106–109, 108*t*
Strategic tropical natural resources management, Covid-19 pandemic, 167–178
Sustainable agricultural bioeconomy, 2–8, 291–292
Sustainable global food order, 20, 21*t*

T

Target stakeholders, 232
TERRA, 109
Third generation biofuels, 251–252
  feedstock for, 258–263
  GHG emissions, 276
Third innovation era (approximately 1900–2000), 215–216, 215*f*
Tissue culture technologies, 156
Topography and soil pattern, 128
Tourism sectors, 15
Trans-esterification, 268
Transcription activator-like effector nucleases (TALEN), 158
Tropical ecosystem, healing of Indonesia, 187–192
Tropical natural resources, 176

U

Umzinyathi district, South Africa, 85–99
UN Department of Economics and Social Affairs, 206
United Nations Conference on Sustainable Development (UNCSD), 87
United States Department of Agriculture's (USDA) National Soils Handbook, 322–323
Universal access to vaccines and pandemic preparedness, 39–67
University education, 289
Unmanned aerial vehicle (UAV) for large-scale disinfection, 113–114
Updated Bioeconomy Strategy, 231
Urban agriculture
  and agro-environment, 315–333
  in Tshwane Municipality, 315–333
Utopian scenarios for world and Turkey, 10–15
Utopias, 21, 22–24*t*

V

Vaccines
  patent protection for, 51–52
  universal access, 39–67
Variant of concern (VOC), 40–42, 43*f*
Variant of high consequence, 40–42
Variant of Interest, 40–42
Vegetables, 150–152
Vocational training, 289

W

Water management, 246–248
Web-o-metric analysis, 30, 30*t*
Weeds, 208
Wet milling, 265
Win-win game, 15
Work from home (WFH) policy, 173
World Farmers Organisation (WFA), 86–87
World food crisis, 183–184

Z

Zinc-finger nucleases (ZFN), 158
ZyCoV-D Vaccine, 56–57

FSC
www.fsc.org
MIX
Papier | Fördert
gute Waldnutzung
FSC® C083411

Kolibri 360 GmbH
Ferdinand-Jühlke-Straße 7
99095 Erfurt, Deutschland
produktsicherheit@kolibri360.de